W9-CJN-235

RYPINS' INTENSIVE REVIEWS

Series Editor

Edward D. Frohlich, MD, MACP, FACC

Alton Ochsner Distinguished Scientist
Vice President for Academic Affairs
Alton Ochsner Medical Foundation
Staff Member, Ochsner Clinic
Professor of Medicine and of Physiology
Louisiana State University of Medicine
Adjunct Professor of Pharmacology and
Clinical Professor of Medicine
Tulane University School of Medicine
New Orleans, Louisiana

RYPINS' INTENSIVE REVIEWS

Physiology

John E. Hall, PhD

Arthur C. Guyton Professor and Chairman
Department of Physiology and Biophysics
University of Mississippi Medical Center
Jackson, Mississippi

Thomas H. Adair, PhD

Professor
Department of Physiology and Biophysics
University of Mississippi Medical Center
Jackson, Mississippi

Lippincott - Raven
PUBLISHERS
Philadelphia • New York

Acquisitions Editor: Richard Winters
Developmental Editor: Mary Beth Murphy
Managing Editor: Susan E. Kelly
Manufacturing Manager: Dennis Teston
Associate Managing Editor: Kathleen Bubbeo
Production Editor: P. M. Gordon Associates
Cover Designer: William T. Donnelly
Interior Designer: Susan Blaker
Indexer: M. L. Coughlin
Compositor: Lippincott–Raven Electronic Production
Printer: Courier/Kendallville

9 8 7 6 5 4 3 2 1

Library of Congress Cataloging-in-Publication Data
Hall, John E. (John Edward), 1946–
    Physiology / John E. Hall, Thomas H. Adair.
        p.   cm.—(Rypins' intensive reviews)
    Includes index.
    ISBN 0–397–51549–9
    1. Human physiology—Examinations, questions, etc.   I. Adair,
Thomas H.   II. Title.   III. Series.
    [DNLM: 1. Physiology—examination questions.   QT 18.2   H177p 1997]
    QP40.H27   1998
    612'.0076—dc21
    DNLM/DLC   97-21928
    for Library of Congress   CIP

Dedicated to Arthur C. Guyton,
our mentor and friend

Who Was "Rypins"?

Dr. Harold Rypins (1892–1939) was the founding editor of what is now known as the RYPINS' series of review books. Originally published under the title *Medical State Board Examinations,* the first edition was published by J. B. Lippincott Company in 1933. Dr. Rypins edited subsequent editions of the book in 1935, 1937, and 1939 before his death that year. The series that he began has since become the longest-running and most successful publication of its kind, having served as an invaluable tool in the training of generations of medical students. Dr. Rypins was a member of the faculty of Albany Medical College in Albany, New York, and also served as Secretary of the New York State Board of Medical Examiners. His legacy to medical education flourishes today in the highly successful *Rypins' Basic Sciences Review* and *Rypins' Clinical Sciences Review,* now in their 17th editions, and in the *Rypins' Intensive Reviews* series of subject review volumes. We at Lippincott–Raven Publishers take pride in this continuing success.

—*The Publisher*

Series Preface

These are indeed very exciting times in medicine. Having made this statement, one's thoughts immediately reflect about the major changes that are occurring in our overall healthcare delivery system, utilization-review and shortened hospitalizations, issues concerning quality assurance, ambulatory surgical procedures and medical clearances, and the impact of managed care on the practice of internal medicine and primary care. Each of these issues has had a considerable impact on the approach to the patient and on the practice of medicine.

But even more mind-boggling than the foregoing changes are the dramatic changes imposed on the practice of medicine by fundamental conceptual scientific innovations engendered by advances in basic science that no doubt will affect medical practice of the immediate future. Indeed, much of what we thought of as having a potential impact on the practice of medicine of the future has already been perceived. One need only take a cursory look at our weekly medical journals to realize that we are practicing "tomorrow's medicine today." And consider that the goal a few years ago of actually describing the human genome is now near reality.

Reflect, then, for a moment on our current thinking about genetics, molecular biology, cellular immunology, and other areas that have impacted upon our current understanding of the underlying mechanisms of the pathophysiological concepts of disease. Moreover, paralleling these innovations have been remarkable advances in the so-called "high tech" and "gee-whiz" aspects of how we diagnose disease and treat patients. We can now think with much greater perspective about the dimensions of more specific biologic diagnoses concerned with molecular perturbations; gene therapy not only affecting genetic but oncological diseases; more specific pharmacotherapy involving highly specific receptor inhibition, alterations of intracellular signal transduction, manipulations of cellular protein synthesis; immunosuppresive therapy not only with respect to organ transplantations but also of autoimmune and other immune-related diseases; and therapeutic means for manipulating organ remodeling or the intravascular placement of stents. Each of these concepts has become inculcated into our everyday medical practice within the past decade. The reason why these changes have so rapidly promoted an upheaval in medical practice is continuing medical education, a constant awareness of the current medical literature, and a thirst for new knowledge.

To assist the student and practitioner in the review process, the publisher and I have initiated a new approach in the publication of *Rypins' Basic Sciences Review* and *Rypins' Clinical Sciences Review.* Thus, when I assumed responsibility to edit this long-standing board review series with the 13th edition of the textbook (first published in 1931), it was with a feeling of great excitement. I perceived that great changes would be coming to medicine, and I believed that this would be one ideal means of not only facing these changes head on but also for me personally to cope and keep up with these changes. Over the subsequent editions, this confidence was reassured and rewarded. The presentation for the updating of medical information was tremendously enhanced by the substitution of new authors, as the former authority "standbys" stepped down or retired from our faculty. Each of the authors who continue to be selected for maintaining the character of our textbook is an authority in his or her respective area and has had considerable pedagogic and formal examination experience. One dramatic recent example of the changes in author replacement just came about with the 17th edition. When I invited Dr. Peter Goldblatt to participate in the authorship of the pathology chapter of the textbook, his answer was "what goes around, comes around." You see, Dr. Goldblatt's father, Dr. Harry Goldblatt, a major contributor to the history of hypertensive disease, was the first author of the pathology chapter in 1931. What a satisfying experience for me personally. Other less human changes in our format came with the establishment of two soft cover volumes, the current basic and clinical sciences review volumes, replacing the single volume text of earlier years. Soon, a third supplementary volume concerned with questions and answers for the basic science volume appeared. Accompanying these more obvious changes was the constant updating of the knowledge base of each of the chapters, and this continues on into the present 17th edition.

And now we have introduced another major innovation in our presentation of the basic and clinical sciences reviews. This change is evidenced by the introduction of the *Rypins' Intensive Reviews* series, along with the 17th edition of *Rypins' Basic Sciences Review, Rypins' Clinical Sciences Review,* and the *Questions and Answers* third volume. These volumes are written to be used separately from the parent textbook. Each not only contains the material published in their respective chapters of the textbook, but is considerably "fleshed out" in the discussions, tables, figures, and questions and answers. Thus, the *Rypins' Intensive Reviews* series serves as an important supplement to the overall review process and also provides a study guide for those already in practice in preparing for specific specialty board certification and recertification examinations.

Therefore, with continued confidence and excitement, I am pleased to present these innovations in review experience for your consideration. As in the past, I look forward to learning of your comments and suggestions. In doing so, we continue to look forward to our continued growth and acceptance of the *Rypins'* review experience.

Edward D. Frohlich, MD, MACP, FACC

Preface

Physiology is an integrative discipline that includes the study of molecules and subcellular components of the body, cells, tissues, organ systems, and interactions between all of these parts in health and in disease. Physiology, therefore, provides a link between basic sciences and clinical medicine, and an understanding of its basic principles is essential for the effective practice of medicine.

Because human physiology covers a broad scope and is a rapidly expanding discipline, the amount of information that is potentially applicable for the practice of medicine can be overwhelming. Our primary goal in writing this book was to distill this vast amount of information into a study guide for medical students preparing for the United States Medical Licensure Examination (USMLE), for physicians preparing for specialty board examinations, and for other health care professionals who wish to rapidly and efficiently review a large amount of physiology in a short time. The book also will be useful to students who wish to grasp overall concepts and key physiological principles before studying more detailed reference textbooks.

An important feature of the book is that the various section headings state succinctly the primary concepts in the accompanying paragraphs. The book contains over 300 questions, presented in the current format of USMLE Step 1. It also includes a series of "must-know topics" for each major section that will assist students in outlining subject areas and testing their knowledge of physiology.

We have, by necessity, omitted many details and clinical examples that enrich the basic physiological principles described. More detailed reference textbooks should be consulted once the most important concepts are mastered. We have tried to make the text as accurate as possible but wish to issue an invitation to all readers to inform us of any inaccuracies or of ways that the book may be improved.

Human physiology is an exciting discipline because it seeks to understand how the body's components interact in ways that permit us to function as living beings and how these mechanicisms are altered in disease. We hope this excitement is conveyed to you and that this book will contribute to your lifelong study of physiology.

John E. Hall, PhD
Thomas H. Adair, PhD

Introduction

Preparing for USMLE

In August 1991 the Federation of State Medical Boards (FSMB) and the National Board of Medical Examiners (NBME) agreed to replace their respective examinations, the FLEX and NBME, with a new examination, the United States Medical Licensing Examination (USMLE). This examination will provide a common means for evaluating all applicants for medical licensure. It appears that this development in medical licensure will at last satisfy the needs for state medical boards licensure, the national medical board licensure, and licensure examinations for foreign medical graduates. This is because the 1991 agreement provides for a composite committee that equally represents both organizations (the FSMB and NBME) as well as a jointly appointed public member and a representative of the Educational Council for Foreign Medical Graduates (ECFMG).

As indicated in the USMLE announcement, "It is expected that students who enrolled in U.S. medical schools in the fall of 1990 or later and foreign medical graduates applying for ECFMG examinations beginning in 1993 will have access only to USMLE for purposes of licensure." The phaseout of the last regular examinations for licensure was completed in December 1994.

The new USMLE is administered in three steps. Step 1 focuses on fundamental basic biomedical science concepts, with particular emphasis on "principles and mechanisms underlying disease and modes of therapy." Step 2 is related to the clinical sciences, with examination on material necessary to practice medicine in a supervised setting. Step 3 is designed to focus on "aspects of biomedical and clinical science essential for the unsupervised practice of medicine."

Today Step 1 and Step 2 examinations are set up and scored as total comprehensive objective tests in the basic sciences and clinical sciences, respectively. The format of each part is no longer subject-oriented, that is, separated into sections specifically labeled Anatomy, Pathology, Medicine, Surgery, and so forth. Subject labels are therefore missing, and in each part questions from the different fields are intermixed or integrated so that the subject origin of any individual question is not immediately apparent, although it is known by the National Board office. Therefore, if necessary, individual subject grades can be extracted.

Step 1 is a two-day written test including questions in anatomy, biochemistry, microbiology, pathology, pharmacology, physiology, and the behavioral sciences. Each subject contributes to the examination a large number of questions designed to test not only knowledge of the subject itself but also "the subtler qualities of discrimination, judgment, and reasoning." Questions in such fields as molecular biology, cell biology, and genetics are included, as are questions to test the "candidate's recognition of the similarity or dissimilarity of diseases, drugs, and physiologic, behavioral, or pathologic processes." Problems are presented in narrative, tabular, or graphic form, followed by questions designed to assess the candidate's knowledge and comprehension of the situation described.

Step 2 is also a two-day written test that includes questions in internal medicine, obstetrics and gynecology, pediatrics, preventive medicine and public health, psychiatry, and surgery. The questions, like those in Step 1, cover a broad spectrum of knowledge in each of the clinical fields. In addition to individual questions, clinical problems are presented in the form of case histories, charts, roentgenograms, photographs of gross and microscopic pathologic specimens, laboratory data, and the like, and the candidate must answer questions concerning the interpretation of the data presented and their relation to the clinical problems. The questions are "designed to explore the extent of the candidate's knowledge of clinical situation, and to test his [or her] ability to bring information from many different clinical and basic science areas to bear upon these situations."

The examinations of both Step 1 and Step 2 are scored as a whole, certification being given on the basis of performance on the entire part, without reference to disciplinary breakdown. The grade for the examination is derived from the total number of questions answered correctly, rather than from an average of the grades in the component basic science or clinical science subjects. A candidate who fails will be required to repeat the entire examination. Nevertheless, as noted above, in spite of the interdisciplinary character of the examinations, all of the traditional disciplines are represented in the test, and separate grades for each subject can be extracted and reported separately to students, to state examining boards, or to those medical schools that request them for their own educational and academic purposes.

This type of interdisciplinary examination and the method of scoring the entire test as a unit have definite advantages, especially in view of the changing curricula in medical schools. The former type of rigid, almost standardized, curriculum, with its emphasis on specific subjects and a specified number of hours in each, has been replaced by a more liberal, open-ended curriculum, permitting emphasis in one or more fields and corresponding deemphasis in others. The result has been rather wide variations in the totality of education in different medical schools. Thus, the scoring of these tests as a whole permits accommodation to this variability in the curricula of different schools. Within the total score, weakness in one subject that has received relatively little emphasis in a given school may be balanced by strength in other subjects.

The rationale for this type of comprehensive examination as replacement for the traditional department-oriented examination in the basic sciences and the clinical sciences is given in the National Board Examiner:

The student, as he [or she] confronts these examinations, must abandon the idea of "thinking like a physiologist" in answering a question labeled "physiology" or "thinking like a surgeon" in answering a question labeled "surgery." The one question may have been written by a biochemist or a pharmacologist; the other question may have been written by an internist or a pediatrician. The pattern of these examinations will direct the student to thinking more broadly of the basic sciences in Step 1 and to thinking of patients and their problems in Step 2.

Until a few years ago, the Part I examination could not be taken until the work of the second year in medical school had been completed, and the Part II test was given only to students who had completed the major part of the fourth year. Now students, if they feel they are ready, may be admitted to any regularly scheduled Step 1 or Step 2 examination during any year of their medical course without prerequisite completion of specified courses or chronologic periods of study. Thus, emphasis is placed on the acquisition of knowledge and competence rather than the completion of predetermined periods.

Candidates are eligible for Step 3 after they have passed Steps 1 and 2, have received the M.D. degree from an approved medical school in the United States or Canada, and subsequent to the receipt of the M.D. degree, have served at least six months in an approved hospital internship or residency. Under certain circumstances, consideration may be given to other types of graduate training provided they meet with the approval of the National Board. After passing the Step 3 examination, candidates will receive their Diplomas as of the date of the satisfactory completion of an internship or residency program. If candidates have completed the approved hospital training prior to completion of Step 3, they will receive certification as of the date of the successful completion of Step 3.

The Step 3 examination, as noted above, is an objective test of general clinical competence. It occupies one full day and is divided into two sections, the first of which is a multiple-choice examination that relates to the interpretation of clinical data presented primarily in pictorial form, such as pictures of patients, gross and microscopic lesions, electrocardiograms, charts, and graphs. The second section, entitled Patient Management Problems, utilizes a programmed-testing technique designed to measure the candidate's clinical judgment in the management of patients. This technique simulates clinical situations in which the physician is faced with the problems of patient management presented in a sequential programmed pattern. A set of some four to six problems is related to each of a series of patients. In the scoring of this section, candidates are given credit for correct choices; they are penalized for errors of commission (selection of procedures that are unnecessary or are contraindicated) and for errors of omission (failure to select indicated procedures).

All parts of the National Board examinations are given in many centers, usually in medical schools, in nearly every large city in the United States as well as in a few cities in Canada, Puerto Rico, and the Canal Zone. In some cities, such as New York, Chicago, and Baltimore, the examination may be given in more than one center.

The examinations of the National Board have become recognized as the most comprehensive test of knowledge of the medical sciences and their clinical application produced in this country.

THE NATIONAL BOARD OF MEDICAL EXAMINERS

For years the National Board examinations have served as an index of the medical education of the period and have strongly influenced higher educational standards in each of the medical sciences. The Diploma of the National Board is accepted by 47 state licensing authorities, the District of Columbia, and the Commonwealth of Puerto Rico in lieu of the examination usually required for licensure and is recognized in the American Medical Directory by the letters DNB following the name of the physician holding National Board certification.

The National Board of Medical Examiners has been a leader in developing new and more reliable techniques of testing, not only for knowledge in all medical fields but also for clinical competence and fitness to practice. In recent years, too, a number of medical schools, several specialty certifying boards, professional medical societies organized to encourage their members to keep abreast of progress in medicine, and other professional qualifying agencies have called upon the National Board's professional staff for advice or for the actual preparation of tests to be employed in evaluating medical knowledge, effectiveness of teaching, and professional competence in certain medical fields. In all cases, advantage has been taken of the validity and effectiveness of the objective, multiple-choice type of examination, a technique the National Board has played an important role in bringing to its present state of perfection and discriminatory effectiveness.

Objective examinations permit a large number of questions to be asked, and approximately 150 to 180 questions can be answered in a 2½-hour period. Because the answer sheets are scored by machine, the grading can be accomplished rapidly, accurately, and impartially. It is completely unbiased and based on percentile ranking. Of long-range significance is the facility with which the total test and individual questions can be subjected to thorough and rapid statistical analyses, thus providing a sound basis for comparative studies of medical school teaching and for continuing improvement in the quality of the test itself.

QUESTIONS

Over the years, many different forms of objective questions have been devised to test not only medical knowledge but also those subtler qualities of discrimination, judgment, and reasoning. Certain types of questions may test an individual's recognition of the similarity or dissimilarity of diseases, drugs, and physiologic or pathologic processes. Other questions test judgment as to cause and effect or the lack of causal relationships. Case histories or patient problems are used to simulate the experience of a physician confronted with a diagnostic problem; a series of questions then tests the individual's understanding of related aspects of the case, such as signs and symptoms, associated laboratory findings, treatment, complications, and prognosis. Case-history questions are set up purposely to place emphasis on correct diagnosis within a context comparable with the experience of actual practice.

It is apparent from recent certification and board examinations that the examiners are devoting more attention in their construction of questions to more practical means of testing basic and clinical knowledge. This greater realism in testing relates to an increasingly interdisciplinary approach toward fundamental material and to the direct relevance accorded practical clinical problems. These more recent approaches to questions have been incorporated into this review series.

Of course, the new approaches to testing add to the difficulty experienced by the student or physician preparing for board or certification examinations. With this in mind, the author of this review is acutely aware not only of the interrelationships of fundamental information within the basic science disciplines and their clinical implications but also of the necessity to present this material clearly and concisely despite its complexity. For this reason, the questions are devised to test knowledge of specific material within the text and identify areas for more intensive study, if necessary. Also, those preparing for examinations must be aware of the interdisciplinary nature of fundamental clinical material, the common multifactorial characteristics of disease mechanisms, and the necessity to shift back and forth from one discipline to another in order to appreciate the less than clear-cut nature separating the pedagogic disciplines.

The different types of questions that may be used on examinations include the completion-type question, where the individual must select one best answer among a number of possible choices, most often five, although there may be three or four; the completion-type question in the negative form, where all but one of the choices is correct and words such as *except* or *least* appear in the question; the true-false type of question, which tests an understanding of cause and effect in relationship to medicine; the multiple true-false type, in which the question may have one, several, or all correct choices; one matching-type question, which tests association and relatedness and uses four choices, two of which use the word, *both* or *neither;* another matching-type question that uses anywhere from three to twenty-six choices and may have more than one correct answer; and, as noted above, the patient-oriented question, which is written around a case and may have several questions included as a group or set.

Many of these question types may be used in course or practice exams; however, at this time the most commonly used types of questions on the USMLE exams are the completion-type question (one best answer), the completion-type negative form, and the multiple matching-type question, designating specifically how many choices are correct. Often included within the questions are graphic elements such as diagrams, charts, graphs, electrocardiograms, roentgenograms, or photomicrographs to elicit knowledge of structure, function, the course of a clinical situation, or a statistical tabulation. Questions then may be asked in relation to designated elements of the same. As noted above, case histories or patient-oriented questions are more frequently used on these examinations, requiring the individual to use more analytic abilities and less memorization-type data.

For further detailed information concerning developments in the evolution of the examination process for medical licensure (for graduates of both U.S. and foreign medical schools), those interested should contact the National Board of Medical Examiners at 3750 Market Street, Philadelphia, PA 19104, USA; telephone number 215–590–9500.

FIVE POINTS TO REMEMBER

In order for the candidate to maximize chances for passing these examinations, a few common sense strategies or guidelines should be kept in mind.

First, it is imperative to prepare thoroughly for the examination. Know well the types of questions to be presented and the pedagogic areas of particular weakness, and devote more preparatory study time to these areas of weakness. Do not use too much time restudying areas in which there is a feeling of great confidence and do not leave unexplored those areas in which there is less confidence. Finally, be well rested before the test and, if possible, avoid traveling to the city of testing that morning or late the evening before.

Second, know well the format of the examination and the instructions before becoming immersed in the challenge at hand. This information can be obtained from many published texts and brochures or directly from the testing service (National Board of Medical Examiners, 3750 Market Street, Philadelphia, PA 19104; telephone 215–590–9500). In addition, many available texts and self-assessment types of examination are valuable for practice.

Third, know well the overall time allotted for the examination and its components and the scope of the test to be faced. These may be learned by a rapid review of the examination itself. Then, proceed with the test at a careful, deliberate, and steady pace without spending an inordinate amount of time on any single question. For example, certain questions such as the "one best answer" probably should be allotted 1 to 1½ minutes each. The "matching" type of question should be allotted a similar amount of time.

Fourth, if a question is particularly disturbing, note appropriately the question (put a mark on the question sheet) and return to this point later. Don't compromise yourself by so concentrating on a likely "loser" that several "winners" are eliminated because of inadequate time. One way to save this time on a particular "stickler" is to play your initial choice; your chances of a correct answer are always best with your first impression. If there is no initial choice, reread the question.

Fifth, allow adequate time to review answers, to return to the questions that were unanswered and "flagged" for later attention, and check every *n*th (e.g., 20th) question to make certain that the answers are appropriate and that you did not inadvertently skip a question in the booklet or answer on the sheet (this can happen easily under these stressful circumstances).

There is nothing magical about these five points. They are simple and just make common sense. If you have prepared well, have gotten a good night's sleep, have eaten a good breakfast, and follow the preceding five points, the chances are that you will not have to return for a second go-around.

Edward D. Frohlich, MD, MACP, FACC

Series Acknowledgments

In no other writing experience is one more dependent on others than in a textbook, especially a textbook that provides a broad review for the student and fellow practitioner. In this spirit, I am truly indebted to all who have contributed to our past and current understanding of the fundamental and clinical aspects related to the practice of medicine. No one individual ever provides the singular "breakthrough" so frequently attributed as such by the news media. Knowledge develops and grows as a result of continuing and exciting contributions of research from all disciplines, academic institutions, and nations. Clearly, outstanding investigators have been credited for major contributions, but those with true and understanding humility are quick to attribute the preceding input of knowledge by others to the growing body of knowledge. In this spirit, we acknowledge the long list of contributors to medicine over the generations. We also acknowledge that in no century has man so exceeded the sheer volume of these advances than in the twentieth century. Indeed, it has been said by many that the sum of new knowledge over the past 50 years has most likely exceeded all that had been contributed in the prior years.

With this spirit of more universal acknowledgment, I wish to recognize personally the interest, support, and suggestions made by my colleagues in my institution and elsewhere. I specifically refer to those people from my institution who were of particular help and are listed at the outset of the internal medicine volume. But, in addition to these colleagues, I want to express my deep appreciation to my institution and clinic for providing the opportunity and ambience to maintain and continue these academic pursuits. As I have often said, the primary mission of a school of medicine is that of education and research; the care of patients, a long secondary mission to ensure the conduct of the primary goal, has now also become a primary commitment in these more pragmatic times. In contrast, the primary mission of the major multidisciplinary clinics has been the care of patients, with education and research assuming secondary roles as these commitments become affordable. It is this distinction that sets the multispecialty clinic apart from other modes of medical practice.

Over and above a personal commitment and drive to assure publication of a textbook such as this is the tremendous support and loyalty of a hard-working and dedicated office staff. To this end, I am tremendously grateful and indebted to Mrs. Lillian Buffa and Mrs. Caramia Fairchild. Their long hours of unselfish work on my behalf and to satisfy their own interest in participating in this major educational effort is appreciated no end. I am personally deeply honored and thankful for their important roles in the publication of the Rypins' series.

Words of appreciation must be extended to the staff of the Lippincott–Raven Publishers. It is more than 25 years since I have become associated with this publishing house, one of the first to be established in our nation. Over these years, I have worked closely with Mr. Richard Winters, not only with the Rypins' editions but also with other textbooks. His has been a labor of commitment, interest, and full support—not only because of his responsibility to his institution, but also because of the excitement of publishing new knowledge. In recent years, we discussed at length the merits of adding the intensive review supplements to the parent textbook and together we worked out the details that have become the substance of our present "joint venture." Moreover, together we are willing to make the necessary changes to assure the intellectual success of this series. To this end, we are delighted to include a new member of our team effort, Ms. Susan Kelly. She joined our cause to ensure that the format of questions, the reference process of answers to those questions within the text itself, and the editorial process involved be natural and clear to our readers. I am grateful for each of these facets of the overall publication process.

Not the least is my everlasting love and appreciation to my family. I am particularly indebted to my parents who inculcated in me at a very early age the love of education, the respect for study and hard work, and the honor for those who share these values. In this regard, it would have been impossible for me to accomplish any of my academic pursuits without the love, inspiration, and continued support of my wife, Sherry. Not only has she maintained the personal encouragement to initiate and continue with these labors of love, but she has sustained and supported our family and home life so that these activities could be encouraged. Hopefully, these pursuits have not detracted from the development and love of our children, Margie, Bruce, and Lara. I assume that this has not occurred; we are so very proud that each is personally committed to education and research. How satisfying it is to realize that these ideals remain a familial characteristic.

Edward D. Frohlich, MD, MACP, FACC
New Orleans, Louisiana

Contents

CONTENTS

Chapter 1

Basic Organization of the Body, Homeostasis, and Cell Function

Physiology is the study of the function of living organisms and their parts. In human physiology, we are concerned with the characteristics of the human body that allow us to sense our environment, to move about, to think and communicate, to reproduce and to perform all of the functions that enable us to survive and thrive as living beings.

The very broad subject of human physiology includes the functions of molecules and subcellular components of the human body; functions of organs such as the heart; organ systems such as the cardiovascular system, as well as the interaction and communication between the various organ systems. A distinguishing feature of this scientific discipline is that it seeks to integrate the large number of individual physical and chemical events occurring at all levels of organization to understand the function of the whole organism.

Cells are the living units of the body. The basic living unit of the body is the cell, and each organism is an aggregate of many different cells held together by intercellular supporting structures. The entire body contains about 75 to 100 trillion cells, each of which is adapted to perform special functions. Although the many cells of the body differ from each other in their special functions, all of them have certain basic characteristics. For example, in all cells oxygen combines with breakdown products of fat, carbohydrate, or protein to release the energy required by the cells for normal function. Most cells have the ability to reproduce, and whenever cells are destroyed, the remaining cells often regenerate new cells until the appropriate number is restored. Finally, all cells are bathed in **extracellular fluid,** the constituents of which are very exactly controlled.

Homeostasis is the maintenance of a stable internal environment. A large part of our discussion of physiology will focus on the mechanisms that regulate the constituents of the extracellular fluid. This process is called **homeostasis,** which means simply the maintenance of constant conditions in the internal environment of the body. Essentially, all organs and tissues of the body perform functions that help to maintain these stable conditions.

HOMEOSTATIC FUNCTIONS OF THE BODY SYSTEMS

The body can be divided into several major functional systems, each of which performs a particular task in maintaining homeostasis as follows:

- The **cardiovascular system** transports fluid and solutes, including nutrients and waste products, through all parts of the body. It keeps the fluids of the internal environment continually mixed by pumping blood through the vascular system. As the blood passes through the capillaries, a large portion of its fluid diffuses back and forth into the interstitial fluid that lies between the cells, allowing continuous exchange of substances between the cells and interstitial fluid and between the interstitial fluid and the blood.

- The **respiratory system** provides oxygen for the body and removes carbon dioxide.

- The **gastrointestinal system** digests food and absorbs different nutrients, including carbohydrates, fatty acids, and amino acids, into the extracellular fluid.

- The **kidneys** regulate the extracellular fluid composition by controlling excretion of salts, water, and waste products of the chemical reactions of the cells. By controlling body fluid volumes and composition, the kidneys also regulate blood volume and blood pressure.

- The **nervous system** directs the activity of the muscular system, thereby providing locomotion. It also controls the function of many internal organs through the autonomic nervous system, and it allows us to sense our external and internal environments and to be intelligent beings so that we can attain the most advantageous conditions for survival.

- The **endocrine glands** secrete hormones that control many of the metabolic functions of the cells, such as growth, rate of metabolism, and special activities associated with reproduction.

- The **musculoskeletal system** consists of skeletal muscle, bones, tendons, joints, cartilage, and ligaments. This system provides protection of internal organs as well as support and movement of the body.

- The **integumentary system,** composed mainly of skin, provides protection against injury and defense against foreign invaders as well as dehydration of underlying tissue. In addition, the skin acts as an important means of maintaining a constant temperature in the body.

- The **immune system** also acts as one of the body's chief defense mechanisms, providing protection against foreign invaders, such as bacteria and viruses, that the body is exposed to daily.

- The **reproductive system** provides for formation of new beings like ourselves; even this can be considered a homeostatic function for it generates new bodies in which trillions of additional cells can exist in a well-regulated internal environment.

THE CELL AND ITS FUNCTION

Cells are highly organized structures. The cell is not merely a bag of fluid and chemicals; it also contains highly organized physical structures called organelles and is surrounded by a **cell membrane** (Fig. 1-1). Inside the cell, there are two major structures: the **nucleus** and the **cytoplasm.** Surrounding the nucleus is a **nuclear membrane,** which is highly permeable, and surrounding the cytoplasm is the cell membrane, which is also permeable but much less so than the nuclear membrane. Both the nucleus and cytoplasm are filled with highly viscous fluid containing water, proteins, carbohydrates, electrolytes, and many other substances. The nucleus holds 23 pairs of **chromosomes,** each of which contains several thousand sequences of **deoxyribose nucleic acid** (DNA) molecules, which are the **genes** that regulate reproduction and the characteristics of the protein enzymes of the cytoplasm.

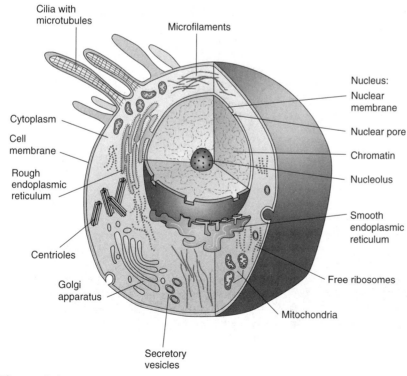

Figure 1-1.
Organization of a typical cell with its organelles in the cytoplasm and in the nucleus. (Adapted from Bullock BL: *Pathophysiology: Adaptations and Alterations in Function,* 4th ed. Philadelphia: Lippincott-Raven Publishers, 1996.)

The cytoplasm contains several organelles, including the **mitochondria,** which are often called the "powerhouses" of the cell. The mitochondria contain large quantities of oxidative and other enzymes that are responsible for supplying energy to the cells as will be discussed below. Other important structures of the cytoplasm include the **lysosome, Golgi apparatus, centrioles,** and the **endoplasmic reticulum,** which is a system of tubes and vesicles that connects with the nucleus through the nuclear membrane and spreads throughout the cytoplasm.

The cell membrane is a lipid bilayer with inserted proteins. Figure 1-2 shows a cell membrane that is composed of a lipid bilayer interspersed with large globular protein molecules. The lipid bilayer is almost entirely made up of phospholipids and cholesterol. Phospholipids have a water-soluble part (hydrophilic) and a part that is soluble only in fats (hydrophobic). The hydrophobic parts of the phospholipids face each other, whereas the hydrophilic parts face the two surfaces of the membrane in contact with the surrounding water.

The lipid bilayer membrane is highly permeable to lipid-soluble substances such as oxygen, carbon dioxide, and alcohol, but acts as a major barrier to water-soluble substances such as ions and glucose. A special feature of the lipid bilayer is that it is a fluid and not a solid.

Floating in the fluid lipid bilayer membrane are proteins, most of which are **glycoproteins** (proteins in combination with carbohydrates). There are two types of membrane proteins: the **integral proteins,** which protrude all the way through the membrane, and the **peripheral proteins,** which are attached to the inner surface of the membrane and do not penetrate. Many of the integral proteins provide **structural channels** (pores) through which water-soluble sub-

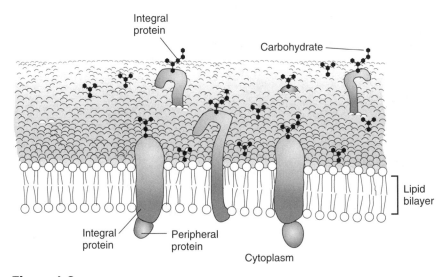

Figure 1-2.
The structure of the cell membrane showing its lipid bilayer with protein molecules protruding through the layer. Also, carbohydrate moieties are attached to the protein molecules on the outside of the membrane and additional protein molecules on the inside.

stances, especially ions, can diffuse. Other integral proteins act as **carrier proteins** for transporting substances, sometimes against their natural gradient for diffusion. The peripheral proteins are normally attached to one of the integral proteins and usually function as enzymes that catalyze chemical reactions in the cell.

The mitochondria release energy in the cell. An adequate supply of energy must always be available to fuel the chemical reactions of the cells. This is provided principally by the chemical reaction of oxygen with any one of the three different types of foods: glucose derived from carbohydrates, fatty acids derived from fats, and amino acids derived from proteins. After entering the cell, the foods are split into smaller molecules, which in turn enter the mitochondria where other enzymes remove carbon dioxide and hydrogen ions in the process called the **citric acid cycle.** Then an oxidative enzyme system, also in the mitochondria, causes progressive oxidation of the hydrogen atoms. The end-products of the reactions of the mitochondria are water and carbon dioxide, and the energy liberated is used by the mitochondria to synthesize still another substance, **adenosine triphosphate** (ATP), a highly reactive chemical that can diffuse through the cell to release its energy whenever it is needed for performing cellular functions.

The endoplasmic reticulum synthesizes multiple substances in the cell. The large network of tubules and vesicles, called the **endoplasmic reticulum** (ER), penetrates almost all parts of the cytoplasm. The membrane of the ER provides an extensive surface area for manufacturing multiple substances that are used inside the cells and released from some cells. These substances include proteins, carbohydrates, lipids, and structures such as lysosomes, peroxisomes, and secretory granules.

Lipids are made in the structure of the ER wall. For the synthesis of proteins, **ribosomes** attach to the outer surface of the ER. These function in association with **messenger RNA** (see below) to synthesize many protein molecules that then enter the ER where the molecules are further modified before release or use in the cell.

The Golgi apparatus functions in association with the ER. The Golgi apparatus has membranes similar to those of the **agranular,** or **smooth,** ER. The Golgi apparatus is prominent in secretory cells and is located on the side of the cell from which the secretory substances are extruded. As shown in Figure 1-3, small "transport vesicles," also called ER vesicles, continually pinch off from the ER and then fuse with the Golgi apparatus. In this way, substances entrapped in the ER vesicles are transported from the ER to the Golgi apparatus. The substances are then processed in the Golgi apparatus to form lysosomes, secretory vesicles, or other cytoplasmic components.

Lysosomes provide an intracellular digestive system. An organelle found in great numbers in cells is the lysosome. This is a small spherical vesicle surrounded by a membrane that contains digestive enzymes that allow lysosomes to digest intracellular substances and structures, especially damaged cell structures, food particles that have been ingested by the cell, and unwanted materials such as bac-

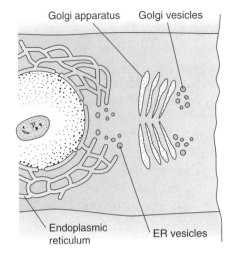

Figure 1-3.
A typical Golgi apparatus and its relationship to the endoplasmic reticulum (ER) in the nucleus.

Golgi apparatus   Golgi vesicles

Endoplasmic
reticulum   ER vesicles

teria. Ordinarily, the membranes surrounding the lysosome prevent the enclosed enzymes from coming in contact with other substances in the cell and therefore prevent their digestive action. However, when these membranes are damaged, enzymes are released, which then split the organic substances with which they come in contact into highly diffusible substances such as amino acids and glucose.

The nucleus acts as a control center of the cell. The nucleus contains large amounts of DNA, also called **genes,** which determines the characteristics of the cell's proteins, including the enzymes of the cytoplasm. Genes also control reproduction, first reproducing themselves through a process called **mitosis** in which two daughter cells are formed, each of which receives one of the two sets of genes.

Separating the nucleus from the cytoplasm is a **nuclear membrane,** also called the **nuclear envelope.** This is actually two separate membranes: The outer membrane is continuous with the ER, and the space between the two nuclear membranes is also continuous with the compartment inside the ER. Both layers of the membrane are penetrated by several thousand **nuclear pores,** almost 100 nanometers in diameter.

The nuclei of most cells contain one or more structures called **nucleoli,** which, unlike many of the organelles, do not have a surrounding membrane. The nucleoli contain large amounts of RNA and proteins of the type found in ribosomes. A nucleolus becomes enlarged when a cell is actively synthesizing proteins. Ribosome RNA is stored in the nucleolus and transported through the nuclear membrane pores to the cytoplasm, where it is used to form mature ribosomes that play an important role in the formation of proteins.

Cell genes control protein synthesis. Proteins play a key role in almost all functions of the cell for two reasons: (1) all of the enzymes that catalyze the chemical reactions of the cells are proteins, and (2) most of the important physical structures of the cell contain structural proteins.

The genes control **protein synthesis** in the cell and in this way control cell function. Each gene is a double-stranded helical molecule of DNA composed of multiple units of the sugar **deoxyribose, phosphoric acid,** and four **nitrogenous bases,** including two purines, **adenine** and **guanine,** and two pyrimidines, **thymine** and **cytosine** (Fig. 1-4).

The first stage in the formation of DNA is the combination of one molecule of phosphoric acid, one molecule of deoxyribose, and one of the four bases to form a **nucleotide.** Four separate nucleotides can be formed, corresponding to four different nitrogenous bases. Multiple nucleotides combine to form DNA in such a way that phosphoric acid and deoxyribose alternate with each other in the two separate strands, and these strands are held together by loose bonds between the purine and pyrimidine bases. The purine-base adenine always bonds with the pyrimidine-base thymine, and the purine-base guanine always bonds with the pyrimidine-base cytosine. The sequence of bases is different for each type of gene, and it is the specific sequence of bases in one of the two strands of DNA molecules that controls the type of protein synthesized.

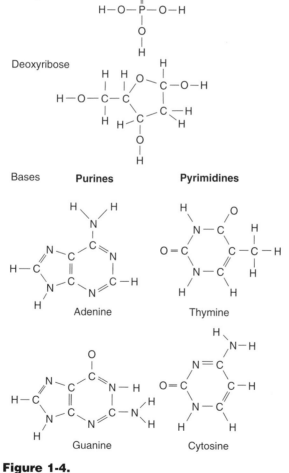

Figure 1-4.
The basic building blocks of deoxyribonucleic acid (DNA)—phosphoric acid, deoxyribose, and purine and pyrimidine bases.

The genetic code consists of triplets of bases. Three successive bases in the DNA strand are each called a **code word.** These code words control the sequence of amino acids in the protein to be formed in the cytoplasm. One code word might be composed of a sequence of adenine, thymine, and guanine, while the next code word might have a sequence of cytosine, guanine, and thymine. These two code words have entirely different meanings because their bases are different. The sequence of successive code words on the DNA strand is known as the **genetic code.**

The process of transcription transfers the DNA code to RNA. Because DNA is located in the nucleus of the cell and many of the functions of the cell are carried out in the cytoplasm, there must be some means for the genes of the nucleus to control the chemical reactions of the cytoplasm. This is achieved through another type of nucleic acid, **ribose nucleic acid** (RNA), the formation of which is controlled by the DNA of the nucleus. In this process called transcription, the code of DNA is transferred to RNA. The RNA diffuses from the nucleus through the nuclear pores into the cytoplasm, where it controls protein synthesis.

Each **RNA molecule** is composed of the sugar ribose, phosphoric acid, and one of the four different pyrimidine and purine bases (the same bases as those found in DNA except that thymine is replaced by uracil). Thus, the RNA strand is similar to the DNA strand. When the DNA strand of the gene causes formation of the RNA strand, it transfers its code to the RNA strand by controlling the sequence of bases in the RNA strand. This control is called transcription because the code of the DNA is "transcribed" onto the RNA.

During the transcription process, the two strands of DNA that make up the chromosome pull apart from each other. One of these strands then serves as the gene and attracts to it the necessary chemicals that form the RNA strand. The four separate bases that are part of the building blocks of the DNA strand are mutually attracted to four complementary bases that subsequently become part of the RNA strand (guanine attracts cytosine; cytosine attracts guanine; adenine attracts uracil; thymine attracts adenine). Thus, the code formed in the RNA strand is complementary to the code of the gene DNA strand. Once the RNA is formed in the nucleus, it diffuses outward into the cytoplasm, where it functions in the synthesis of a specific cell protein.

Three different types of RNA strands are formed: (1) **messenger RNA** (mRNA), which carries the genetic code to the cytoplasm for controlling the formation of the proteins; (2) **ribosomal RNA,** which along with other proteins forms the **ribosomes,** the structures in which protein molecules are actually assembled; and (3) **transfer RNA** (tRNA), which transports activated amino acids to the ribosomes to be used in assembling the proteins.

There are twenty separate types of **transfer RNA,** each of which combines specifically with one of the twenty different amino acids and conducts this amino acid to the ribosome, where it is combined into the protein molecule.

Translation—polypeptide synthesis on the ribosomes from the genetic code contained in the mRNA. There are many thousands of different mRNAs, each of which carries the genetic code that determines the sequence in which successive amino acids will be arranged in one specific type of protein molecule. Messenger RNA is a single-stranded long molecule having a succession of **codons** along its axis (Fig. 1-5). These codons are mirror images of the code words in the gene DNA, and they also consist of three successive bases.

To manufacture proteins, one end of the RNA strand enters the ribosome and the entire strand then threads its way to the ribosome, taking just over 1 minute. As it passes through the ribosome, the ribosome "reads" the genetic code and causes the proper succession of amino acids to bind together by chemical bonds called **peptide linkages.** Actually, the mRNA does not recognize the different types of amino acids but instead recognizes the different tRNA molecules. However, each type of tRNA molecule carries only one specific type of amino acid that will be incorporated into the protein.

Thus, as the strand of mRNA passes through the ribosome, each of its codons draws to it a specific transfer RNA that in turn delivers a specific amino acid. This amino acid then combines with the preceding amino acids by forming a peptide linkage, and the sequence continues to build until an entire protein molecule is formed. At this point, a special codon appears, which indicates completion of the process, and the protein is released into the cytoplasm or through the membrane of the ER into the interior of this reticulum. This control of amino acid sequence by the RNA code during protein formation is called **translation.**

It is estimated that there are about 100,000 different types of genes in the nucleus. Therefore, one can understand that a large

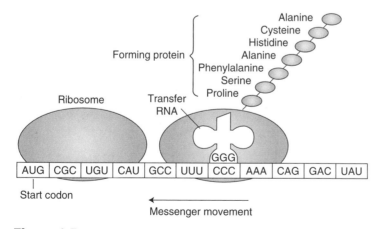

Figure 1-5.
A messenger RNA strand that is moving through two ribosomes. As each codon passes through, an amino acid is added to the growing protein chain, which is shown in the right-hand ribosome. The transfer RNA molecule determines which of the 20 amino acids will be added at each stage of protein formation.

number of different types of proteins can also be formed in each cell. The character of each cell depends on the relative proportion of different types of proteins that are formed. Thus, the genes control the structure of the cell through the types of structural proteins formed, and the genes control the function of the cell mainly through the types of protein enzymes that are formed.

Cellular reproduction occurs by the process of mitosis. Most cells of the body, with the exception of mature red blood cells, striated muscle cells, and neurons in the nervous system, are capable of reproducing other cells of their own type. Ordinarily, if sufficient nutrients are available, each cell grows larger and larger until it automatically divides by the process of **mitosis** to form two new cells. Before mitosis occurs, all of the genes in the nucleus, as well as the chromosomes carrying the genes, are themselves reproduced to create a complete new set of genes. During mitosis, one set of genes enters one of the daughter cells while the other set enters the second daughter cell. Thus, not only are the physical characteristics of the two new cells alike, but they are still controlled by the same types of genes so that their functions will be very similar. If, during the process of reproduction, one or more of the genes fails to be reproduced or becomes suppressed, then the two new cells will not be exactly alike, and the cells will become slightly differentiated from each other.

Cell differentiation allows different cells of the body to perform different functions. As a human being develops from a fertilized ovum, the ovum divides repeatedly until trillions of cells are formed. Gradually, however, the new cells differentiate from each other with certain cells having different genetic characteristics from other cells. This differentiation process occurs as a result of inactivation of certain genes and activation of others during successive stages of cell division. This process of differentiation leads to the ability of different cells in the body to perform different functions.

Chapter 2

Body Fluids, Kidneys, and Acid-Base Regulation

BODY FLUID COMPARTMENTS

Total body water is comprised of extracellular and intracellular fluid. In a normal 70-kg adult person, the total body water averages 60% of the body weight, or about 42 liters. The percentage of total body water is greater in newborns and lean persons, and is lower in adult females, elderly persons, or adults with a large amount of adipose tissue.

The total body water can be divided into two major compartments: the **extracellular fluid,** which is about 20% of total body weight, and **intracellular fluid,** which is 40% of total body weight (Fig. 2-1).

The extracellular fluid can be subdivided into two main subcompartments: the plasma, which makes up almost one-fourth of the extracellular fluid, and the interstitial fluid, which lies between the tissue cells and amounts to more than three-fourths of the extracellular fluid. Because the plasma and interstitial fluids are separated only by highly permeable capillary membranes, their ionic compositions are similar and they are often considered together as one large compartment of homogeneous fluid. The most important difference between plasma and interstitial fluid is the higher concentration of protein in the plasma, which exists because the capillaries have a low permeability to the plasma proteins.

Comparison of extracellular and intracellular fluid solutes. Figure 2-2 shows the concentrations of important substances in extracellular and intracellular fluids. Both of these fluids contain nutrients that are needed by the cells, including glucose, amino acids, oxygen, and other nutrients not shown in the figure. The intracellular fluid is separated from the extracellular fluid by a cell membrane that is highly permeable to water but not to many of the electrolytes in the body; thus, there are important differences between extracellular and intracellular fluids in electrolyte concentrations. Extracellular fluid contains large quantities of sodium and chloride ions, but only small amounts of potassium, magnesium, and phosphate ions.

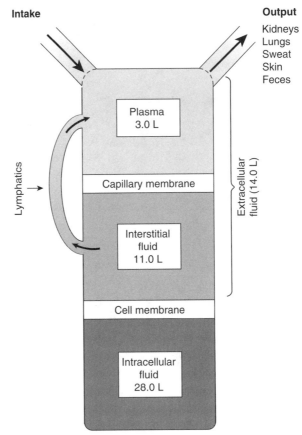

Figure 2-1.
Summary of the body fluids, including the major body fluid compartments and the membranes that separate these compartments. The values shown are for an average 70-kg man.

In contrast, intracellular fluid contains large amounts of potassium and phosphate ions, moderate amounts of magnesium ions, and exceedingly few calcium ions. These differences in the ionic composition of the fluids cause a membrane potential to develop across the two sides of the cell membrane—negative on the inside and positive outside. Later, we shall see how this potential develops and the manner in which it changes during the transmission of nerve and muscle impulses. Because the cell membrane is highly permeable to water, the osmolarity (the concentration of osmotically active solute particles) of the intracellular and extracellular compartments is normally the same, about 285 milliosmoles per liter (285 mOsm/L).

Measurement of Volumes in Different Body Fluid Compartments Using the Indicator-Dilution Principle

The volume of a fluid compartment in the body can be measured by placing a substance in the compartment, allowing it to disperse

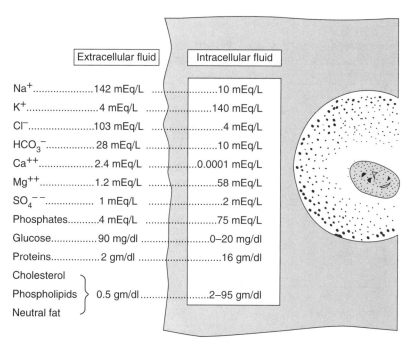

Figure 2-2.
Approximate chemical compositions of extracellular and intracellular fluids.

throughout the compartment's fluid, and then analyzing the extent to which it has become diluted in the compartment. Thus, the unknown volume of a compartment can be calculated by knowing the total mass of substance injected into the compartment (Q) divided by the concentration of the substance after dilution in the compartment:

$$Volume = Q / concentration$$

For this method to be used properly, the substance must be uniformly distributed in the compartment, and only in the compartment that is being measured. Table 2-1 lists some of the different indicators that can be used to measure volumes of the body fluid compartments.

Total body water is measured using substances that disperse throughout the body fluids, such as radioactive water (3H2O) or heavy water (deuterium, 2H2O). **Extracellular fluid** is measured using several substances that disperse in the plasma and interstitial fluid but do not permeate the cell membrane, such as radioactive sodium, inulin, and thiosulfate. **Intracellular volume** cannot be measured directly, but can be calculated as the difference between total body water and extracellular volume. **Plasma volume** is measured by injecting substances, such as radioactive albumin, that do not penetrate capillary membranes and therefore remain in the vascular system. **Interstitial fluid volume** cannot be measured directly but can be calculated as the difference between extracellular fluid volume and plasma volume.

TABLE 2-1.
Measurement of Body Fluid Volumes

Volume	Indicators
Total body water	^3H_2O, ^2H_2O, antipyrine
Extracellular fluid	^{22}Na, ^{125}I-iothalamate, thiosulfate, inulin
Intracellular fluid	(Calculated as total body water – extracellular fluid volume)
Plasma volume	^{125}I-albumin, Evans blue dye (T-1824)
Blood volume	^{51}Cr-labeled red blood cells, or calculated as Blood volume = Plasma volume/1 – hematocrit
Interstitial fluid	(Calculated as extracellular fluid volume – plasma volume)

EXCHANGE OF FLUID AND ELECTROLYTES ACROSS THE CELL MEMBRANE

Diffusion is the net movement of molecules through the cell membrane along chemical or electrical gradients. Substances pass through the cell membrane, back and forth between the extracellular and intracellular fluids, by the process of **diffusion,** which occurs because molecules migrate from a region of high concentration to one of lower concentration as a result of random motion. The rate of diffusion across the cell membrane is directly related to (1) the electrical potential and chemical concentration differences across the membrane, (2) the surface area of the membrane, and (3) the permeability of the membrane for the solute. The permeability of the membrane for a solute is inversely related to the size of the solute and the membrane thickness.

Substances that are highly soluble in lipids, such as carbon dioxide and oxygen, can diffuse directly through the lipid matrix of the cell membrane. The various ions, large molecules such as glucose, and water enter and leave the cells through the membrane channels formed by glycoprotein molecules that extend all the way through the membrane. These channel proteins allow rapid passage of water and other small molecules. However, they are selectively permeable to certain substances such as the ions. Depending on various conditions of the cells, some of these channels are open and allow rapid diffusion of specific substances, whereas at other times the channels are closed and diffusion is greatly decreased. Thus, opening and closing these channels is a means by which movement of many substances through the cell membrane can be controlled.

Facilitated diffusion—carrier-mediated diffusion. Many substances are transported through the cell membrane by combining chemically with a carrier, which is a protein that penetrates through the cell membrane. The carrier combines chemically with the substance and "facilitates" its diffusion through the cell membrane to the opposite side where the substance is released.

Facilitated diffusion differs from simple diffusion in the following important way: Although the rate of diffusion increases proportionally with the concentration of the diffusing substance, there is a **transport maximum,** called V_{max}, with facilitated diffusion. Among the important substances that cross cell membranes by facilitated diffusion are **glucose** and **amino acids.**

Another important characteristic of facilitated diffusion is that it occurs **down an electrochemical gradient** and is, therefore, a passive type of transport. However, because of the protein carriers, it is more rapid than simple diffusion.

Primary active transport can move molecules through the cell membrane against an electrochemical gradient. Active transport is similar to facilitated diffusion in that the substance to be transported combines with a carrier protein and is released from the carrier on the opposite side of the membrane. However, active transport differs from facilitated diffusion in that it can transport the substance in the absence of an electrochemical gradient, or even against an electrochemical gradient. To achieve "uphill" transport against a concentration or electrical gradient, energy is required. This energy is supplied by the high energy compound **adenosine triphosphate** (ATP) or by some other high-energy phosphate compound inside the cell.

One of the most important active transport systems in the body is that for transport of sodium out of the cells and potassium into the cells—the **sodium-potassium (Na⁺-K⁺) pump** (Fig. 2-3). The protein structure of the pump serves both as the carrier and as the energy-transferring mechanism for moving the sodium out of the cell and potassium into the cell. The pump causes the concentration of sodium inside the cell to become very low and at the same time causes the intracellular potassium concentration to build up to a very high level. This sodium-potassium pump maintains much of the ionic concentration difference between intracellular and extracellular fluids and is, therefore, extremely important in con-

Figure 2-3.
Postulated mechanism of the sodium-potassium pump.

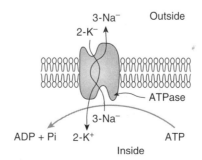

trolling the volume of cells as well as the function of excitable nerve and muscle fibers, as will be described later in the chapter.

Other important primary active transport systems include the following: (1) The **calcium pump** (calcium-ATPase, or Ca^{++}-ATPase), which transports calcium to the outside of the cell membrane as well as into the internal organelles of the cell. This calcium pump maintains a very low intracellular calcium concentration. (2) **Hydrogen pumps** (H$^+$), which are especially important for secreting hydrochloric acid in the stomach and for secreting hydrogen ions into the renal tubules.

Secondary active transport—cotransport and countertransport. When sodium ions are transported out of the cells by the sodium-potassium pump, a large concentration gradient of sodium develops, with very high concentration outside the cell and low concentration inside. This gradient represents a storehouse of energy and, under appropriate conditions, this diffusion energy of sodium can pull other substances along with sodium into the cell. This phenomenon is called **cotransport** and is one form of active transport. Examples of substances cotransported with sodium include glucose and amino acids, especially in the epithelial cells of the intestine and renal tubules.

Another form of secondary active transport is **countertransport,** in which the substance is transported in the opposite direction as sodium, again by means of a carrier protein that binds both sodium and the substance to be transported. Two very important countertransport mechanisms include **sodium-calcium countertransport** and **sodium-hydrogen countertransport.**

OSMOTIC PRESSURE AND OSMOTIC EQUILIBRIA AT THE CELL MEMBRANE

Osmosis is the diffusion of water across a membrane. Cell membranes in the body are highly permeable to water and whenever there is a higher concentration of solute on one side of the membrane, water rapidly diffuses across the membrane toward the region of higher solute concentration. The reason for this is as follows: The greater the concentration of dissolved substances in a water solution, the lower the concentration of water molecules in the fluid. Therefore, if a solution with a low solute is placed on one side of the cell membrane, water will diffuse from the side of the membrane with the low solute concentration (high water concentration) to the side that has the high solute concentration (low water concentration) until the water concentration on both sides becomes equal. This is **osmosis.**

An osmole is equal to a mole of osmotically active particles. The total number of solute particles in a solution is measured in terms of osmoles. One osmole (Osm) is equal to one mole (mol; 6.02×10^{23}) of solute particles. Therefore, a solution containing one mole of glucose in each liter has a concentration of 1 Osm/L. If a molecule dissociates into two ions (giving two particles), such as sodium chloride ionizing to give chloride and sodium ions, then the solution containing 1.0 mol/L will have an osmotic concentration of 2.0 Osm/L. For body fluids, the osmotic activity of solutes is usually expressed in terms of milliosmoles (mOsm; 1/1000 osmoles).

Osmotic pressure is the hydrostatic pressure required to prevent osmosis. If sufficient pressure is applied across the membrane through which osmosis is occurring, but the pressure is applied in the direction opposite the osmotic flow, the osmosis can be stopped. The amount of pressure that is required to stop osmosis is called the **osmotic pressure.** Osmotic pressure, therefore, is an indirect measurement of the water and solute concentrations of a solution. The higher the osmotic pressure of a solution, the higher the solute concentration of the solution.

The osmotic pressure of a solution is directly proportional to the concentration of osmotically active particles, regardless of whether the solute is a large molecule or a small molecule. Expressed mathematically, osmotic pressure (π) can be calculated as

$$\pi = CRT$$

where C = the concentration of solute in Osm/L, R = the ideal gas constant, T = the absolute temperature.

For biological solutions, π calculates to be about 19,300 mmHg for a solution having a concentration of 1.0 Osm/L. This means that for a concentration of 1.0 mOsm/L, π is equal to 19.3 mmHg. Thus, for each mOsm concentration gradient across the cell membrane, 19.3 mmHg osmotic force are exerted. This means that only small differences in solute concentration across the cell membrane can cause rapid osmosis of water.

Sample Calculation. What is the osmolarity of a 0.9% solution of sodium chloride?

A 0.9% solution means that there is 0.9 gm of sodium chloride per 100 ml of solution, or 9 gm/L. Because the molecular weight of sodium chloride is 58.5 gm/mol, the molarity of the solution is 9 gm/L divided by 58.5 gm/mol, or about 0.154 mol/L. Because each molecule of sodium chloride is equal to 2 Osm, the osmolarity of this solution is 2×0.154, or 0.308 Osm/L. Therefore, the osmolarity of this is 308 mOsm/L. The potential osmotic pressure of this solution would be 308 mOsm/L $\times$ 19.3 mmHg/mOsm/L, or 5944 mmHg. (This calculation is only an approximation since sodium and chloride ions do not behave entirely as independent ions in solution because of interionic attractions between these ions.)

Isotonic, hypotonic, and hypertonic fluids. A solution is said to be **isotonic** if no osmotic force develops across the cell membrane when a normal cell is placed in the solution. This means that an isotonic solution has the same osmolarity as the cell and that the cells will not shrink or swell if placed in the solution. Examples of isotonic solutions are a 0.9% sodium chloride solution and a 5% glucose solution.

A solution is said to be **hypertonic** when it contains a higher osmotic concentration of substances than does the cell. In this case, osmotic force develops that causes water to flow out of the cell into the solution, thereby greatly concentrating intracellular fluid and shrinking the cell.

The solution is said to be **hypotonic** if the osmotic concentration of substances in the solution is less than their concentration in the cell. An osmotic force develops immediately when the cell is exposed to the solution, causing water to flow by osmosis into the cell until the intracellular fluid has about the same concentration as the extracellular fluid, or until the cell bursts from excessive swelling.

Volumes and Osmolarities of Extracellular and Intracellular Fluid in Pathophysiologic Conditions

Some of the most common and important clinical problems include abnormalities of the composition and volume of the body fluids. Understanding and treating these disorders requires knowledge of fluid shifts between intracellular and extracellular compartments before and after therapy.

You can approximate the changes in intracellular and extracellular fluid volumes and the types of therapy that must be instituted if the following basic principles are kept in mind:

- **Water moves rapidly across cell membranes.** Therefore, the osmolarities of intracellular and extracellular fluids remain almost exactly equal to each other except for a few minutes after a change in one of the compartments.
- **Cell membranes are almost completely impermeable to most solutes,** especially the electrolytes. Therefore, the number of osmoles in the extracellular and intracellular fluid remains relatively constant unless solutes are added to or lost from the extracellular compartment.

Effect of adding saline solutions to the extracellular fluid. If an *isotonic* saline solution is added to the extracellular fluid compartment, the osmolarity does not change and, therefore, the only effect is an increase in extracellular fluid volume (Fig. 2-4A). The sodium chloride remains largely in the extracellular fluid because the cell membrane behaves as though it were virtually impermeable to sodium chloride.

If a *hypertonic* solution is added to the extracellular fluid, the extracellular osmolarity increases, causing osmosis of water out of the cells into the extracellular compartment (Fig. 2-4B). Almost all of the added sodium chloride remains in the extracellular compartment, and fluid diffuses from the cells into the extracellular space to

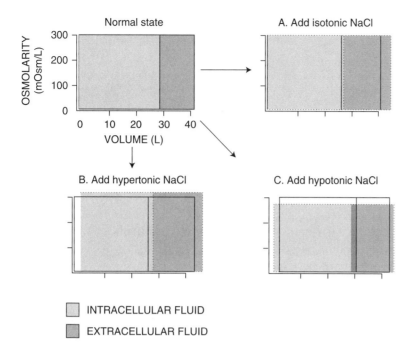

INTRACELLULAR FLUID
EXTRACELLULAR FLUID

Figure 2-4.
Effect of adding isotonic, hypertonic, and hypotonic solutions to the extracellular fluid. The normal state is indicated by the solid lines, and the shifts from normal after osmotic equilibrium are shown by the dashed lines and shaded compartments. The volumes of intracellular and extracellular fluid compartments are shown on the abscissa of each diagram, and the osmolarities of these compartments are shown on the ordinates.

achieve osmotic equilibrium. The net effect is an increase in extracellular fluid volume (greater than the volume of fluid added), a decrease in intracellular volume, and an increase in osmolarity of both compartments.

If a *hypotonic* solution is added to the extracellular fluid, the osmolarity of the extracellular fluid decreases and some of the extracellular water diffuses into the cells until intracellular and extracellular compartments have the same osmolarity (Fig. 2-4*C*). Both the intracellular and extracellular volumes are increased, although the intracellular volume increases to a greater extent.

Administration of glucose and other solutions for nutritive purposes. When solutions of glucose, amino acids, or homogenized fat are administered, their concentrations are adjusted nearly to isotonicity, or they are given slowly enough so that they do not upset the osmotic equilibria of the body fluids. After these nutrients are metabolized, an excess of water often remains, but ordinarily the kidneys excrete this in the form of very dilute urine. The net result is, therefore, addition of only nutrients to the body.

Clinical abnormalities of fluid volume regulation: hyponatremia and hypernatremia. One of the primary measures readily available to the clinician in evaluating a patient's fluid status is plasma sodium concentration. Because sodium and its associated anions (mainly chloride) account for more than 90% of the solute in the extracellular fluid, plasma sodium concentration is a reasonable indicator of plasma osmolarity under many conditions. When plasma sodium concentration is reduced below normal (about 142 mEq/L), a person is said to have hyponatremia. When plasma sodium is elevated above normal, a person is said to have hypernatremia.

Hyponatremia results from either loss of sodium chloride from the extracellular fluid or addition of excess water in the extracellular fluid. A primary loss of sodium chloride usually results in **hypoosmotic dehydration** and is associated with decreased extracellular fluid volume. Some examples of this include sodium chloride loss due to diarrhea, vomiting, overuse of diuretics that inhibit the kidney's ability to conserve sodium, and Addison's disease resulting from decreased secretion of the hormone aldosterone, which impairs the ability of the kidney to reabsorb sodium.

Hyponatremia caused by excess water retention results in **hypoosmotic overhydration.** For example, excess secretion of antidiuretic hormone (syndrome of inappropriate ADH) causes the kidney tubules to reabsorb more water, leading to hyponatremia and overhydration.

Hypernatremia is caused by water loss or excess sodium in the extracellular fluid. Primary loss of water results in **hyperosmotic dehydration.** This condition can occur from decreased secretion of ADH, which is needed for the kidneys to conserve water, or dehydration associated with insufficient water intake or excess water loss from the body, as can occur with sweating during heavy exercise.

Hypernatremia can also occur with an excess of sodium chloride, resulting in **hyperosmotic overhydration.** An example of this is excessive secretion of the sodium-retaining hormone aldosterone, which causes a mild degree of hypernatremia and overhydration.

Thus, when analyzing abnormalities of plasma sodium concentration in deciding a proper therapy, you must first determine whether the abnormality is caused by a primary loss or gain of sodium and a primary loss or gain of water.

EXCHANGE OF FLUIDS AND SOLUTES THROUGH CAPILLARY MEMBRANES

Capillaries are highly permeable to most solutes. Most capillaries are very porous with several million slits, or "pores," between the cells that make up their walls (the widths of the pores are about 8 nm) to each square centimeter of capillary surface. Because of the high permeability of the capillaries for most solutes, as blood flows through the capillaries very large amounts of dissolved substances diffuse in both directions through these pores. In this way, sodium, chloride, potassium, glucose, and almost all other dissolved substances in the plasma, except the plasma proteins, continually mix with the interstitial fluid. In capillaries, the most important means of moving solutes from the blood to the interstitium is by **simple diffusion** across the capillary wall. The rate of diffusion for most solutes is so great that cells as far as 50 microns away from the capillaries can still receive adequate quantities of nutrients.

Bulk flow (ultrafiltration) of fluid across the capillary wall is due to hydrostatic and colloid osmotic forces. Although exchange of nutrients, oxygen, and metabolic end-products across the capillaries occurs almost entirely by diffusion, the distribution of fluid across the capillaries is determined by another process: the bulk flow, or **ultrafiltration,** of protein-free plasma. As discussed above, capillary walls are highly permeable to water and to most plasma solutes, except the plasma proteins. Therefore, hydrostatic pressure differences across the capillary wall push protein-free plasma (ultrafiltrate) through the capillary wall into the interstitium. The rate at which this ultrafiltration occurs depends on the difference in hydrostatic and colloid osmotic pressures of the capillary and interstitial fluid (Fig. 2-5).

The **capillary hydrostatic pressure,** which normally averages about 17 mmHg, tends to push fluid out of the pores of the capillaries into the interstitial fluid. However, the proteins in the plasma are relatively impermeable to the capillary wall and therefore exert an osmotic force that tends to draw fluid back into the capillaries. This osmotic force is often called the **oncotic pressure,** or **colloid osmotic pressure,** and averages about 28 mmHg in most capillaries. The **interstitial fluid hydrostatic pressure** tends to oppose fluid filtration, but in many tissues (such as subcutaneous tissue) the normal interstitial fluid hydrostatic pressure averages –3 mmHg, a negative rather than a positive value. Proteins in the interstitial fluid also exert an **interstitial fluid colloid osmotic pressure** of approximately 8 mmHg, which tends to draw fluid out of the capillaries.

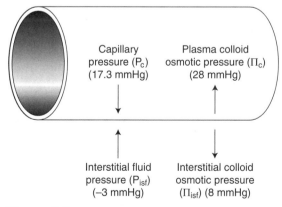

Figure 2-5.
Hydrostatic and colloid osmotic forces acting across the capillary membrane that tend to move fluid either outward or inward through the membrane pores.

Thus, there are opposing forces acting to move fluid across the capillary wall: (1) the difference between capillary hydrostatic pressure and interstitial fluid hydrostatic pressure, which tends to cause filtration out of the capillary; and (2) the difference in colloid osmotic pressure between the plasma and interstitial fluid, which favors movement of fluid into the capillary. Under normal conditions, these forces favor a slight net movement of fluid out of the capillaries, which is what causes formation of tissue fluid and lymph.

The normal values for these different pressures are as follows:

Capillary pressure = 17.3 mmHg
Interstitial fluid hydrostatic pressure = −3 mmHg
Plasma colloid osmotic pressure = 28 mmHg
Tissue fluid colloid osmotic pressure = 8 mmHg

Thus, the sum of forces that tend to move fluid into or out of the capillary wall is about +0.3 mmHg (see Fig. 2-5). These forces are often called **Starling forces.**

> **Sample Calculation of Starling Forces.** If the capillary hydrostatic pressure is 25 mmHg, plasma colloid osmotic pressure is 28 mmHg, interstitial fluid hydrostatic pressure is 0 mmHg, and interstitial fluid colloid osmotic pressure is 5 mmHg, what is the net force across the capillaries? Will filtration or absorption occur?
>
> $$\text{Net pressure} = 25 - 28 - 0 + 5 \text{ mmHg}$$
> $$\text{Net pressure} = +2 \text{ mmHg}$$

Thus, filtration will occur in this circumstance.

Edema Is the Collection of Excess Fluid in the Tissues

Intracellular edema can occur when the cell membrane is damaged or when there is inadequate nutrition to the cells. When this happens, sodium ions are no longer efficiently pumped out of the cells, and the excess sodium ions inside the cells cause osmosis of water into the cells.

Extracellular edema is more common than intracellular edema and occurs with accumulation of fluid in the interstitial spaces. There are two general causes of extracellular edema: (1) abnormal leakage of fluid from the plasma to the interstitial spaces across the capillaries, and (2) failure of the lymphatics to return fluid from the interstitium back to the blood.

Factors that can increase capillary filtration and cause interstitial fluid edema. To understand the cause of excess capillary filtration, it is useful to state the mathematical determinants of capillary filtration:

$$\text{Filtration} = K_f \times (P_C - P_{isf} - \pi_C + p_{isf})$$

where K_f is the capillary filtration coefficient (the product of the permeability and surface area of the capillaries), P_{isf} is the interstitial fluid hydrostatic pressure, π_C is the capillary plasma colloid osmotic pressure, and p_{isf} is the interstitial fluid colloid osmotic pressure. From this equation, you can see that any of the following changes could increase the capillary filtration rate:

- **Increased capillary filtration coefficient,** which allows leakage of fluids and plasma proteins through the capillary membranes; this can occur as a result of allergic reactions, bacterial infections, and toxic substances that injure the capillary membranes.
- **Increased capillary hydrostatic pressure,** which can result from obstruction of a vein, excess flow of blood from the arteries into the capillaries, or failure of the heart to pump blood rapidly out of the veins (heart failure).
- **Decreased plasma colloid osmotic pressure,** which can occur as a result of failure of the liver to produce sufficient quantities of plasma proteins, loss of large amounts of proteins into the urine in certain kidney diseases, or loss of large quantities of proteins through burned areas of the skin or other denuding lesions.
- **Increased interstitial fluid colloid osmotic pressure,** which will draw fluid out of the plasma into the tissue spaces. This results most frequently from lymphatic blockage, which prevents the return of proteins from the interstitial spaces to the blood as will be discussed in the following sections.

THE LYMPHATIC SYSTEM

The Lymphatic System Returns Protein and Fluid from the Interstitium to the Blood

The lymphatics are a system of accessory vessels that accompany blood vessels to almost all parts of the body. Minute **lymphatic capillaries,** like the blood capillaries, are found in almost all tissues. These lead into progressively larger lymphatic channels that finally converge mainly in the **thoracic duct,** which passes upward through the chest and empties into the venous system at the juncture of the left internal jugular and subclavian veins. The lymphatic capillaries are highly permeable so that bacteria, various types of debris in the interstitial fluids, and large proteins can enter the lymph from the interstitial fluid with great ease.

Because of their large molecular size, most proteins that move out of the blood capillaries into the interstitial fluids cannot pass back into the blood capillaries. After these collect in the interstitial fluid and become progressively more and more concentrated, they finally flow into the lymphatic system and are returned to the bloodstream in this way. Approximately one-half of all the protein in the blood leaks into the interstitial spaces each day, and were it not for its return by the lymph to the blood, a person would lose most of the plasma colloid osmotic pressure within a few hours and would, therefore, no longer be able to maintain normal blood volume. Thus, return of protein from the interstitial spaces to the blood is probably the single most important function of the lymphatic system.

Lymph Is Formed from Interstitial Fluid

Lymph is formed from the fluid that flows out of the interstitial spaces into the lymphatic capillaries. Therefore, its constituents are almost identical to those of the interstitial fluid. Most lymph from the peripheral tissues has a protein concentration between 2 and 3 gm/100 ml, while the liver contains a concentration of about 6 gm/100 ml, and the intestines about 3 to 4 gm/100 ml. Since the liver produces much more lymph in proportion to its weight than any other tissue in the body, the thoracic lymph has a protein concentration of about 4 gm/l00 ml.

Increased interstitial fluid pressure and the lymphatic pump increase lymph flow. The two main factors that control the rate of lymph flow are (1) the interstitial fluid hydrostatic pressure and (2) the pumping activity of the lymphatics. The greater the volume of fluid in the tissue spaces, the greater the interstitial fluid hydrostatic pressure. This, in turn, promotes increased flow of fluid into the lymphatic capillaries through their large openings. Once in the lymphatic capillaries, the lymph is pumped by intermittent compression of the lymphatic vessels. Compression results either from repetitive contraction of the lymphatic walls or from external compression of the lymph vessels by contracting muscles, joint movement, or other effects that cause tissue compression. The lymphatics contain valves that allow flow to occur only away from the tissues and toward the bloodstream. Therefore, during each compression cycle of a lymph vessel, fluid is pumped progressively along the vessel toward the circulation.

Bacteria and debris from the tissues are removed by the lymphatic system at lymph nodes. Because of the very high degree of permeability of the lymphatic capillaries, bacteria and other small particulate matter in the tissues can pass into the lymph. However, the lymph passes through a series of nodes on its way to the blood. In these nodes, bacteria and other debris are filtered out, then phagocytized by macrophages in the nodes, and finally digested into amino acids, glucose, fatty acids, and other small molecular substances before being released into the blood.

THE KIDNEYS AND THEIR FUNCTIONS

The multiple functions of the kidneys in maintaining homeostasis include the following:

- Regulation of water and electrolyte balances in the body
- Regulation of body fluid osmolarity and electrolyte concentrations
- Regulation of acid-base balance
- Excretion of metabolic waste products and foreign chemicals
- Regulation of arterial pressure
- Secretion of hormones (e.g., erythropoietin, renin)
- Synthesis of glucose from amino acids (gluconeogenesis)

Urine Formation Results from Glomerular Filtration, Tubular Reabsorption, and Tubular Secretion

The kidneys perform their most important functions by "clearing" unwanted substances from the blood and excreting them in the

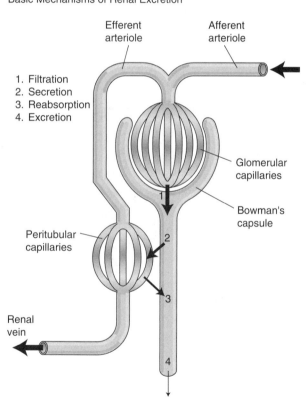

Basic Mechanisms of Renal Excretion

1. Filtration
2. Secretion
3. Reabsorption
4. Excretion

Excretion = Filtration − Reabsorption + Secretion

Figure 2-6.
Basic kidney processes of glomerular filtration, tubular secretion, and tubular reabsorption that determine the urine excretion. Urinary excretion rate of a substance is equal to the rate at which the substance is filtered minus its reabsorption rate plus the rate at which it is secreted from the peritubular capillary blood into the tubules.

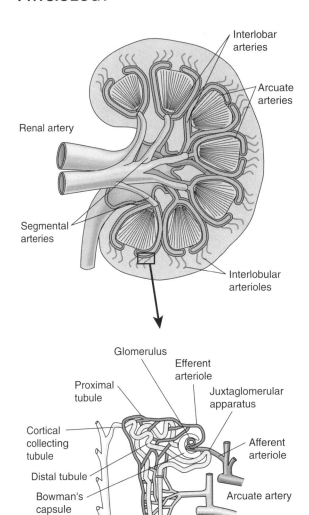

Figure 2-7.
Section of the human kidney showing the major vessels that supply the blood flow to the kidney, and the microcirculation and tubular segments of each nephron.

urine while returning substances that are needed back to the blood. Excretion of urine begins with the filtration of fluid from the glomerular capillaries into the renal tubules, a process called **glomerular filtration** (Fig. 2-6). As glomerular filtrate flows along the renal tubules, the volume of the filtrate is reduced and its composition is altered by **tubular reabsorption** (the movement of water and solutes from the tubules back into the blood) and by **tubular secretion** (the net movement of water and solutes into the tubules). For each substance excreted in the urine, a particular combination of filtration, reabsorption, and secretion occurs:

Urinary excretion rate

$$= \text{filtration rate} - \text{reabsorption rate} + \text{secretion rate}$$

As discussed later in the chapter, changes in urinary excretion rate of different substances that are needed to maintain body fluid and electrolyte homeostasis can occur as a result of changes in glomerular filtration, tubular reabsorption, or in some cases, tubular secretion.

The nephron is the structural and functional unit of the kidney. Each kidney in the human is made up of about one million nephrons, all capable of forming urine. The functional parts of the nephron are shown in Figure 2-7. Each nephron is composed of a tuft of glomerular capillaries, called the glomerulus, through which large amounts of fluid are filtered from the blood, a capsule around the glomerulus called Bowman's capsule, and a long tubule (with multiple segments that differ anatomically and functionally) in which the filtered fluid is converted into urine on its way to the renal pelvis, which receives urine from all of the nephrons. Urine passes from the renal pelvis to the bladder where it is stored until it is eventually expelled from the body by the process of micturition, or urination.

RENAL BLOOD FLOW, GLOMERULAR FILTRATION, AND THEIR CONTROL

Renal Blood Flow Is About 21% of the Cardiac Output

Blood flows to the kidney at a rate of about 1200 ml/min (21% of cardiac output) through the renal artery, which branches progressively to form the **interlobar arterioles, arcuate arteries, interlobular arteries,** and **afferent arterioles,** which lead to the glomerular capillaries (see Fig. 2-7). Blood leaves the glomerular capillaries through an **efferent arteriole** that empties into a second capillary network—the **peritubular capillaries.** The peritubular capillaries surround the renal tubules and empty into the venous system, which leaves the kidney via the renal vein.

Renal blood flow is determined by the pressure gradient across the renal vasculature (the difference between renal artery and renal vein hydrostatic pressure) divided by the total renal vascular resistance:

Renal blood flow = (renal artery pressure − renal vein pressure) / total renal vascular resistance

Although changes in renal artery pressure have some effect on renal blood flow, the kidneys normally have effective mechanisms for maintaining renal blood flow relatively constant over the arterial pressure range between 80 and 170 mmHg, a process called **autoregulation.**

Total renal vascular resistance is determined by the sum of resistances in the individual vascular segments, including the arteries, arterioles, capillaries, and veins. However, the afferent and efferent arterioles appear to be the sites that are most subject to physiologic control, as discussed later. Increased resistance of any of the vascular segments of the kidney tends to reduce renal blood flow, whereas a decrease in vascular resistance increases blood flow if renal artery pressure remains constant.

Glomerular Filtration Is the First Step in Urine Formation

The glomerular filtrate is an ultrafiltrate of plasma that has a composition almost identical to that of plasma except that it has almost no protein (only 0.03%). The glomerular filtration rate (GFR) is normally about 125 ml/min, or about 20% of the renal plasma flow. Thus, the fraction of renal plasma flow that is filtered (**filtration fraction**) averages about 0.2. This means that about 20% of the plasma flowing through the kidney is filtered. The filtration fraction is calculated as

$$\text{Filtration fraction} = \text{GFR} / \text{renal plasma flow}$$

As in other capillaries, the GFR is the product of the **net filtration pressure** and the glomerular **capillary filtration coefficient** (K_f), which is determined by the permeability and surface area of the capillaries:

$$\text{GFR} = \text{net filtration pressure} \times K_f$$

The net filtration pressure represents the sum of hydrostatic and colloid osmotic forces that either favor or oppose filtration across the glomerular capillaries (Fig. 2-8). These forces include (1) the hydrostatic pressure inside the glomerular capillaries (P_G), which is normally about 60 mmHg and promotes filtration; (2) the hydrostatic pressure in Bowman's capsule outside the capillaries (P_B), which is normally 18 mmHg and opposes filtration; (3) the colloid osmotic pressure of the glomerular capillary plasma proteins (π_G), which averages 32 mmHg and opposes filtration; and (4) the colloid osmotic pressure of the proteins in Bowman's capsule (π_B), which is normally near zero and therefore has little effect on filtration. Therefore,

$$\text{net filtration pressure} = P_G - P_B - \pi_G = 10 \text{ mmHg}$$

The net filtration pressure can be decreased in several ways:

- By decreasing the arterial pressure, which decreases the glomerular hydrostatic pressure. In normal kidneys, glomerular hydrostatic pressure changes only 1 to 2 mmHg over a wide range of arterial pressures (from 80 to 160 mmHg) due to

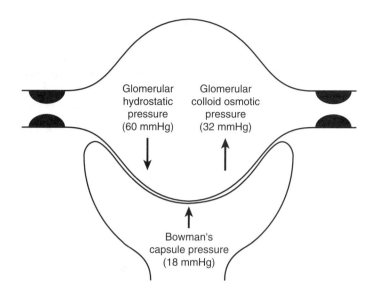

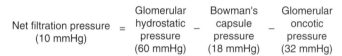

Net filtration pressure (10 mmHg) = Glomerular hydrostatic pressure (60 mmHg) − Bowman's capsule pressure (18 mmHg) − Glomerular oncotic pressure (32 mmHg)

Figure 2-8.
Summary of hydrostatic and colloid osmotic forces causing filtration by the glomerular capillaries. The values shown are estimates for normal humans.

autoregulation, which causes adjustments of renal arteriolar resistance.

- By increasing the resistance of the afferent arterioles, which decreases glomerular hydrostatic pressure.
- By decreasing resistance of the efferent arterioles, which decreases glomerular hydrostatic pressure.
- By increasing the glomerular plasma colloid osmotic pressure.

The GFR can also be reduced by decreasing K_f. Because K_f is the measure of the product of the hydraulic conductivity and surface area of the glomerular capillaries, decreased K_f can occur as a result of loss of capillary surface area because of renal disease or decreased hydraulic conductivity as a result of thickening of the glomerular capillary membrane. Often, both of these changes occur simultaneously. For example, in chronic hypertension there is thickening of the glomerular capillary membrane, glomerular sclerosis, and eventual complete loss of capillary function.

Physiologic Control of Glomerular Filtration

Table 2-2 summarizes the physical determinants of GFR and some of the physiologic and pathophysiologic factors that can influence these determinants, as follows:

- **Increased sympathetic activity** decreases GFR and renal blood flow by causing constriction of renal arterioles and by decreasing the glomerular capillary filtration coefficient.
- **Increased circulating levels of norepinephrine and epinephrine,** released by the adrenal medulla, constrict afferent and efferent arterioles causing reductions in GFR and renal blood flow.
- **Increased levels of endothelin,** a peptide released from damaged vascular endothelial cells, cause constriction of arterioles and reduces GFR and renal blood flow.
- **Increased angiotensin II levels cause preferential constriction of efferent arterioles,** which raises glomerular hydrostatic pressure and reduces renal blood flow. Increased angiotensin II formation usually occurs when there is volume depletion or decreased arterial pressure, which tends to decrease GFR. In these instances, increased angiotensin II, by constricting efferent arterioles, helps to prevent decreases in GFR but at the same time also reduces renal blood flow.
- **Endothelial-derived nitric oxide,** an autacoid released from the blood vessels, decreases renal vascular resistance and increases GFR.
- **Increased renal prostaglandins,** especially PGE_2 and PGI_2, cause decreased renal vascular resistance, increased renal blood flow, and increased GFR. Conversely, blockade of prostaglandin synthesis with nonsteroidal anti-inflammatory drugs (e.g., aspirin) tends to reduce renal blood flow and GFR.

TABLE 2-2.

Summary of Factors That Can Decrease GFR*

Physical Determinants	Physiologic/Pathophysiologic Causes
$\downarrow K_f \rightarrow \downarrow$ GFR	$\uparrow$ Sympathetic activity; renal disease; diabetes mellitus; hypertension
$\uparrow P_B \rightarrow \downarrow$ GFR	Urinary tract obstruction (e.g., kidney stones)
$\uparrow \pi_G \rightarrow \downarrow$ GFR	$\downarrow$ Renal blood flow; increased plasma proteins
$\downarrow P_G \rightarrow \downarrow$ GFR	
• $\downarrow A_P \rightarrow \downarrow P_G$	$\downarrow$ Arterial pressure (has only small effect due to autoregulation)
• $\downarrow R_E \rightarrow \downarrow P_G$	$\downarrow$ Angiotensin II (drugs that block angiotensin II formation)
• $\uparrow R_A \rightarrow \downarrow P_G$	Sympathetic activity; vasoconstrictor hormones (e.g., norepinephrine, endothelin)

*Opposite changes in the above determinants/factors usually increase GFR. K_f, glomerular-filtration coefficient; P_B, Bowman's capsule hydrostatic pressure; π_G, glomerular capillary colloid osmotic pressure; P_G, glomerular capillary hydrostatic pressure; A_p, systemic arterial pressure; R_A, afferent arteriolar resistance; R_E, efferent arteriolar resistance.

- **Autoregulation of GFR and renal blood flow** occurs through mechanisms that are intrinsic to the kidney. The autoregulatory mechanisms keep renal blood flow and GFR relatively constant during variations in renal artery pressure between 80 and 100 mmHg in normal kidneys.

TUBULAR SECRETION AND TUBULAR REABSORPTION

As the glomerular filtrate enters the renal tubules, it flows sequentially through the successive parts of the tubule before it is excreted in the urine. Along this course, some substances are selectively reabsorbed from the tubule back into the blood, whereas others are secreted from the blood into the tubular lumen. Eventually the urine that is formed and all of the substances in the urine represent the sum of three basic renal processes as follows (see Fig. 2-6):

Urinary excretion
= glomerular filtration − tubular reabsorption + tubular secretion

Tubular Secretion Moves Solutes from the Peritubular Capillaries into the Tubules

Some substances are not only filtered by the glomerular capillaries but are also secreted from the peritubular capillaries into the tubules. The first step of tubular secretion is simple diffusion of the substance from the peritubular capillaries into the interstitium. The next step is movement of the substance into the tubular cell or between the tubular cells into the lumen of the tubule by active or passive transport. Substances that are actively secreted into the tubules include potassium and hydrogen ions (which will be discussed in more detail later in this chapter). Certain organic acids and bases are also highly secreted into the renal tubules, leading to rapid removal of these substances from the blood and excretion in the urine.

Reabsorption of Solutes and Water from the Tubules to the Peritubular Capillaries

As the glomerular filtrate enters the renal tubules, it flows sequentially through four major sections of the tubule: (1) **proximal tubules,** (2) **loops of Henle,** (3) **distal tubules,** and (4) **collecting tubules and ducts**. From the 125 ml/min of glomerular filtrate, approximately 124 ml/min are normally reabsorbed by the tubules into the peritubular capillary blood, leaving only 1 ml/min to pass into the urine. Approximately 65% is reabsorbed in the proximal tubules, 15% in the loop of Henle, 10% in the distal tubule, and 9.3% in the collecting tubules and ducts. Thus, most of the reabsorption occurs in the early part of the renal tubular system.

For a substance to be reabsorbed, it must first be transported across the renal tubular epithelial membranes into the renal interstitial fluid, and then through the peritubular capillary membrane back to the blood (Fig. 2-9). Reabsorption across the tubular epithelium may include active and passive transport mechanisms, as in other cell membranes of the body. For example, water and solutes can be transported either through the cell membranes themselves (**transcellular route**) or through the junctional spaces between the cells (**paracellular route**). Then after absorption into the interstitial fluid, water and solutes are transported the rest of the way through the peritubular capillaries into the blood by **ultrafiltration (bulk flow)** that is mediated by hydrostatic and colloid osmotic forces. In contrast to the glomerular capillaries, which filter large amounts of fluid, peritubular capillaries have a large reabsorptive force that moves fluid and solutes from the interstitium into the blood.

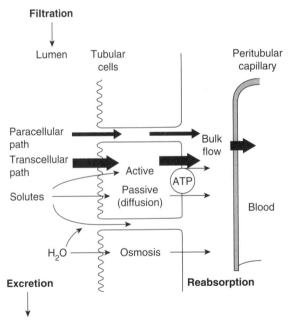

Figure 2-9.
Reabsorption of filtered water and solutes from the tubular lumen across the tubular epithelial cells, through the renal interstitium, and back into the blood of the peritubular capillaries. Solutes are transported through the cells (transcellular route) by passive diffusion or active transport, or between the cells (paracellular route) by diffusion. Water is transported through the cell and between the tubular cells by osmosis. Transport of water and solutes from the interstitial fluid into the peritubular capillaries occurs by ultrafiltration (bulk flow). ATP, adenosine triphosphate.

Some substances that are filtered, such as glucose and amino acids, are almost completely reabsorbed by the tubules so that urinary excretion rate is essentially zero (Table 2-3). Most of the ions in the plasma—such as sodium, chloride, and bicarbonate—are also highly reabsorbed, but their rates of reabsorption and urinary excretion are variable, depending upon the needs of the body. Certain waste products, such as urea and creatinine, are poorly reabsorbed and excreted in relatively large amounts. Therefore, *tubular reabsorption is highly selective, allowing the kidneys to regulate excretion of solutes independently of one another.* This allows precise control of the composition of the body fluids.

The proximal tubules have a high capacity for reabsorption. Approximately two-thirds of the water and solutes filtered are reabsorbed in the proximal tubules. Thus, one important function of the proximal tubules is to conserve substances needed by the body, including glucose, amino acids, and proteins, as well as water and electrolytes—such as sodium, potassium, and chloride. On the other hand, the proximal tubules are relatively impermeable to the waste products of the body and therefore reabsorb a much smaller percentage of the filtered load of these substances.

The loop of Henle consists of three functionally distinct segments: the descending thin segment, the ascending thin segment, and the ascending thick segment. The three distinct segments of the loop of Henle dip into the inner part of the kidney, the **renal medulla,** and play an important role in allowing the kidney to form a concentrated urine. Normally, there is a graded increase in the osmolarity of the renal medullary interstitium, with the inner renal medulla being about 1200 mOsm/L, more than four times as great as that of the plasma.

The descending loop of Henle is highly permeable to water and therefore water is rapidly reabsorbed from the tubular fluid into the hyperosmotic interstitium. Approximately 20% of the glomerular filtrate volume is reabsorbed in the thin descending loop of Henle,

TABLE 2-3.

Filtration, Reabsorption, and Excretion Rates of Different Substances by the Kidneys

	Amount Filtered	Amount Reabsorbed	Amount Excreted	Filtered Load Reabsorbed (%)
Glucose (gm/day)	180	180	0	100
Bicarbonate (mEq/day)	4320	4318	2	>99.9
Sodium (mEq/day)	25,560	25,410	150	99.4
Chloride (mEq/day)	19,440	19,260	180	99.1
Urea (gm/day)	46.8	23.4	23.4	50
Creatinine (gm/day)	1.8	0	1.8	0

causing the tubular fluid to become hyperosmotic as it moves toward the inner part of the renal medulla.

In the thin and thick segments of the ascending loop of Henle, water permeability is virtually zero but large amounts of sodium, chloride, and potassium are reabsorbed, causing the tubular fluid to become dilute (hypotonic) as it flows back up toward the cortex. At the same time, active transport of sodium chloride out of the ascending loop of Henle into the interstitium causes a very high concentration of these ions in the interstitial fluid of the renal medulla. This mechanism for trapping sodium chloride and building up a high osmolality of the renal medullary interstitium is called the **countercurrent multiplier,** which is essential for forming a highly concentrated urine, as discussed later.

Reabsorption in the distal and collecting tubules is highly variable. In the distal tubules and collecting tubules of the nephron, fluid and solute reabsorption can vary greatly depending on the body's needs. If the amount of sodium in the body fluids is high, very little sodium is reabsorbed in these sections of the tubules so that large amounts of sodium are lost in the urine. It is in these parts of the renal tubule where final processing of the urine takes place that several hormones influence reabsorption. For example, **aldosterone** stimulates sodium reabsorption in these segments and **ADH** increases water reabsorption. Normally, about 5% of the glomerular filtrate is reabsorbed in the distal tubules and less than 10% in the collecting tubules and ducts.

Active and passive reabsorption by the renal tubules. Reabsorption of some electrolytes, such as sodium, occurs by active transport, especially by the sodium-potassium ATPase pump (Fig. 2-10).

Many of the nutrients, such as glucose and amino acids, are reabsorbed by **cotransport** with sodium. In most instances, reabsorption of these substances displays a **transport maximum,** which refers to the maximum rate of reabsorption (Fig. 2-11). When the filtered load of these substances exceeds the transport maximum, the excess amount is excreted. The **threshold** is the tubular load at which the transport maximum is exceeded in one or more nephrons, resulting in the appearance of that solute in the urine. Some substances, such as water and urea, are absorbed through the tubular membrane by passive diffusion and osmosis, a process called **passive reabsorption.**

Tubular reabsorption is highly regulated. Because it is essential to maintain precise balance between tubular reabsorption and glomerular filtration, multiple nervous, hormonal, and local control mechanisms regulate the rate of tubular reabsorption as well as GFR. An important feature of tubular reabsorption is that some solutes and water can be independently regulated. Some of the mechanisms that control tubular reabsorption are the following:

- **Glomerulotubular balance** refers to the ability of the tubules to increase reabsorption rate in response to increased tubular load (e.g., caused by increased glomerular filtration).

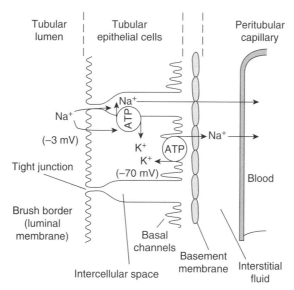

Figure 2-10.
Basic mechanism for active transport of sodium through the tubular epithelial cell. The sodium-potassium pump transports sodium from the interior of the cell through the basolateral membrane, creating a low intracellular sodium concentration and a negative intracellular electrical potential. The low intracellular sodium concentration and the negative electrical potential cause sodium ions to diffuse from the tubular lumen into the cell through the brush border.

- Increased **arterial pressure** reduces tubular reabsorption and increases renal excretion of sodium and water; this is called **pressure natriuresis** and **pressure diuresis.** Conversely, decreased arterial pressure increases tubular reabsorption and decreases sodium and water excretion.

- **Aldosterone,** secreted by the adrenal cortex, increases sodium reabsorption and potassium secretion by the tubules. Therefore, aldosterone reduces sodium excretion and increases potassium excretion.

- **Angiotensin II** increases sodium reabsorption directly and indirectly by stimulating aldosterone secretion and greatly decreases sodium excretion.

- **ADH,** secreted by the posterior pituitary gland, increases water reabsorption in the distal and collecting tubules and greatly decreases water excretion. This, in turn, increases the concentration of solutes in the urine.

- **Parathyroid hormone** increases calcium reabsorption and decreases phosphate reabsorption, leading to decreased calcium excretion and increased phosphate excretion.

- **Atrial natriuretic peptide** decreases sodium reabsorption and increases renal excretion of sodium and water.

- **Sympathetic nervous activity** increases sodium reabsorption and decreases sodium excretion.

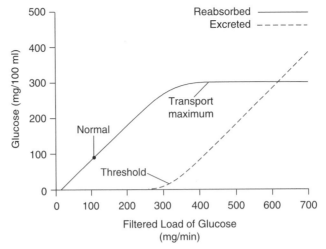

Figure 2-11.
Relations among plasma glucose concentration, the filtered load of glucose, the rate of glucose reabsorption by the renal tubules, and the rate of glucose excretion in the urine. The transport maximum is the maximum rate at which glucose can be reabsorbed from the tubules. The threshold for glucose is the filtered load of glucose at which glucose first begins to appear in the urine.

USE OF RENAL CLEARANCE TO QUANTIFY KIDNEY FUNCTION

Renal clearance is the volume of plasma that is completely cleared of a substance each minute. The concept of renal clearance is useful in quantitating the efficiency of the kidneys in removing a substance from the plasma. For a given substance X, renal clearance is defined as the ratio of the excretion rate of substance X to its concentration in the plasma:

$$C_X = (U_X V) / P_X$$

where C_X is renal clearance of the substance in ml/min, $U_X V$ is the excretion rate of substance X (U_X is the concentration of X in urine, and V is urine flow rate in ml/min), and P_X is plasma concentration of X.

Renal clearances can be used to quantitate several aspects of kidney function, including GFR, tubular reabsorption, and tubular secretion of different substances.

Renal clearance of creatinine or inulin can be used to estimate GFR. Creatinine, a by-product of skeletal muscle metabolism, is filtered at the glomerulus but is not reabsorbed or secreted to a great extent by the tubule. Therefore, the entire 125 ml of plasma that filters into the tubules each minute (GFR) is cleared of creatinine (e.g., creatinine clearance is 125 ml/min, which is equal to the GFR). For this reason, creatinine clearance is often used as an index of GFR. However, a more accurate measure of GFR is the clearance of **inulin,** a polysaccharide that is not reabsorbed or secreted by the renal tubules.

For substances that are completely reabsorbed from the tubules (e.g., amino acids, glucose), the clearance rate is zero because urinary excretion rate is zero. For substances that are highly reabsorbed (e.g., sodium), the clearance rate is usually less than 1% of the GFR, or less than 1 ml/min. In general, waste products of metabolism, such as urea, are poorly reabsorbed and have relatively high clearance rates.

Renal clearance of *para*-aminohippuric acid (PAH) can be used to estimate renal plasma flow. Some substances, such as *para*-aminohippuric acid, are filtered and not reabsorbed by the tubules, but are secreted into the tubules. Therefore, the renal clearance of these substances is greater than GFR. In fact, about 90% of the plasma flowing through the kidney is completely cleared of PAH and the renal clearance of PAH (C_{PAH}) can be used to estimate the renal plasma flow:

$$C_{PAH} = (U_{PAH} V) / P_{PAH} = \text{renal plasma flow}$$

where U_{PAH} and P_{PAH} are urine and plasma concentrations of PAH, respectively, and V is urine flow rate.

URINE CONCENTRATION AND DILUTION

Formation of a dilute urine depends on decreased ADH secretion. The kidneys can produce urine with a range of osmolarity from 50 to 1400 mOsm/L. Because ADH normally increases the permeability of the distal and collecting tubules to water, decreased ADH secretion reduces water permeability and therefore prevents water from being reabsorbed in these parts of the nephron. In the absence of ADH, solutes continue to be reabsorbed from the tubules, but water cannot be effectively reabsorbed in the distal and collecting tubules, making the urine very dilute.

Formation of a concentrated urine—the role of ADH and the countercurrent multiplier. When there is a water deficit in the body, the kidney decreases urine volume and forms a concentrated urine by increasing water reabsorption and continuing to excrete solutes at a normal rate. Concentration of the urine takes place as tubular fluid flows through the medullary collecting ducts, which are normally surrounded by a very hyperosmotic interstitial fluid (Fig. 2-12). When ADH levels are high, water diffuses out of the medullary collecting ducts into the interstitial fluid, leaving behind a concentrated urine.

Thus, the formation of a concentrated urine depends on (1) high ADH levels, which increase permeability of the distal and

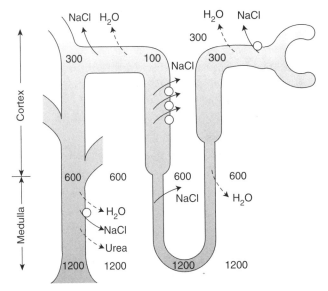

Figure 2-12.
Formation of a concentrated urine when ADH levels are high. Fluid leaving the loop of Henle is dilute, but the tubular fluid becomes concentrated as water is reabsorbed from the distal tubules and collecting tubules. With high ADH levels, the osmolarity of the urine is about the same as the osmolarity of the renal medullary interstitial fluid in the papilla, which is about 1200 mOsm/L. (Numerical values are in mOsm/L.)

collecting tubules to water; and (2) a high osmolarity of the renal medulla. Tubular fluid flowing out of the loop of Henle is dilute, with an osmolarity of only about 50 to 100 mOsm/L. As fluid flows into the distal and collecting tubules, it can equilibrate with the surrounding interstitium when high levels of ADH are present and the tubules are permeable to water. The medullary interstitium outside the collecting tubules is normally very concentrated with sodium and urea due to the operation of the **countercurrent multiplier,** which depends on the special permeability characteristics of the loop of Henle discussed previously. Thus, as fluid flows into the distal tubules and finally into the collecting tubules and ducts, water is reabsorbed until the tubule osmolarity equilibrates with medullary interstitial fluid osmolarity. This leads to a highly concentrated urine with an osmolarity of 1200 to 1400 mOsm/L when high levels of ADH are present.

Major factors that contribute to the buildup of a high solute concentration in the renal medulla. These factors include:

- Active transport of sodium ions and cotransport of potassium, chloride, and other ions out of the thick ascending limb of the loop of Henle into the medullary interstitium
- Active transport of ions from the collecting ducts into the medullary interstitium
- Diffusion of large amounts of urea from the inner medullary collecting ducts into the medullary interstitium

- Diffusion of only small amounts of water from the medullary tubules into the interstitium, far less than the reabsorption of solutes into the medullary interstitium

Factors that can impair urine-concentrating ability. The **vasa recta** are specialized peritubular capillaries that act as a countercurrent exchange system to preserve solute concentration in the renal medulla. However, excessively high blood flow in the vasa recta can "wash out" the medullary interstitium and reduce urine-concentrating ability.

The inability to form a highly concentrated urine can also be caused by (1) decreased secretion of ADH (**"central" diabetes insipidus**); (2) a lack of responsiveness of the renal tubules to ADH (**nephrogenic diabetes insipidus**); or (3) **impaired operation of the countercurrent multiplier** (e.g., certain diuretics or drugs that interfere with the countercurrent multiplier).

The osmoreceptor-ADH system controls extracellular sodium concentration and osmolarity. There are two primary mechanisms for controlling the osmolarity of the body fluids: (1) the **hypothalamic-pituitary ADH feedback system,** and (2) the **thirst mechanism.**

When extracellular fluid osmolarity is increased above normal, both of these systems are activated, causing increased thirst and therefore increased intake of water, and stimulation of ADH secretion, which increases water reabsorption in the renal tubules and decreases water excretion. Therefore, the quantity of water in the body increases, diluting the concentration of solutes. The opposite effect occurs with low plasma osmolarity, which reduces both ADH secretion and thirst.

REGULATION OF SODIUM BALANCE AND EXTRACELLULAR FLUID VOLUME

Sodium and chloride are the most abundant ions in the extracellular fluid. Consequently, changes in sodium chloride content of the extracellular fluid usually cause parallel changes in extracellular fluid volume, provided the ADH-thirst mechanisms are operative. When the ADH-thirst mechanisms are functioning normally, a change in the amount of sodium chloride in the extracellular fluid is matched by a similar change in the amount of extracellular water, so that osmolarity and sodium concentration are maintained relatively constant.

The total amount of sodium in the extracellular fluid is determined by the balance between sodium intake and sodium excretion by the kidneys. Although sodium intake and excretion must be precisely balanced under steady-state conditions, temporary imbalances can lead to changes in extracellular fluid sodium content and consequently extracellular fluid volume. Because sodium intake is usually governed more by one's eating and drinking habits than by physiologic control mechanisms, the regulation of extracellular fluid volume is vested largely in the control of renal sodium excretion. The two variables that control renal sodium excretion ($U_{Na}V$) are (1) the rate of sodium filtration (GFR $\times$ P_{Na}), and (2) the rate of tubular sodium reabsorption:

$$U_{Na}V = (GFR \times P_{Na}) - (\text{tubular sodium reabsorption})$$

where P_{Na} is the plasma (extracellular fluid) sodium concentration. As discussed previously, glomerular filtration and tubular reabsorption are regulated by multiple factors including hormones, sympathetic activity, and arterial pressure.

Pressure natriuresis is a key regulator of body fluid volume and arterial pressure. Perhaps the most powerful mechanism for control of blood volume and extracellular fluid volume, as well as for the maintenance of sodium balance, is the effect of blood pressure on sodium excretion, called **pressure natriuresis,** as discussed previously. When extracellular fluid volume and blood volume increase, arterial pressure also increases. The increased pressure, in turn, has a potent effect to increase the rate of sodium excretion by the kidneys, thereby reducing extracellular fluid volume back toward the normal level. This mechanism for control of extracellular fluid volume and blood volume is called the **renal-body fluid feedback mechanism** (see Chapter 5).

Integrated responses to changes in sodium intake and increased extracellular fluid volume. One important aspect of the renal-body fluid feedback is that nervous and hormonal factors act in concert with pressure natriuresis to make this mechanism more effective in minimizing changes in blood volume, extracellular fluid volume, and blood pressure in response to day-to-day challenges. For example, with increased sodium intake, there are reductions in **angiotensin II** and **aldosterone** formation, which, in turn, decrease sodium reabsorption and add to the direct effect of blood pressure to raise sodium excretion.

Increased extracellular fluid volume also reduces **sympathetic nervous system activity** and increases secretion of **atrial natriuretic factor** (ANF; a hormone secreted by the cardiac atria), which both decrease sodium reabsorption, causing the kidneys to excrete increased amounts of sodium and water and reducing extracellular fluid volume and blood volume back toward normal. Thus, combined activation of natriuretic systems and suppression of sodium- and water-retaining systems leads to increased sodium excretion when sodium intake and extracellular fluid volume are increased. The opposite changes take place when sodium intake and extracellular fluid volume are reduced below normal.

Potassium Excretion Is Regulated Mainly by the Distal and Collecting Tubules

Extracellular potassium concentration normally is regulated precisely at about 4.2 mEq/L. A special difficulty in regulating extracellular potassium concentration is the fact that over 98% of the total body potassium is contained in the cells and only about 2% in the extracellular fluid. Therefore, it is extremely important to maintain a precise balance between potassium intake and potassium excretion, which occurs mainly via the kidneys. Renal potassium excretion, in turn, is determined by the sum of three processes: (1) potassium filtration at the glomerulus, (2) potassium reabsorption by the tubules, and (3) potassium secretion by the tubules.

Most of the daily variation in potassium excretion occurs by changes in potassium secretion in the distal and collecting tubules, rather than by changes in glomerular filtration or tubular reabsorption of potassium. The cells in the late distal tubules and cortical collecting tubules that secrete potassium are called **principal cells.** Secretion of potassium from the peritubular capillary blood into the tubular lumen is a three-step process (Fig. 2-13), beginning with passive diffusion of potassium from the blood to the

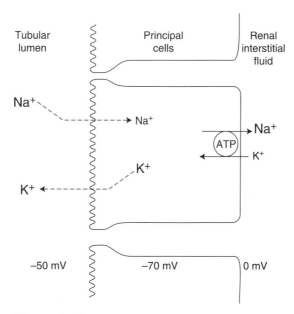

Figure 2-13.
Mechanisms of potassium secretion and sodium reabsorption by the principal cells of the distal tubules and collecting tubules.

interstitium, active transport of potassium from the interstitium into the cell by the sodium-potassium-ATPase pump at the basolateral membrane, and finally passive diffusion of potassium from the cell interior into the tubular fluid.

The primary factors that control potassium secretion by the principal cells are listed below:

- Increased extracellular potassium concentration increases potassium secretion.
- Increased aldosterone concentration increases potassium excretion.
- Increased tubular flow rate increases potassium secretion.
- Increased hydrogen ion concentration (acidosis) decreases potassium secretion.

Feedback control of plasma potassium concentration by aldosterone. Increased extracellular potassium concentration stimulates potassium excretion directly as well as indirectly by stimulating aldosterone secretion. Aldosterone, a hormone secreted by the adrenal cortex, increases sodium reabsorption and increases potassium secretion by the distal and collecting tubules, which reduces urine excretion of sodium and increases potassium excretion.

Aldosterone is the primary hormonal mechanism for regulating potassium ion concentration because there is a direct feedback by which aldosterone and potassium ion concentration are linked. This feedback mechanism operates as follows: Whenever extracellular fluid potassium concentration rises above normal, this stimulates secretion of aldosterone, which then increases renal excretion of potassium, returning extracellular fluid potassium concentration toward normal. The opposite changes take place when potassium concentration is too low.

Regulation of other extracellular fluid electrolytes. Other extracellular fluid electrolytes that are also closely regulated include calcium, magnesium, chloride, and bicarbonate. For example, a decrease in calcium ion concentration stimulates **parathyroid hormone** secretion, which in turn increases calcium reabsorption by the renal tubules and decreases loss of calcium in the urine. (This is discussed in more detail in Chapter 10.) Chloride and bicarbonate ion concentration are adjusted by the renal and respiratory mechanisms for control of acid-base balance as explained in the following paragraphs.

REGULATION OF ACID-BASE BALANCE (HYDROGEN ION CONCENTRATION)

Hydrogen ion concentration $[H^+]$ in the extracellular fluid is precisely regulated, averaging only 0.00004 mEq/L (40 nanoequivalents/L). Normally, $[H^+]$ is expressed in terms of pH, which is the logarithm of the reciprocal of the $[H^+]$:

$$pH = \log (1/[H^+])/ = - \log [H^+]$$

Arterial blood has a normal pH of 7.4. A pH of 7.8 is considered to be highly alkaline, while a pH of 7.0 is considered to be highly acidic.

The body has three primary lines of defense against changes in $[H^+]$: (1) **Acid-base buffer systems,** which react within seconds to prevent changes in $[H^+]$ concentration. (2) The **lungs,** which eliminate CO_2 and therefore carbonic acid (H_2CO_3, sometimes called volatile acid); this mechanism operates within seconds to minutes and acts as a second line of defense. (3) The **kidneys,** which can secrete H^+, reabsorb bicarbonate (HCO_3^-), and produce new HCO_3^-; this mechanism is a third line of defense that operates slowly but very powerfully in regulating acid-base balance.

ACID-BASE BUFFER SYSTEMS

Among the most important buffer systems of the body are the **proteins** of the cells, and to a lesser extent, the proteins of the plasma and interstitial fluids. The **phosphate buffer system** ($HPO_4^=/H_2PO_4^-$) is not a major buffer in extracellular fluid but is important as an intracellular buffer and as a buffer in renal tubular fluid. The most important extracellular fluid buffer is the **bicarbonate buffer system** (HCO_3^-/P_{CO_2}), primarily because the components of this system, CO_2 and HCO_3^-, are closely regulated by the lungs and the kidneys, respectively.

The pH of extracellular fluid can be expressed as a function of the concentration of the components of the bicarbonate buffer system according to the **Henderson-Hasselbalch equation:**

$$pH = 6.1 + \log [HCO_3^- / (0.03 \times P_{CO_2})]$$

In this equation, HCO_3^- is expressed in mmol/L and P_{CO_2} is expressed as mmHg. The greater the P_{CO_2}, the lower the pH; the greater the HCO_3^-, the higher the pH.

THE LUNGS CONTROL $[H^+]$ BY ELIMINATING CO_2

Because the lungs expel CO_2 from the body, rapid ventilation by the lungs decreases the concentration of CO_2 in the blood, which in turn reduces carbonic acid (H_2CO_3) and $[H^+]$ in the blood. Conversely, decreasing pulmonary ventilation increases the concentration of CO_2 and $[H^+]$ in the blood.

Because increased $[H^+]$ stimulates respiration and alveolar ventilation, which in turn decreases $[H^+]$, the respiratory system acts as a negative feedback controller of $[H^+]$. Thus, when $[H^+]$ increases above normal, the respiratory system is stimulated and alveolar ventilation increases, thereby reducing P_{CO_2} and $[H^+]$ back toward normal. Conversely, if $[H^+]$ falls below normal, the respiratory center becomes depressed, alveolar ventilation decreases, and $[H^+]$ increases back toward normal. The respiratory system can return $[H^+]$ and pH about two-thirds of the way back toward normal within a few minutes after a sudden disturbance of acid-base regulation.

RENAL CONTROL OF ACID-BASE BALANCE

The Kidneys Control Acid-Base Balance by H^+ Secretion, HCO_3^- Reabsorption, and HCO_3^- Production

When the respiratory system fails to completely restore $[H^+]$ to normal, the kidneys are capable of bringing it back toward normal within 12 to 24 hours in most cases. If the kidneys have sufficient time to function, they are many times as effective as either the buffers or the respiratory mechanism in returning the pH of the body fluids to normal.

The kidneys readjust $[H^+]$ by excreting either an acidic or basic urine. Excreting an acidic urine reduces the amount of acid in the extracellular fluid, whereas excreting a basic urine removes base from the extracellular fluid, which is the same as adding H^+.

The overall mechanism by which the kidneys excrete acid or basic urine is as follows: Large amounts of HCO_3^- are filtered continuously into the tubules and if they are excreted in the urine, this removes base from the blood. On the other hand, large numbers of H^+ are also secreted into the tubular lumen by the epithelial cells, thus removing acid from the blood. If more H^+ are secreted than HCO_3^- are filtered, there will be a net loss of acid from the extracellular fluids. Conversely, if more HCO_3^- is filtered than H^+ is secreted, there will be a net loss of base.

When there is a reduction in extracellular fluid hydrogen ion concentration (**alkalosis**), the kidneys fail to reabsorb all of the filtered bicarbonate, thereby increasing excretion of bicarbonate. Because HCO_3^- normally buffers H^+ in the extracellular fluid, the loss of HCO_3^- is the same as adding a H^+ to the extracellular fluid. Thus, in alkalosis the removal of HCO_3^- raises the extracellular fluid $[H^+]$ and decreases pH back toward normal.

When [H⁺] in the extracellular fluid increases (**acidosis**), the kidneys do not excrete bicarbonate in the urine, but instead reabsorb all of the filtered bicarbonate and produce new bicarbonate, which is added back to the extracellular fluid. This reduces extracellular fluid [H⁺] toward normal.

Secretion of hydrogen ions and reabsorption of bicarbonate ions by the renal tubule. Bicarbonate is not reabsorbed directly by the tubules. Instead, bicarbonate is reabsorbed as a result of the combination of secreted H⁺ with filtered HCO_3^- in the tubular fluid under the influence of carbonic anhydrase in the tubular epithelium. The proximal tubule, loop of Henle, and early distal tubule, all secrete H⁺ into the tubular fluid by sodium-hydrogen countertransport as shown in Figure 2-14. Secreted H⁺ is consumed by reaction with HCO_3^-, forming H_2CO_3, which dissociates to CO_2 and H_2O. The CO_2 diffuses into the cell and is used to reform H_2CO_3 and eventually HCO_3^-, which is then reabsorbed across the basolateral membranes of the tubules.

In the distal and collecting tubules, the same basic mechanisms are used for HCO_3^- reabsorption except that H⁺ are secreted by primary active transport. Normally, over 99% of the filtered HCO_3^- is reabsorbed by the renal tubules, with about 95% of this occurring in the proximal tubules, loops of Henle, and early distal tubule.

Combination of excess H⁺ with phosphate and ammonia buffers in the tubule—a mechanism for generating new HCO_3^-. The H⁺ remaining in excess of that which reacts with HCO_3^- can react with other urinary buffers, especially NH_3 and HPO_4^-, and is then

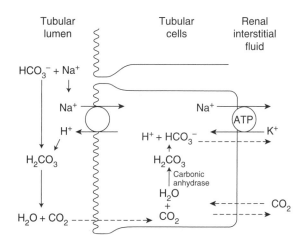

Figure 2-14.
The cellular mechanisms for (1) active secretion of hydrogen ions into the renal tubules; (2) tubular reabsorption of bicarbonate by combination with hydrogen ions to form carbonic acid, which dissociates to form carbon dioxide and water; (3) sodium ion reabsorption in exchange for hydrogen ions secreted. This pattern of hydrogen ion secretion occurs in the proximal tubule.

excreted as buffer salts (e.g., NH_4Cl and Na_2HPO_4). For each H^+ secreted that combines with a non-HCO_3^- buffer, a new HCO_3^- is formed within the renal tubular cells and added to the body fluids.

The urinary phosphate buffer that combines with the secreted H^+ is called **titratable acid.** The amount of titratable acid that forms is limited by the supply of urinary phosphate buffer, since about 75% of filtered phosphate is reabsorbed. Therefore, the other buffer, ammonia, is more effective in buffering large amounts of secreted H^+. Under conditions of chronic acidosis, most of the urinary buffer is NH_3, which is converted to NH_4^+ by the addition of H^+.

Quantification of Renal Tubular Acid Secretion

H^+ secretion rate = HCO_3^- reabsorption rate
                      + titratable acid excretion rate + NH_4^+ excretion rate

Net acid excretion rate = urinary titratable acid excretion rate
                      + NH_4^+ excretion rate − HCO_3^- excretion rate

Virtually all of the H^+ secreted is used to reabsorb HCO_3^- or to react with urinary buffers, especially phosphate and ammonia. Normally, about 4400 mEq/day of hydrogen ions are secreted. Most of the secreted H^+ (4320 mEq/day) combines with the filtered bicarbonate ions to allow almost complete reabsorption of the tubular bicarbonate ions, except for about 2 mEq/day, which is excreted. About 80 mEq of the secreted H^+ is used to combine with ammonia or phosphate and therefore to rid the body of non-volatile acids.

The total rate of hydrogen ion excretion as free H^+ is negligible; almost all of the hydrogen ions excreted are buffered by phosphate (titratable acid) or by ammonia. Therefore, the total rate of hydrogen ions excreted can be approximated as the sum of titratable acid excretion and NH_4^+ excretion.

Acid-Base Disturbances

The condition of **acidosis** occurs when arterial pH is below 7.4, whereas **alkalosis** occurs when arterial pH is above 7.4. An abnormality of pH that results primarily from a change in HCO_3^- is called **metabolic** acidosis or alkalosis, whereas a disturbance that results primarily from a change in P_{CO_2} is called **respiratory** acidosis or alkalosis:

- **Respiratory acidosis results from high P_{CO_2}** due to the failure of the lungs to eliminate CO_2 adequately. As a compensation, increased P_{CO_2} stimulates H^+ secretion by the renal tubular cells, causing increased HCO_3^- reabsorption. The excess H^+ remaining in the renal tubular cells combines with buffers, especially NH_3 in chronic acidosis, which leads to generation

TABLE 2-4.

Characteristics of Primary Acid-Base Disturbances

	pH	[H+]	P_{CO_2}	[HCO$_3^-$]
Respiratory acidosis	↓	↑	⇑	↑*
Respiratory alkalosis	↑	↓	⇓	↓*
Metabolic acidosis	↓	↑	↓†	⇓
Metabolic alkalosis	↑	↓	↑†	⇑

The primary event is indicated by the double arrows (⇑ or ⇓). Note that respiratory acid-base disorders are initiated by an increase or a decrease in P_{CO_2}, whereas metabolic disorders are initiated by an increase or a decrease in [HCO$_3^-$].
*Compensatory changes in [HCO$_3^-$] caused by the kidneys.
†Compensatory changes in P_{CO_2} by the lungs.

of new HCO$_3^-$ that is added back to the blood. These changes return plasma pH toward normal.

- **Respiratory alkalosis is caused by decreased P_{CO_2}** associated with excessive ventilation. Renal compensation for this includes a reduction in H+ secretion, which results in incomplete HCO$_3^-$ reabsorption and a loss of HCO$_3^-$ in the urine, thereby restoring plasma pH toward normal.

- **Metabolic acidosis is caused by decreased HCO$_3^-$,** resulting from a loss of buffer such as HCO$_3^-$ or too much acid in the body fluids. The compensatory responses include stimulation of the respiratory centers, which eliminates CO_2 and returns pH toward normal. At the same time, renal compensation increases reabsorption of bicarbonate and excretion of buffer salts, which leads to new HCO$_3^-$ formation and a return of H+ and HCO$_3^-$ concentrations toward normal.

- **Metabolic alkalosis is caused by increased HCO$_3^-$,** resulting from excessive loss of H+ (e.g., loss of HCl) or excessive intake or retention of bases. This raises HCO$_3^-$ concentration, thereby decreasing H+ concentration and increasing pH. The compensatory responses include a reduction in respiration rate causing retention of CO_2 and an increased renal excretion of HCO$_3^-$, both of which help return plasma pH toward normal. Table 2-4 shows the different types of acid-base disturbances and the characteristic changes in plasma pH, [H+], P_{CO_2}, and [HCO$_3^-$].

MICTURITION—THE PROCESS OF URINATION

The urinary bladder empties when it becomes filled through two main steps: (1) The bladder fills progressively until the tension in its walls rises above a threshold level, which elicits the second step; (2) a nervous reflex called the micturition reflex occurs and empties the bladder or, if this fails, at least causes a conscious desire to urinate.

The micturition reflex is a spinal cord reflex. The micturition reflex is a complete cycle of a progressive and rapid increase in bladder pressure, a period of sustained increase in pressure, and a return of the pressure to the basal tone of the bladder, described as follows:

- Sensory signals from the stretch receptors in the bladder wall and are conducted to sacral segments of the spinal cord through the **pelvic nerves** and then reflexly back to the bladder through the **parasympathetic nerves** by way of the pelvic nerves.
- Once the micturition reflex is powerful enough, it causes another reflex that passes through the **pudendal nerves** to the **external sphincter** to inhibit this. If this inhibition is more potent than the voluntary constrictor signals to the external sphincter, urination will occur.
- The micturition reflex is an autonomic spinal cord reflex, but it can be inhibited or facilitated by centers in the brain stem, mainly the **pons,** and several centers in the **cerebral cortex,** which are mainly excitatory.

Chapter 3

Blood, Hemostasis, and Immunity

Red Blood Cells

Red blood cells (erythrocytes) are anuclear, biconcave disks. Red blood cells average 7 μm in diameter, but their extreme pliability allows them to squeeze through capillaries less than 5 μm in diameter. The human body contains about 25 trillion red blood cells in an average concentration of about 5 million per μL of blood. The percentage of the total blood volume comprised of red blood cells is called the **hematocrit,** and this is normally about 40% in women and about 45% in men.

The main function of red blood cells is to transport hemoglobin. In turn, hemoglobin transports oxygen from the lungs to the tissues and transports carbon dioxide from the tissues back to the lungs. Also, hemoglobin is an excellent acid-base buffer (as is true of most proteins), providing most of the buffering power of whole blood.

Tissue hypoxia stimulates red blood cell formation. The mass of red blood cells in the circulation is regulated within narrow limits to provide adequate oxygenation of the tissues, but is not so concentrated as to impede blood flow to the tissues. The general mechanism of this is shown in Figure 3-1. Tissue hypoxia causes the kidneys to release a hormone called **erythropoietin,** which then flows in the blood to the bone marrow where it stimulates **erythropoiesis.** Once red blood cells are formed, their average life span is 120 days in the circulatory system.

Many factors decrease tissue oxygenation and thus stimulate red blood cell production. Some of these factors include low blood volume, anemia, low hemoglobin, poor blood flow, and pulmonary disease. In persons residing at very high altitudes where oxygen concentration is low, the concentration of red blood cells sometimes becomes as much as 50% greater than normal. The condition is called **polycythemia.**

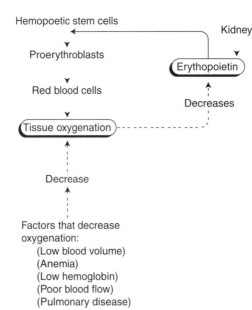

Hemopoetic stem cells

Proerythroblasts

Red blood cells

Tissue oxygenation

Kidney

Erythopoietin

Decreases

Decrease

Factors that decrease
oxygenation:
(Low blood volume)
(Anemia)
(Low hemoglobin)
(Poor blood flow)
(Pulmonary disease)

Figure 3-1.
The erythropoietin mechanism
for stimulating red blood cell
production. (Adapted from
Guyton AC and Hall JE: *Textbook of Medical Physiology,*
9th ed. Philadelphia: WB Saunders, 1995, p 428.)

Vitamin B_{12} and folic acid are necessary for final maturation of red blood cells. The process of red blood cell formation can be divided into two principal processes: formation of the cell structure itself and formation of hemoglobin. The red cell is formed in the bone marrow by a series of divisions from the **hemocytoblast**. Two vitamins, vitamin B_{12} and folic acid, are essential for the synthesis of DNA. Lack of either vitamin leads to decreased DNA and thus failure of nuclear maturation and division. The erythroblastic cells of the bone marrow become larger than normal and are called **megaloblasts**. The adult cell has a flimsy membrane and is often large, with an ovoid shape rather than the usual biconcave disk shape. Therefore, vitamin B_{12} or folic acid deficiency causes maturation failure in the process of erythropoiesis.

The Anemias

Blood loss anemia can be acute or chronic. The plasma can be replaced within 1 to 3 days following acute blood loss, but about 3 to 4 weeks are required for the red blood cell concentration to return to normal. When chronic blood loss occurs, the individual cannot absorb enough iron from the gastrointestinal tract to produce adequate amounts of hemoglobin. Red blood cells with too little hemoglobin give rise to a condition called **microcytic hypochromic anemia**.

Lack of iron can cause hypochromic anemia. Iron is a primary nutritive factor necessary for formation of hemoglobin. It is present in the diet in only very small quantities and even then is rather poorly absorbed from the gastrointestinal tract. Therefore, many persons fail to form sufficient quantities of hemoglobin to fill the red blood cells as they are being produced. This causes hypochromic anemia, in which the number of cells may be normal but the amount of **hemoglobin** in each cell is far **below normal.**

Aplastic anemia can occur when the bone marrow has been damaged. The damage may be caused by irradiation, excessive x-ray treatment, reactions to drugs or chemicals, immunologic disturbances (e.g., following infections), and defects of the bone marrow microenvironment.

Megaloblastic anemia results from deficiency of vitamin B_{12} or folic acid. As discussed above, lack of vitamin B_{12} or folic acid leads to decreased DNA and thus failure of nuclear maturation and division. The erythroblastic cells of the bone marrow grow too large, have odd shapes, and are called **megaloblasts.**

Hemolytic anemia occurs when red blood cells are fragile and easily ruptured. The production of red blood cells may be normal or even greater than normal, but the life span of the red blood cell is so short that anemia occurs. Examples of hemolytic anemias include the following:

* **Hereditary spherocytosis,** in which the red blood cells are spherical and easily ruptured.
* **Sickle cell anemia,** in which an abnormal type of hemoglobin, called **hemoglobin S,** precipitates when it is exposed to hypoxic conditions. The precipitated hemoglobin forms long crystals that give the cells a sickle shape, making them more susceptible to rupture as they pass through the splenic pulp.
* **Erythroblastosis fetalis** occurs when anti-Rh antibodies from the mother diffuse into the fetus causing red blood cell agglutination, as discussed below.

White Blood Cells

White blood cells (leukocytes) protect the body from infectious agents. The number of white cells in the blood is normally only 1/600 the number of red blood cells, about 7000 per μL of blood. They perform the very important function of protecting or helping to protect the body from invasion by infectious agents. The various types of white cells include **polymorphonuclear neutrophils, polymorphonuclear eosinophils, polymorphonuclear basophils,** and **monocytes,** all formed in the bone marrow, and lymphocytes formed in the lymph nodes, as shown in Table 3-1. There are also large numbers of **platelets,** which are fragments of **megakaryocytes,** also formed in the bone marrow. The polymorphonuclear cells have multiple nuclei and are also called **granulocytes** because they have a granular appearance.

The neutrophils are the most numerous of the white blood cells. They represent about 62% of the total white cells in the blood. They are highly motile, highly phagocytic, and are attracted out of the blood into tissue areas where tissue destruction is occurring by a process called **chemotaxis,** which means attraction by the destruc-

TABLE 3-1.		
Normal White Blood Cell Values		

	Number/μL	%
Polymorphonuclear neutrophils	4340	62.0
Polymorphonuclear eosinophils	161	2.3
Polymorphonuclear basophils	28	0.4
Monocytes	371	5.3
Lymphocytes	2100	30.0

tion products from the damaged tissues. Once in the tissue area, the neutrophils phagocytize bacteria and small amounts of dead tissue debris.

The monocyte-macrophage system comprises the "reticuloendothelial system." The **monocytes** are much larger cells than the neutrophils. Normally large numbers of them wander continually through the capillary membranes and into the tissues. Many of them become attached to tissue cells and become very large; they are then called **macrophages.** They can phagocytize 5 to 10 times as many bacteria and much larger particles of tissue debris than can the neutrophils. Almost all tissues of the body contain macrophages, but they are especially abundant in those tissues that are routinely exposed to bacteria, such as the lung alveoli, the sinusoids of the liver, the sinusoids of the bone marrow, the sinusoids of the lymph nodes, and the subcutaneous tissue. This extensive monocyte-macrophage system is frequently also called the **reticuloendothelial system.**

Macrophages and neutrophils respond to inflammation with four lines of defense. The **macrophages** already present in a tissue provide the first line of defense against infection. Within a few hours, **neutrophils** present in the blood invade the inflamed area, providing the second line of defense. At the same time, the number of neutrophils in the blood can increase severalfold, resulting in **neutrophilia.** Neutrophilia is caused by products from the infected or inflamed tissues that enter the bloodstream and are transported to the bone marrow where they mobilize stored neutrophils. Several days later the concentration of **macrophages** in the inflamed tissue becomes very **high,** the macrophages replacing most of the neutrophils. This is the third line of defense. The fourth line of defense consists of **increased production of white blood cells** by the bone marrow, which is stimulated by products released from the infected or inflamed tissues.

Eosinophils combat parasitic infections. Eosinophils are similar to the neutrophils except that they are less chemotactic and less phagocytic. **Eosinophils** are produced in large numbers in persons with parasitic infections. The parasites are usually too large to be phagocytized, but the eosinophils attach themselves to the surface and release lethal substances that can kill many of the parasites. Large numbers of eosinophils also appear in the blood in allergic conditions and may help to detoxify toxins that are released by allergic reactions.

Mast cells and basophils are important for allergic reactions. The type of antibody, the IgE type, that causes allergic reactions binds to **mast cells** and **basophils,** causing them to release various inflammatory products that in turn cause many of the manifestations of allergic reactions. Also, basophils and mast cells liberate heparin into the blood, a substance that can prevent blood coagulation.

Platelets are fragments of megakaryocytes. Like heparin, they too are important in the blood coagulation process but instead cause coagulation, as will be discussed below.

BLOOD COAGULATION

Mechanism of Coagulation

Blood clots can patch a ruptured blood vessel. When a blood vessel ruptures, a blood clot develops within a few minutes to fill the gap and stop the bleeding—if the rupture is not too large. This process is caused by polymerization of plasma **fibrinogen** molecules into

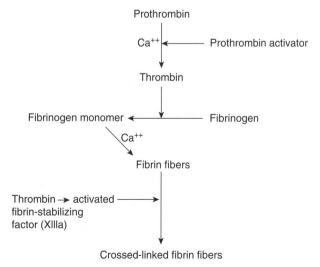

Figure 3-2.
Conversion of prothrombin to thrombin and polymerization of fibrinogen to form fibrin threads. (Adapted from Guyton AC and Hall JE: *Textbook of Medical Physiology,* 9th ed. Philadelphia: WB Saunders, 1995, p 465.)

long **fibrin threads** that entrap large numbers of red blood cells, white blood cells, platelets, and plasma to form a soft gelatinous mass—the blood clot. The fibrin threads gradually contract, expressing most of the plasma from the clot, which leaves a reasonably solid barrier in the opening of the blood vessel.

Blood clotting is initiated by conversion of prothrombin into thrombin. Prothrombin is a plasma protein continually formed by the liver that can be split into two smaller molecules, one of which is thrombin. **Thrombin** is an enzyme that then enzymatically causes polymerization of the plasma fibrinogen molecules into the fibrin threads that lead to blood clotting, as shown in Figure 3-2. In the normal circulation, very little prothrombin is converted into thrombin, and blood clotting does not occur. Two principal conditions that can lead to blood clotting are damage to the vessel wall and damage to the blood itself.

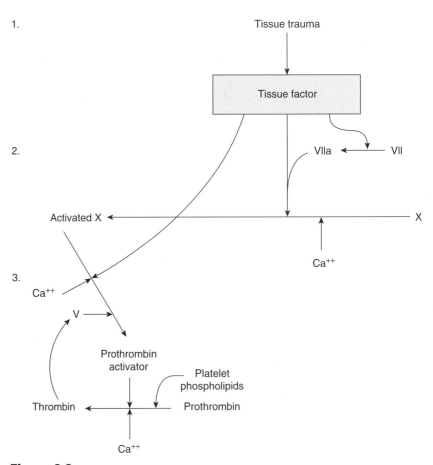

Extrinsic Pathway

Figure 3-3.
The extrinsic pathway for initiating blood clotting. (Adapted from Guyton AC and Hall JE: *Textbook of Medical Physiology,* 9th ed. Philadelphia: WB Saunders, 1995, p 467.)

Damage to the vessel wall initiates the extrinsic pathway for pro-thrombin activation. The damaged tissues in the wall of the vessel release a substance called **tissue factor** or **tissue thromboplastin.** This is composed mainly of phospholipids from the damaged tissues. The thromboplastin in turn catalyzes a series of enzymatic reactions among multiple blood plasma proteins called **blood coagulation factors.** These reactions eventually form **prothrombin activator,** which converts prothrombin to thrombin and thereby initiates blood coagulation. The extrinsic pathway for initiating blood clotting is shown in Figure 3-3.

Damage to the blood initiates the intrinsic pathway for pro-thrombin activation. Damage to the blood causes direct activation of special protein blood coagulation factors. Also, damage to the platelets of the blood causes release of **platelet thromboplastin,** which has effects similar to those of tissue thromboplastin released by torn blood vessels. The combined activation of the protein coagulation factors and release of the platelet thromboplastin leads eventually to the formation of prothrombin activator. Then the prothrombin activator converts prothrombin to thrombin, which subsequently causes blood clotting. The intrinsic pathway for initiating blood clotting is shown in Figure 3-4.

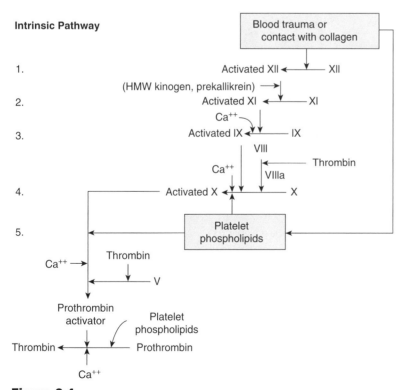

Figure 3-4.
The intrinsic pathway for initiating blood clotting. (Adapted from Guyton AC and Hall JE: *Textbook of Medical Physiology,* 9th ed. Philadelphia: WB Saunders, 1995, p 468.)

Clotting Disorders

Hemophilia is a bleeding tendency caused by deficiency of factor VIII. Hemophilia occurs exclusively in males, usually because of a deficiency in factor VIII, or **antihemophilic factor,** which is required for function of the intrinsic pathway of blood coagulation. Excessive bleeding will usually not occur without trauma, but the amount of trauma required to cause severe and prolonged bleeding is hardly noticeable.

Liver disease and vitamin K deficiency can lead to excessive bleeding. Most of the coagulation factors are formed in the liver. Therefore, hepatitis, cirrhosis, and other diseases of the liver can depress the normal coagulation system, causing a person to bleed excessively. Another cause of decreased coagulation factor production by the liver is vitamin K deficiency. Vitamin K is necessary for formation of prothrombin, factor VII, factor IX, and factor X.

Thrombocytopenia causes petechial hemorrhages. Thrombocytopenia means that the number of platelets in the blood is greatly reduced. Platelets, aside from their capability to induce blood clotting, also have the ability to attach themselves to very minute rupture points in blood vessels and thereby close these holes even without causing actual blood coagulation. In thrombocytopenia this function is lost, and as a result, the person develops many minute bleeding spots, called **petechial hemorrhages,** throughout all organs and beneath the skin.

IMMUNITY

Immunity is the resistance of the body to invasion by bacteria, viruses, or other infectious agents or toxins.

A person is born with innate immunity. Each person is born with a certain amount of innate immunity that results from several special mechanisms: (1) the reticuloendothelial system and the white blood cells (previously discussed); (2) resistance of the intact skin to invasion by microorganisms; (3) destruction of bacterial organisms by the digestive enzymes in the stomach; and (4) substances circulating in the blood.

Acquired immunity occurs following exposure to invading agents or foreign substances. In addition to the natural immunity that normally exists in all persons, a person can develop acquired immunity to many destructive agents to which he is not naturally immune. Most destructive agents, such as bacteria, viruses, or toxins, are mainly composed of protein molecules. On entering the body, these proteins act as antigens and cause two types of immunity: One is called **humoral immunity** and the other **cell-mediated immunity.**

Humoral Immunity

In humoral immunity, the body develops circulating antibodies. The foreign antigens first enter the lymphoid tissue, especially the lymph nodes. There they cause plasma cells, which are derived from lymphocytes, to produce large quantities of antibodies that are specifically reactive for the type of protein (or other chemical) that initiated their production. Once these antibodies have been formed and released into the body fluids, which usually requires 1 week to several weeks, they then destroy the specific invader that had caused their formation and can also destroy any future invader of this same type.

Antibodies destroy an invading agent or make it more susceptible to phagocytosis. Antibodies are large protein molecules, usually γ-globulins, that attach to the surfaces of bacteria or viruses, or they combine directly with toxins. They either destroy the invading agent or make it more susceptible to phagocytosis by the tissue macrophages or by white blood cells. Or, in the case of toxins, the antibodies can simply neutralize these agents by combining chemically with them.

Lymphocytes processed in the liver become antibody-producing plasma cells. In early fetal development of the lymph nodes, no lymphocytes are present in the nodes. Instead, the early lymphocytes are formed and processed in the liver and the thymus gland. After processing, the lymphocytes are released into the circulating blood and eventually become entrapped in the lymph nodes. Once in these nodes, those lymphocytes processed in the liver eventually are converted into plasma cells that become part of the humoral immune process to form antibodies.

Cell-Mediated Immunity

Cell-mediated immunity involves activation of lymphocytes processed in the thymus gland. The lymphocytes that are processed in the thymus gland also end up in the lymph nodes and other lymphoid tissues of the body. However, instead of forming antibodies when the lymph node is exposed to antigens, these cells form so-called **sensitized lymphocytes,** also called **T cells** because of their earlier processing in the thymus. The T cells form chemical substances that are similar to antibodies, but these remain attached to the cell membranes of the lymphocytes. Large numbers of T cells are then released into the circulating blood, and they spread throughout the body.

There are three major types of sensitized T cells. The types of sensitized T cells include:

1. **Cytotoxic T cells** that combine directly with antigens on the surfaces of invading organisms and can therefore destroy the organisms

2. **Helper T cells** that function mainly in association with the plasma cells in the lymph nodes. They multiply manyfold the capability of the plasma cells to produce humoral antibodies in response to antigens.
3. **Suppressor T cells** that suppress some of the immune reactions, in this way preventing the immune system from running wild and becoming destructive to normal tissues

Immune Tolerance

Normally the immune system does not attack the body's own tissues. The immune process of the normal human body does not develop antibodies or sensitized lymphocytes that can destroy the body's own tissues, although the body tissues are to a great extent like bacteria in their chemical composition. This phenomenon is called **tolerance** to the body's own proteins and tissues. This results mainly from destruction during fetal life of those primordial lymphocytes in the thymus and liver that are capable of forming antibodies or sensitized lymphocytes against the body's own proteins and tissues. It is likely that special suppressor T cells also develop to help cause tolerance.

Failure of immune tolerance leads to autoimmune disease. Autoimmunity occurs particularly in older age or after some disease causes destruction of large amounts of body tissue with release of tissue antigens into the circulating body fluids. Once the immune process has caused production of antibodies or sensitized lymphocytes that can attack the body's own tissues, these will then react against specific tissues and cause serious debility. Examples of autoimmune disease include rheumatic heart disease, rheumatoid arthritis, thyroiditis, acute glomerulonephritis, myasthenia gravis, and lupus erythematosus.

ALLERGY

There are a number of different types of allergy, some of which can occur under appropriate conditions in normal persons and some only in persons who have specific allergic tendencies.

Delayed-reaction allergy can occur in the normal person. It is caused by sensitized lymphocytes and not by antibodies. A typical example is the reaction to poison ivy. The toxin of poison ivy becomes deposited in the skin, and a small portion of it finds its way to the lymph nodes, which, over a period of several days to 1 week or more, form sensitized lymphocytes. These then are carried by the blood back to the original site of entry of the poison ivy toxin. Reaction of the lymphocytes with the toxin in direct association with the cells produces severe local tissue damage, the well-known rash and blisters associated with poison ivy.

Allergy in the allergic person is characterized by excess IgE antibodies. These IgE antibodies are called **reagins.** They have a different protein structure from that of normal IgG antibodies. Also, the reagins tend to attach themselves to basophils and mast cells throughout the tissues. When the specific antigen (called the **allergen**) that reacts with the reagin enters the tissue, it combines with the reagin. This combination, occurring on the cell surface of the basophils and mast cells, causes cellular damage with release of histamine and proteolytic enzymes from the affected cells. Severe local tissue damage can result. Examples of allergies of this type are hay fever, asthma, urticaria (called also hives), and some types of anaphylaxis.

BLOOD GROUPS AND TRANSFUSION

Successful transfusion of blood from the donor to the recipient is mainly a problem of immunity. The recipient may already be immune to the transfused blood, or he may develop immunity, which then causes damage or death to the red blood cells in the transfusion.

Blood Groups

The four major A-B-O blood groups are based on the presence or absence of two agglutinogens (cell antigens). The membranes of the red blood cells in about 60% of all people contain one or both of two very important antigens, called **group A** or **group B agglutinogens,** that frequently cause transfusion reactions. The bloods of different persons are generally **typed,** as illustrated in Table 3-2, on the basis of the presence or the absence of these agglutinogens in the blood cells. Thus, the four major blood groups of humans are **group A,** which contains type A agglutinogen; **group B,** which contains type B agglutinogen; **group AB,** which contains both A and B agglutinogens; and **group O,** which contains neither.

Agglutinins (antibodies) directed against the missing agglutinogen are almost always present in the plasma. The agglutinins, like other antibodies, are γ-globulins that develop in response to small numbers of group A and B antigens that enter the body in food, bacteria, and in other ways. Antibodies that agglutinate the A agglutinogen are called anti-A agglutinins, while those that agglutinate the B agglutinogen are called anti-B agglutinins. Thus, type A blood contains **anti-B agglutinins;** type B blood contains **anti-A agglutinins;** type AB blood contains neither of the agglutinins; and type O blood

TABLE 3-2.

The Blood Groups Showing Their Genotypes and Their Agglutinogens and Agglutinins

Genotypes	Blood Groups	Agglutinogens*	Agglutinin†
OO	O		Anti-A and anti-B
OA or AA	A	A	Anti-B
OB or BB	B	B	Anti-A
AB	AB	A and B	

*Cell antigens
†Serum antibodies

contains both anti-A and anti-B agglutinins. Therefore, mixing bloods of different types will often cause agglutination of at least some of the cells and can result in a transfusion reaction.

The Rh antigen is often present in red blood cells. The blood cells of about 85% of all white persons, 95% of American blacks, and virtually 100% of African blacks contain another antigen, called the **Rh antigen,** which exists in several different forms. Those persons who have the Rh antigen are said to be Rh positive, while those who do not have any Rh antigen are said to be Rh negative.

Transfusion Reaction

A transfusion reaction consists mainly of an attack on the red blood cells by the recipient's antibodies. A transfusion reaction is likely to occur if the host is already immune to the transfused blood or becomes immune soon after the transfusion. The antibodies, or agglutinins, attach themselves to the surfaces of the red blood cells and make them **agglutinate** with each other, which means that the cells stick together in clumps. Occasionally the antibodies are powerful enough to cause the cells to rupture. However, even if the cells do not rupture, the clumped cells become caught in the capillaries of the circulatory system and during the next few hours become ruptured because of progressive trauma or attack by white blood cells or by tissue macrophages. Thus, the final result in all transfusion reactions is rupture of the red cells, called **hemolysis,** with release of hemoglobin and other intracellular substances into the blood.

Acute renal shutdown can occur following a transfusion reaction. Much of the free hemoglobin in the blood resulting from a transfusion reaction filters through the glomerular membrane into the renal tubules. Then water is reabsorbed from the tubules, allowing the hemoglobin to become so concentrated that it precipitates.

If large amounts of blood are hemolyzed, this process can block many or most of the tubules of the kidneys, causing either oliguria or anuria. As a result, a person occasionally dies within 1 week or later of **uremia** rather than as the immediate result of the transfusion reaction.

A transfusion reaction can occur when Rh-positive blood is transfused into an Rh-negative person. However, a reaction will not occur unless the Rh-negative person has been exposed previously to Rh-positive blood because, unlike the anti-A and anti-B agglutinins, anti-Rh antibodies do not occur "spontaneously" in the blood. Yet, if the Rh-negative person has been exposed previously to Rh-positive blood, that person will have developed anti-Rh antibodies against the Rh factor, and a subsequent transfusion with Rh-positive blood can cause a transfusion reaction equally as severe as that which occurs with the A-B-O blood groups.

Erythroblastosis fetalis occurs when anti-Rh antibodies from the mother diffuse into the fetus causing red blood cell agglutination. If the mother is Rh negative, and the baby inherits the Rh-positive trait from the father, some of the Rh-positive antigens from the baby can sometimes cause the mother to develop anti-Rh antibodies. These antibodies then diffuse through the placenta into the baby and cause agglutination of the baby's circulating red cells. Although this effect rarely occurs with the first Rh-positive baby, it does occur frequently during subsequent pregnancies. The reason is that immunity develops in the mother after birth of the baby in response to antigens entering her blood from degenerating products of the placental tissues. If the mother is given antiserum against the Rh factor immediately after each delivery, most instances of immunization can be prevented. But without such preventive measures, death will occur in large numbers of newborn or unborn children. The clinical condition, when present, is called erythroblastosis fetalis because the fetus responds with an erythroblastic reaction to form more blood cells.

Chapter 4

Nerve and Muscle

FUNCTION OF THE NERVE FIBER

Diffusion and Equilibrium Potentials

An electrical potential can be established across a cell membrane by diffusion of ions through the membrane. Figure 4-1 shows the establishment of a diffusion potential for sodium and potassium. The concentration of potassium is very high inside the cell and very low outside the cell, creating a strong tendency for potassium ions to diffuse out of the cell. As positively charged potassium ions move to the cell exterior, a state of electronegativity is created inside the cell. This negative membrane potential prevents the further diffusion of positively charged potassium ions out of the cell despite the continued existence of a high potassium concentration inside the cell. The membrane potential required to balance the tendency for diffusion of potassium caused by the concentration difference is called the **equilibrium potential** and is about −94 millivolts (mV) for potassium in the normal, large mammalian nerve cell.

Figure 4-1*A* shows a similar phenomenon for sodium ions. However, this time the concentration is very low inside the cell and very high outside the cell, creating a large concentration gradient for diffusion of sodium ions into the cell. Diffusion of sodium ions into the cell creates positivity in the cell interior, which tends to repel the further diffusion of ions into the cell. The **equilibrium potential** for sodium is about +61 mV.

The Nernst equation is used to calculate equilibrium potentials. The Nernst equation shown below is used to calculate the equilibrium potential for any ion at a given concentration difference across the membrane. The greater the concentration difference, the greater the tendency for diffusion through the membrane, and therefore the greater the Nernst potential for that ion.

EMF (mV) = ± 61 log (concentration inside / concentration outside)

The sign of the potential is positive when the ion is negative and negative when the ion is positive. For example, when the concentration of a positive ion such as sodium is 10 times higher on the out-

Diffusion Potentials

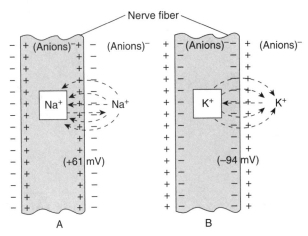

A   B

Figure 4-1.
Establishment of diffusion potentials across the cell membrane for sodium and potassium ions. *A:* When the membrane is permeable only to sodium ions, the diffusion potential is +61 mV; *B:* when the membrane is permeable only to potassium ions, the diffusion potential is –94 mV. (Adapted from Guyton AC and Hall JE: *Textbook of Medical Physiology,* 9th ed. Philadelphia: WB Saunders, 1995, p 58.)

side compared to the inside, the Nernst potential calculates to be –61 × –1.0 = +61 mV.

Resting Membrane Potential

The negative membrane potential is caused by the relative absence of positively charged ions from inside the cell. A membrane potential exists across the membranes of all cells. In large nerve and muscle cells, this potential amounts to about 90 mV during resting conditions, with the inside of the cell negatively charged with respect to the outside. The development of this **resting membrane potential** occurs as follows: All cell membranes contain sodium-potassium pumps that transport sodium to the outside of the cell and potassium to the inside. Because more sodium is transported outward than potassium inward, the net effect of these pumps is principally to transport sodium outward. This active transport of sodium, which is a positively charged ion, to the outside of the membrane causes loss of positive charges from inside the membrane and gain of positive charges on the outside, thus creating negativity inside the membrane and positivity outside.

More important, the membrane is far more permeable to potassium compared to sodium. If the membrane were permeable only to potassium ions and not to sodium ions, the membrane potential would be equal to –94 mV, the equilibrium potential for potassium. On the other hand, if the membrane were permeable only to sodium ions, the membrane potential would be equal to the equilibrium potential for sodium, +61 mV. Therefore, the relatively high permeability of the membrane to potassium causes the resting membrane potential (−90 mV) to be relatively close to the equilibrium potential for potassium (−94 mV).

The Action Potential

The depolarization phase of the action potential is caused by increased sodium permeability. The action potential, shown in Figure 4-2, is a sequence of changes in the membrane potential that occurs within a small fraction of a second when a nerve or muscle membrane impulse spreads over its surface. Any factor that makes the cell membrane suddenly excessively permeable to sodium ions can elicit an action potential. Such factors include

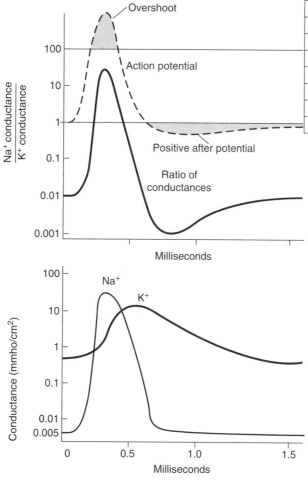

Figure 4-2.
An action potential with the associated changes in sodium and potassium conductances (permeabilities). (Adapted from Guyton AC and Hall JE: *Textbook of Medical Physiology,* 9th ed. Philadelphia: WB Saunders, 1995, p 64.)

passing an electrical current through the membrane, pinching the nerve fiber, pricking it with a pin, crushing it, or applying a drug such as acetylcholine. When the membrane becomes very permeable to sodium (i.e., sodium conductance increases), the positive sodium ions on the outside of the membrane now flow rapidly to the inside. Therefore, the negative potential inside the membrane suddenly becomes greatly reduced or even reversed with positivity on the inside and negativity on the outside. This process is called **depolarization** because the polarity across the cell membrane decreases.

Repolarization of the action potential is caused by increased potassium permeability and return of the sodium permeability to normal. When depolarization has occurred and the negative potential on the inside of the membrane has been lost, the movement of sodium ions to the interior of the cell decreases, and the permeability to sodium returns to its resting state, as shown in Figure 4-2. At the same time, the permeability to potassium increases greatly. This low permeability to sodium and high permeability to potassium permits only a few sodium ions to pass to the inside, but large numbers of potassium ions to pass to the outside, since these are present inside the cell in high concentration. Because potassium ions are also positively charged, this once again causes loss of positive charges from the inside of the membrane, reestablishing the normal negative membrane potential. This process is called **repolarization,** and the total sequence of depolarization and repolarization is called the **action potential.**

Action potentials can be propagated in both directions along a nerve fiber. An action potential elicited at one point on an excitable membrane can excite adjacent portions of the membrane, thus causing the action potential to spread in both directions. The mechanism of this is as follows: Each time an action potential begins at any given point on the membrane, the normal negative charge on the inside of the membrane is lost, causing an electrical current to flow to adjacent and as yet unexcited regions, where an **electrotonic potential** is generated. This potential reaches the **threshold value** and serves as the stimulus for exciting these adjacent regions of the membrane. As a result, an action potential then occurs at each of these points, and the process is repeated again and again until the action potential has spread to both ends of the fiber.

Conduction velocity is greatest in large, myelinated nerve fibers. The velocity of conduction in very large myelinated nerve fibers can be as great as 100 meters per second (m/sec) and as little as only 0.5 m/sec in very small unmyelinated fibers.

The refractory period limits the frequency of action potentials. The refractory period of the action potential is the duration of time between the beginning of depolarization and the end of repolarization. In large nerve fibers, the repolarization process follows about 1/2500 second after the depolarization process. Therefore, as many as 2500 nerve impulses (action potentials) can be transmitted along a large nerve fiber per second. On the other hand, the process of repolarization is so slow in some smooth muscle fibers that only one impulse can be transmitted every few seconds.

The action potential follows the all-or-nothing law. A very important aspect of nerve function is that a stimulus usually causes a nerve fiber either to transmit a complete impulse (action potential) or, if the stimulus is too weak, none at all. This is called the **all-or-nothing law of excitation.** Furthermore, once an impulse begins, it normally travels in all directions over the fiber—either backward or forward and into all branches of the fiber—until each portion of the neuronal membrane has become depolarized.

The action potential is a physical phenomenon and does not consume energy. The energy required to reestablish the sodium and potassium concentrations across the membrane following a large number of action potentials is derived from the metabolic processes of the cell. However, the number of sodium and potassium ions that move across the cell membrane with each action potential is so infinitesimal that tens of thousands of action potentials can be transmitted by a nerve fiber after the metabolic machinery of the cell has been poisoned. Therefore, nerve metabolism can be blocked for as long as several hours at a time without greatly affecting transmission of the impulse; once the differences in ion concentration across the cell membrane have been built up, impulse transmission is purely a physical phenomenon without involving the enzymatic metabolic processes of the cells.

NEUROMUSCULAR JUNCTION

Skeletal muscles are innervated by large, myelinated nerve fibers called motor neurons or **motoneurons.** The neuromuscular junction is the point of connection between a motor nerve fiber and a skeletal muscle fiber. The nerve ending, where it lies on the skeletal muscle fiber, branches to form a complex of fifty or more nerve terminals and the entire structure is called the **motor endplate.** The nerve terminals in this motor endplate synthesize and store a chemical transmitter substance—acetylcholine.

Acetylcholine is released at the neuromuscular junction of skeletal muscle. When a nerve impulse reaches the motor endplate, portions of acetylcholine stored in small vesicles are released into the **synaptic cleft** between the nerve terminals and the muscle membrane. The acetylcholine binds to receptors on the muscle membrane, increasing its permeability to sodium and thus depolarizing the muscle fiber. Within another 1/500 sec, a protein enzyme called **acetylcholinesterase** (AChE) destroys the acetylcholine. Thus, the net result of a nerve impulse reaching the motor endplate is the release of a small pulse of acetylcholine that lasts only a few milliseconds at most, but still long enough to stimulate the muscle fiber.

Acetylcholine initiates the endplate potential. When the short pulse of acetylcholine increases the permeability of the muscle membrane, sodium ions immediately begin to pour to the inside of the fiber, causing a depolarization in the immediate vicinity of the endplate. This change in potential is called the **endplate potential.** The endplate potential is nearly always above the **threshold value** of the membrane, and the action potential that follows spreads over the entire muscle fiber, as described above for nerve fibers.

Myasthenia gravis can cause muscle paralysis. Myasthenia gravis is an autoimmune disease in which the immune system has produced antibodies against acetylcholine receptors on the muscle membrane. The endplate potentials are often too small to stimulate the muscle fibers, and the result is muscle weakness. In severe cases, the patient dies following paralysis of the respiratory muscles. When treated with a drug, such as neostigmine, which can inactivate the enzyme that normally inactivates acetylcholine, the levels of acetylcholine build and thereby enhance neuromuscular transmission.

SKELETAL MUSCLE

Physiologic Anatomy

Figure 4-3 shows the structure of **skeletal muscle** from the gross to the molecular level. Skeletal muscle is composed of muscle fibers that run the entire length of the muscle. The cytoplasm of the muscle fiber contains myofibrils that consist of **actin filaments** and **myosin filaments** arranged linearly in the muscle fiber so that the myosin filaments alternate with the actin filaments. The portion of a myofibril that lies between two successive **Z disks** is called the **sarcomere,** which is the basic functional unit of skeletal muscle. The resting length of the typical sarcomere is about 2 micrometers (μm). The interdigitating actin and myosin filaments within a sarcomere cause myofibrils to have alternate light and dark bands, which give skeletal and cardiac muscle their striated appearance. The light bands contain only actin filaments and are called **I bands,** and the dark bands contain both actin and myosin filaments and are called **A bands** (Table 4-1).

Molecular characteristics of contractile filaments. Although the myosin filaments are composed entirely of myosin molecules, the actin filament is composed of long threads of **actin** and **tropomyosin** wrapped around each other in a helical manner. Also, attached periodically to the tropomyosin is a protein complex called **troponin.** The troponin and tropomyosin control muscle contraction.

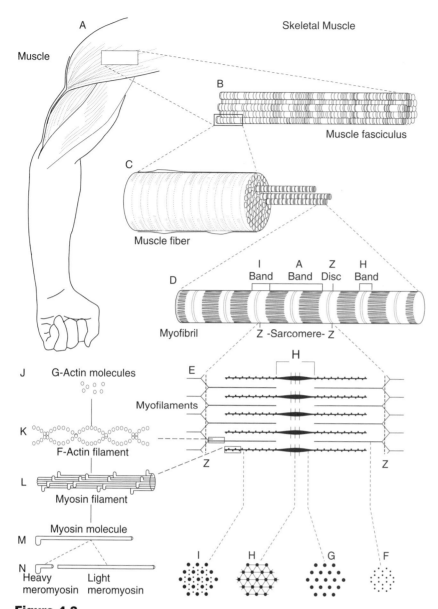

Figure 4-3.
Structure of skeletal muscle from the gross to the molecular level. (Adapted from Guyton AC and Hall JE: *Textbook of Medical Physiology,* 9th ed. Philadelphia: WB Saunders, 1995, p 74.)

Mechanism of Contraction

Actin and myosin filaments slide past each other causing muscle contraction. When an action potential passes over the muscle fiber membrane, the actin and myosin filaments interact with each other so that they slide past each other, thus shortening the fiber. Figure 4-4 shows the relaxed and contracted states of three successive sarcomeres. During the relaxed state, the actin filaments connected to the Z disks barely overlap each other, but they overlap the myosin filaments completely. In the contracted state, the actin filaments have

TABLE 4-1.

Comparison of Cardiac, Skeletal, and Smooth Muscle

Muscle Characteristic or Attribute	Cardiac Muscle	Skeletal Muscle	Smooth Muscle
Striated	Yes	Yes	No
Myosin	Yes	Yes	Yes
Actin connects to Z disks	Yes	Yes	No
Actin connects to dense bodies	No	No	Yes
Sarcomere	Yes	Yes	Functional
Troponin	Yes	Yes	No
Calmodulin	No	No	Yes
Sliding filament model of contraction	Yes	Yes	Yes
Ca^{++} required	Yes	Yes	Yes
Ca^{++} from sarcoplasmic reticulum	Yes	Yes	Minimal
Ca^{++} from T tubules	Great amount	Minimal	None
Volume of T-tubule system	25X	1X	None
Effect of Ca^{++} concentration in extracellular fluid on strength of contraction	Great effect	Little or no effect	Great effect
Major ion causing depolarization*	Na^+	Na^+	Ca^{++}
Mitochondria	Numerous	Fewer	?
Capillarity	Higher	Lower	?
Type of contraction	Phasic	Phasic	Tonic/phasic

*Ca^{++} is important for maintenance of the action potential plateau in cardiac muscle.

been pulled inward among the myosin filaments and the two successive Z disks are now much closer together. As soon as the action potential is over, the interaction between the actin and myosin filaments disappears so that the alternating sets of filaments then slide away from each other.

The "walk-along" theory of muscle contraction. The nature of the forces that cause the actin and myosin filaments to slide along each other during the contractile process is not entirely understood. One theory, however, known as the **"walk-along" theory of muscle contraction,** is as follows (Fig. 4-5): Each myosin filament has approximately 200 arms, called **crossbridges,** that extend anglewise to the sides. At the end of each crossbridge is a hinged head that makes contact with one of the actin filaments. The head can attach temporarily to an active site on the actin filament, and it can also rock back and forth where it is hinged to the crossbridge. It is believed that when the head attaches to the actin filament, the molecular forces in the head change, causing it to bend in a forward direction. This pulls the actin filament forward, but the forward bending also causes the head to detach from the actin filament and

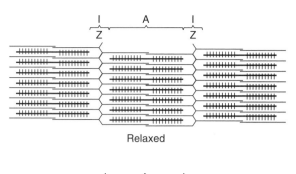

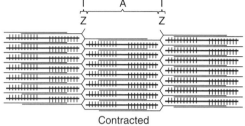

Figure 4-4.

Relaxed and contracted states of a myofibril showing sliding of the actin filaments into spaces between myosin filaments. Note that actin filaments are connected to Z disks and that myosin filaments have crossbridges. (Adapted from Guyton AC and Hall JE: *Textbook of Medical Physiology,* 9th ed. Philadelphia: WB Saunders, 1995, p 76.)

to bend back to its original position. Next, it attaches to another active site farther along the actin filament. And the head bends again and pulls the actin filament still another short distance. Thus, by a process of attachment, bending, detachment, attachment farther along the myosin, and so forth, the actin and myosin filaments are pulled together, causing muscle contraction.

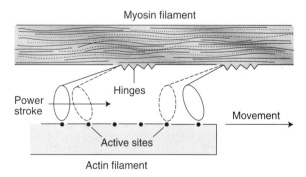

Figure 4-5.

The "walk-along" theory for contraction of muscle. (Adapted from Guyton AC and Hall JE: *Textbook of Medical Physiology,* 9th ed. Philadelphia: WB Saunders, 1995, p 78.)

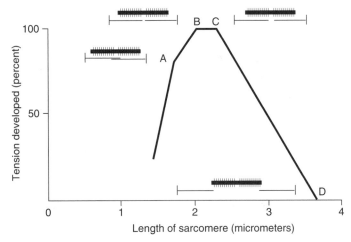

Figure 4-6.
Length-tension diagram for a single sarcomere, illustrating maximum strength of contraction when the sarcomere is 2.0 to 2.2 μm in length. The degree of actin and myosin overlap is shown at different points along the curve. (Adapted from Guyton AC and Hall JE: *Textbook of Medical Physiology,* 9th ed. Philadelphia: WB Saunders, 1995, p 79.)

Length-Tension Relationship

The degree of actin and myosin overlap determines tension development by contracting muscle. Figure 4-6 illustrates the effect of sarcomere length (actin and myosin overlap) on tension development in contracting muscle. The degree of actin and myosin overlap at different points along the curve is shown. Note at points B and C that maximum tension develops in association with maximum overlap of actin and myosin filaments at this ideal sarcomere length. Any increase or decrease in sarcomere length from this ideal position decreases the overlap of actin and myosin filaments and thus decreases the development of tension. Point D indicates that zero tension is developed when zero overlap occurs. This diagram indicates that maximum contraction occurs with maximum overlap of actin and myosin filaments and supports the idea that the number of myosin crossbridges pulling on actin filaments controls the strength of a muscle contraction.

Excitation-Contraction Coupling

Transmission of an action potential over the muscle fiber membrane causes the fiber to contract a few milliseconds later, a process called **excitation-contraction coupling.** The mechanism by which this occurs is described below.

Transverse tubules conduct the action potential to the interior of a muscle fiber. Minute tubules, called transverse tubules or T tubules, pass transversely all the way through the muscle fiber from one side of the fiber membrane to the other side. The action potential, on reaching one of these tubules, travels to the interior of the muscle fiber along the membranes of the tubules. Thus, electrical current from the action potential is distributed to the inner substance of the muscle fiber as well as along its surface. The T tubules make physical contact inside the muscle fiber with many additional very fine tubules called **longitudinal tubules.** These are part of the endoplasmic reticulum and are collectively called the **sarcoplasmic reticulum.** They lie parallel to the myofibrils that cause muscle contraction. Also, they contain large amounts of highly concentrated calcium ions.

Calcium released from longitudinal tubules binds to the troponin complex and initiates muscle contraction. When the action potential travels along the T tubules, electrical currents spread from these to the longitudinal tubules and cause a pulse of calcium ions to be released from the longitudinal tubules into the fluid surrounding the actin and myosin filaments of the myofibrils. The calcium ions then combine with the troponin complex, and in some way not yet understood, this changes the physical relationship of the tropomyosin and actin molecules. The net result is exposure of active electrical sites on the actin filament that attract the heads of the myosin filament, causing contraction of the muscle. Within another small fraction of a second the calcium ions are actively transported back into the longitudinal tubules, which decreases the calcium ion concentration back to a very low value and thereby inactivates the myosin, allowing relaxation of the muscle.

Energetics

Energy from adenosine triphosphate (ATP) returns the head of a myosin crossbridge to its "cocked" position. Contraction of the muscle requires energy. Each time a muscle fiber contracts, a certain amount of adenosine triphosphate, the energy-rich compound that is synthesized during the metabolism of food, is destroyed. It is believed that one molecule of ATP is degraded each time the head of a crossbridge bends to cause movement of the actin filament. The energy derived from this ATP is used to return the head back to its original "cocked" position. In this way the energy is stored in the cocked head, and it is the release of this energy that provides the force exerted by the head to pull the actin filament forward.

The more times a muscle fiber contracts and the greater the load against which the muscle contracts, the greater is the quantity of ATP that is degraded. During extreme muscle activity, the rate of metabolism in individual muscles sometimes rises to more than 50 times the metabolic rate of the resting muscle.

Force Summation

The strength of a muscle contraction is controlled by multiple motor unit summation and frequency summation. A skeletal muscle is composed of many muscle fibers connected in parallel with each other. The strength of contraction of the entire muscle can be increased by (1) **multiple motor unit summation,** in which many parallel muscle fibers contract simultaneously, and (2) **frequency summation,** in which successive contractions of each muscle fiber occur so close together that they actually fuse into one long continuous contraction rather than many individual twitches. This latter effect is called **tetanization.**

CARDIAC MUSCLE

The basic structure of cardiac muscle is similar to that of skeletal muscle. **Cardiac muscle** is striated with actin and myosin filaments almost identical to those found in skeletal muscle. The sarcomere runs from Z disk to Z disk, and alternating thick (myosin) filaments and thin (actin) filaments slide past each other during the process of contraction. Contraction of cardiac muscle is also triggered by depolarization of the cell membranes, which leads to increased concentration of calcium in the cell interior.

Unlike skeletal muscle fibers, cardiac muscle fibers are connected in series by low-resistance intercalated disks that allow the fibers to contract as a syncytium. The intercalated disks contain gap junctions that are very permeable, allowing ions to diffuse freely from one cell to the next. Thus, an action potential can pass from one cell to the next, allowing the cardiac muscle fibers to contract in a syncytial manner (i.e., to contract almost all at once).

Large quantities of calcium ions diffuse from the T tubules into the cardiac muscle fibers during the action potential. The volume of the T tubule system is about 25 times greater in cardiac muscle compared to skeletal muscle. Because the T tubule system is open to the extracellular fluid, the strength of contraction of cardiac muscle depends to a large extent upon the concentration of calcium in the extracellular fluid. In contrast, the strength of skeletal muscle contraction is hardly affected by the extracellular calcium concentration.

SMOOTH MUSCLE

Excitation-contraction coupling is different in smooth muscle compared to skeletal and cardiac muscle. Smooth muscle does not contain troponin; instead, another regulatory protein called **calmodulin** reacts with calcium ions (see Table 4-1). The following steps occur in excitation-contraction coupling in smooth muscle:

1. Depolarization of the cell membrane opens voltage-gated calcium channels, allowing calcium to move down its electrochemical gradient into the cell.
2. Calcium ions bind to calmodulin, and the calmodulin-calcium combination binds with and activates myosin kinase.
3. One of the myosin chains becomes phosphorylated, allowing it to bind with actin. Muscle contraction follows.
4. Contraction is terminated when myosin is dephosphorylated by myosin phosphatase.

Multiunit smooth muscle is composed of discrete smooth muscle fibers. Each fiber is often innervated by a single nerve ending, allowing it to operate independently of the other muscle fibers. Contraction is controlled almost entirely by the nervous system and is rarely initiated spontaneously. Some examples of multiunit smooth muscle are the smooth muscle fibers of the ciliary muscle of the eye, the pilo-erector muscles that cause hairs to erect, the iris of the eye, and the vas deferens.

Single-unit smooth muscle is so named because many smooth muscle fibers contract as a single unit. This is the most common type of smooth muscle, also called **visceral smooth muscle,** and is present in the gastrointestinal tract, uterus, ureter, bladder, and many blood vessels. Blood vessels also have properties of single-unit smooth muscle. The cell membranes are joined by **gap junctions** that enable ions to move from cell to cell, thus allowing an action potential initiated on one cell to spread rapidly to neighboring cells.

Smooth muscle can exhibit a continuous contraction called tone. An important distinguishing characteristic of smooth muscle is that it can contract continuously in addition to contracting rhythmically. The continuous contraction is called **tone.** The degree of tone can change from time to time, becoming almost negligible or increasing to a truly strong contraction. The intermittent rhythmic contractions are then superimposed on the basic tone. Thus, in the gastrointestinal tract, tonic contraction maintains a basal amount of pressure in the lumen of the gut, while rhythmic contractions superimposed on this cause propulsion of food along the gastrointestinal tract.

Chapter 5

Cardiovascular System

FUNCTIONAL ANATOMY OF THE HEART

The heart is actually two separate pumps: a right heart, which pumps blood through the lungs, and a left heart, which pumps blood through the peripheral organs (Fig. 5-1). Each of these two pumps is, in turn, comprised of two chambers, an **atrium** and a **ventricle.** Located between the atria and ventricles on both sides of the heart are the **atrioventricular (A-V) valves,** which normally allow blood to flow from the atrium to the ventricle but prevent backward flow from the ventricles to the atria. The right A-V valve is called the **tricuspid valve,** and the left A-V valve is called the **mitral valve.** Blood exits from the right ventricle through the **pulmonary valve** into the pulmonary artery, and from the left ventricle through the **aortic valve** into the aorta. The pulmonary and aortic valves allow blood to flow into the arteries during ventricular contraction (**systole**) but prevent blood from moving in the opposite direction during ventricular relaxation (**diastole**).

The walls of the heart, the myocardium, are composed primarily of cardiac muscle cells called myocytes. The inner surface of the myocardium, which comes in contact with the blood within the cardiac chambers, is lined with a layer of cells called **endothelial cells,** providing a smooth surface in the heart and throughout the cardiovascular system to help prevent blood clotting.

The cardiac muscle cells are arranged in layers that completely encircle the chambers of the heart. When the walls of the chambers contract, this exerts pressure on the blood that the chambers enclose and propels the blood forward. Cardiac muscle cells are striated and have typical myofibrils containing **actin** and **myosin** filaments, similar to those found in skeletal muscle, which slide along each other during the process of contraction. Adjacent cardiac muscle cells are joined end to end at structures called **intercalated disks**—actually cell membranes that have very low electrical resistance. This permits ions and, therefore, action potentials to move with ease from one cardiac muscle cell to another. Thus, cardiac muscle is a **syncytium** of many myocytes that are interconnected. When one of these cells becomes excited, the action potential spreads rapidly throughout the interconnections.

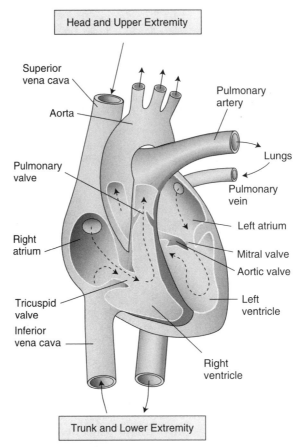

Figure 5-1.
Structure of the heart and flow of blood through the
heart chambers.

RHYTHMIC EXCITATION OF THE HEART

Action Potential

Action potentials in cardiac muscle are triggered by depolarization of the cell membrane. Contraction of cardiac muscle, similar to contraction of other muscles, is triggered by the depolarization of the cell membranes and the development of action potentials that spread from one cell to another. The resting membrane potential of normal ventricular muscle is approximately –90 millivolts (mV).

As in skeletal muscle, depolarization and development of action potentials in cardiac muscle are due mainly to a sudden increase in sodium permeability caused by the opening of voltage-sensitive **fast sodium channels** that allow large numbers of sodium ions to enter the muscle fibers. However, in cardiac muscle, unlike skeletal muscle, the action potential has a plateau and is prolonged because of opening of **slow calcium channels** (also called **calcium-sodium channels**), which remain open for several tenths of a second, allowing large amounts of calcium and sodium ions through these channels to the interior of the cardiac cell. This maintains a prolonged period of depolarization, causing a plateau in the action potential. The calcium ions that enter the cardiac muscle during this action potential play an important role in helping to excite the muscle contractile process. Eventually, repolarization occurs when the permeabilities of calcium and potassium return to their original state and the slow calcium channels close.

The action potentials of atrial cells have a similar shape except that the duration of their plateau phase is shorter. In contrast, the action potentials in the conducting system of the heart (described later) differ considerably from that of the ventricular cells just described. In some parts of the conducting system of the heart there is a gradual, progressive depolarization, resulting in a rhythmic cell excitation of the cardiac tissue.

Excitation-contraction coupling in cardiac muscle—function of calcium ions and transverse (T) tubules. When an action potential passes over the cardiac muscle membrane, the action potential also spreads to the interior of the muscle fiber along the membranes of penetrating **T tubules,** similar to skeletal muscle. The T-tubule action potentials, in turn, act on the membranes of the longitudinal sarcoplasmic tubules to cause the release of very large amounts of calcium ions into the muscle sarcoplasm from the sarcoplasmic reticulum. These calcium ions diffuse very quickly into the myofibrils and catalyze the chemical reactions that promote sliding of the actin in the myosin filaments along each other, thereby producing muscle contraction.

In addition to the calcium ions released from the sarcoplasmic reticulum, large quantities of calcium ions also diffuse into the sarcoplasm from the T tubules during the action potential. This contrasts with skeletal muscle in which virtually all of the calcium comes from the sarcoplasmic reticulum. Thus, the strength of contraction of cardiac muscle depends to a great extent on the concentration of calcium ions in the extracellular fluid. The reason for this is that the ends of the T tubules of cardiac muscle open directly to the outside of the cardiac fibers, allowing calcium in the extracellular fluid to percolate through the T tubules.

At the end of the plateau of the action potential, the influx of calcium ions to the interior of the muscle is suddenly stopped, and the calcium ions in the sarcoplasm are rapidly pumped back into both the sarcoplasmic reticulum and the T tubules. As a result, the contraction ceases until a new action potential occurs.

The sinoatrial node serves as the pacemaker of the heart. The heart is also endowed with a specialized system that initiates rhythmic contraction of the heart and rapidly conducts signals throughout the heart, thus controlling the heart contraction (Fig. 5-2). Although many cardiac fibers have the capability of self-excitation, initial depolarization normally arises in a small group of specialized cardiac cells, called the **sinoatrial (S-A) node,** located in the wall of the right atrium near the opening of the superior vena cava. The

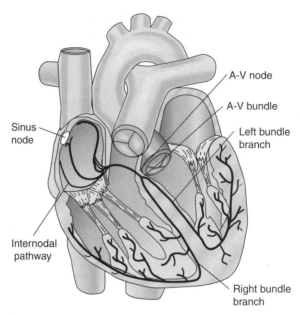

Figure 5-2.
Sinus node and the Purkinje system of the heart, also
showing the A-V node, atrial internodal pathways,
and ventricular bundle branches.

S-A node has a natural rate of rhythmicity of about 72 times per
minute, whereas the natural rate of rhythmicity of atrial muscle, if
separated from the S-A node, is only 40 to 60 times per minute and
that of the ventricular muscle is about 15 to 30 times per minute.
Thus, the more rapid basic rhythmicity of the S-A node usually initiates the heartbeat, and for this reason the S-A node normally serves
as the **pacemaker** of the heart.

The pacemaker function of the S-A node is accomplished as follows: Every time the muscle fibers of the S-A node contract, an
impulse is transmitted to the rest of the atrial muscle and from
there to the ventricular muscle. Thus, contraction of the S-A node
causes contraction of the entire heart. Then, long before the atria
or the ventricles can recover enough to contract again spontaneously, another excitatory impulse arrives from the S-A node.
Therefore, the more slowly contracting parts of the heart are never
allowed to contract at their natural rates of rhythm under normal
conditions.

**The Purkinje system rapidly conducts impulses throughout the
heart.** The **Purkinje fibers** are modified cardiac muscle cells that
allow a single impulse to rapidly spread over the entire heart muscle
mass (see Fig. 5-2). In the ventricles, the Purkinje fibers conduct
impulses at a velocity of approximately 2 m/sec, which is 4 to 6 times
the velocity in normal cardiac muscle. This rapid conduction allows
all portions of the ventricular muscle to contract almost at the same
time, rather than one part contracting ahead of another part. This,
in turn, allows effective and forceful pumping by the heart.

The Purkinje system begins in the S-A node and from here passes through the atria by way of the **internodal pathways** to the **atrioventricular (A-V) node,** which lies in the posterior wall of the right atrium near the tricuspid valve. From the A-V node, a large bundle of Purkinje fibers called the **A-V bundle** (also called the **bundle of His**) passes into the ventricular wall. Then the bundle divides into the **left and right bundle branches,** which spread, respectively, around the endocardial surfaces of the left and right ventricles. The atria and the ventricles are separated from each other by fibrous tissue everywhere, except where the A-V bundle passes into the ventricles. Therefore, in the normal heart, the only pathway by which the impulse can travel from the atria to the ventricles is through the A-V bundle.

Another very important feature of the Purkinje system is the presence of **junctional fibers** in the A-V node. These fibers are very small and have extremely slow velocity of impulse transmission, only about one-tenth the velocity of transmission in normal cardiac muscles. This slow velocity allows the prolonged delay of the impulse so that the atria contract 0.1 to 0.2 sec ahead of the ventricles. This allows the atria to pump blood into the ventricles before the ventricles begin their pumping cycle.

Conduction disorders of the heart are caused by damage to the A-V node or A-V bundle. Occasionally, pathologic conditions destroy or damage the A-V bundle or A-V node so that impulses no longer pass from the atria to the ventricles. When this occurs, the atria continue to beat at the normal rate of the S-A node, while the ventricles establish their own rate of rhythm. This condition is called **heart block.** Usually the Purkinje fibers in the A-V bundle or one of the bundle branches of the ventricles become the **ventricular pacemaker** because these fibers have a higher rate of rhythm than the muscle fibers of the ventricles. The natural rate of rhythm in the ventricles after heart block can be as slow as 15 beats/min or as rapid as 60 beats/min. In heart block, the atrial contractions are not coordinated with ventricular contractions, which prevents the ventricles from becoming as well filled before contraction as in the normal heart. Loss of this coordinated atrial function impairs maximal heart pumping about 30%. However, a person with heart block can live for many years because the normal heart has tremendous reserve capacity for pumping blood.

Cardiac flutter and fibrillation—the circus movement. Occasionally, an impulse in the heart continues all the way around the heart and, when arriving back at the starting point, reexcites the heart muscle to cause still another impulse that goes around the heart, continuing indefinitely. This **circus movement** does not occur in the normal heart for two reasons: First, normal cardiac muscle has a long refractory period, usually about 0.25 sec, which means the muscle fiber cannot be reexcited during this time. Second, the impulse of the normal heart travels so rapidly that it will normally pass over the entire atrial or ventricular muscle mass in about 0.06 sec and, therefore, disappears before the heart muscle becomes reexcitable.

In the *abnormal heart,* however, the circus movement can occur in the following conditions:

1. *When the refractory period of cardiac muscle becomes much less than 0.25 sec*
2. *When the Purkinje system is destroyed* so that impulses take a far longer amount of time to travel through the ventricles because of the slow conduction in the muscle fibers
3. *When the atria or the ventricles become dilated* so that the length of the pathway around the heart is greatly increased, thereby increasing the time required for the impulse to travel around the heart
4. *When the impulse does not travel directly around the heart,* but instead travels in a zigzag direction, which lengthens the pathway to as much as tenfold the direct distance around the heart. This obviously prolongs the time for transmission of the impulse and can result in reexcitation of the cardiac muscle.

In the atria, a regular circus movement around the atria causes **atrial flutter,** whereas zigzag impulses cause **atrial fibrillation.** The zigzag impulses in fibrillation also divide into multiple impulses so that there may be as many as 5 to 10 impulses traveling in different directions at the same time. As a result, the atria remain partially contracted all the time, but they never contract rhythmically to provide any pumping action.

Flutter only very rarely occurs in the ventricles, but **ventricular fibrillation,** with many zigzag impulses spreading in all directions at once, is a major cause of cardiac failure and death. Either an electric shock to the ventricles or ischemia of the ventricular muscle as a result of coronary thrombosis can initiate ventricular fibrillation.

THE ELECTROCARDIOGRAM—INDIRECT RECORDING OF ELECTRICAL POTENTIALS OF THE HEART

The principles and clinical use of electrocardiography obviously cannot be presented in this chapter, but the normal electrocardiogram (ECG), illustrated in Figure 5-3, can be related to the rhythmic excitation of the heart. The normal ECG consists of

- The **P wave,** which is caused by the depolarization process in the atria
- The **QRS complex** of waves, which is caused by the depolarization process of the ventricles
- The **T wave,** which is caused by repolarization of the ventricles

The PQ interval is the time between atrial and ventricular depolarization. Normally, depolarization begins in the atria approximately 0.16 sec before it begins in the ventricles. Therefore the length of

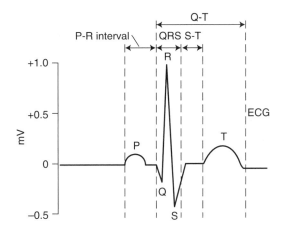

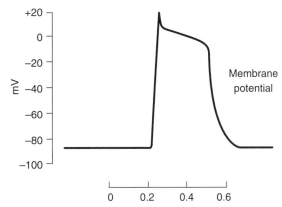

Figure 5-3.
Relationship of ventricular membrane potential and events of the electrocardiogram.

time between the beginning of the P wave and the Q wave, called the **PQ interval** (or PR interval when the Q wave is absent), is normally 0.16 sec. However, increased heart rate reduces the PQ interval.

Abnormally slow conduction of the impulse through the A-V bundle, as occurs when the bundle becomes ischemic following **coronary thrombosis** or when it becomes inflamed in the acute phase of **rheumatic fever,** prolongs the PQ interval to as much as 0.25 to 0.4 sec. The process of rheumatic inflammation of the heart can be assessed by following the changes in the PQ interval. As the disease becomes worse, the PQ interval often increases; as it becomes better, the PQ interval decreases toward normal. When the PQ interval becomes very long, conduction to the A-V bundle will eventually cease, causing heart block (as explained previously).

Abnormal QRS waves indicate aberrant impulse transmission through the ventricles. Since the QRS wave represents passage of the depolarization process through the ventricles, any condition that causes abnormal impulse transmission will alter the shape, the voltage, or the duration of the QRS complex. For example, hypertrophy of one ventricle will cause increased voltage and is likely to increase predominantly the R or the S wave, depending on the electrocardiographic lead and the ventricle affected. Also, damage to any portion of the Purkinje system will delay transmission of the impulse through the heart and therefore cause an abnormal shape of the QRS complex as well as prolongation of the complex.

Abnormal T wave indicates altered repolarization of the ventricular muscle. Some diseases damage the ventricular muscle just enough that it becomes difficult for the muscle to reestablish normal membrane potentials after each heart beat. As a result, some of the ventricular fibers may continue to emit electrical current longer than usual, which causes a bizarre pattern to the T wave, such as a biphasic pattern or sometimes inversion of the T wave. Thus, an abnormal T wave usually means mild to severe damage to at least a portion of the ventricular muscle.

Elevated or depressed S-T segment—current of injury. Occasionally, the ECG segment between the S and T waves is displaced either above or below the major level of the ECG. This is caused by failure of some of the cardiac muscle fibers to repolarize between each two heartbeats. As a result, between heart beats these fibers continue to emit large quantities of electrical current, called **current of injury,** which causes an elevated or depressed S-T segment. Therefore, when an elevated or depressed S-T segment is observed, one can be certain that at least some portion of the ventricular muscle is severely damaged, as occurs after acute heart attacks.

Abnormal heart rhythms can be diagnosed with the ECG. For example, in heart block the P waves are completely dissociated from the QRS and T waves because of blocked conduction from the atria to the ventricles through the A-V bundle. In atrial fibrillation, no true P wave can be discerned, but many very fine waves continue indefinitely in the ECG. Finally, in extrasystoles of the heart, occasional QRS and T waves appear in the record at points completely out of rhythm with the remaining portions of the ECG.

It should be emphasized that the ECG does not record changes in membrane potential across the individual cardiac muscle cells, but is a measure of electrical activity of many cells. In some cases, changes in mechanical and pumping ability of the heart can occur without measured changes in electrical activity, and therefore without major changes in the electrocardiogram. However, despite this limitation, the ECG can be a powerful tool for diagnosing many types of heart disease.

MECHANICS AND REGULATION OF HEART PUMPING

The cardiac cycle includes the events in the heart that occur from the beginning of one heart beat to the beginning of the next. The two major phases of the cardiac cycle are both named for events occurring in the ventricles: (1) a period of ventricular relaxation called **diastole,** lasting for about 0.5 sec, in which the ventricles fill with blood, and (2) a period of ventricular contraction and blood ejection, called **systole,** lasting about 0.3 sec. Thus, at a normal heart rate of about 72 beats/min, the entire cardiac cycle lasts about 0.8 sec. As the heart rate increases, the fraction of the cardiac cycle in diastole decreases, which means that the heart beating very fast may not remain relaxed long enough to allow complete filling of the ventricles before the next contraction.

Figure 5-4 shows the pressure changes in the heart during the cardiac cycle and the relationship to the ECG, the heart sounds, and ventricular volume changes. Important events during the transition from diastole to systole include closure of the A-V (mitral and tricuspid) valves, which causes the first heart sound, a brief period of isovolumic contraction, and opening of the aortic and pulmonary valves. This transition period begins with the heart at the largest volume seen during the cardiac cycle—the **end diastolic volume.** At the end of ventricular contraction, the heart has its smallest volume during the cardiac cycle—the **end systolic volume.** During this period, the aortic and pulmonary valves close, causing the second heart sound and preventing backflow from the aorta into the ventricles. Other events occurring during this transition from systole to diastole include **isovolumic relaxation** (the initial period of ventricular

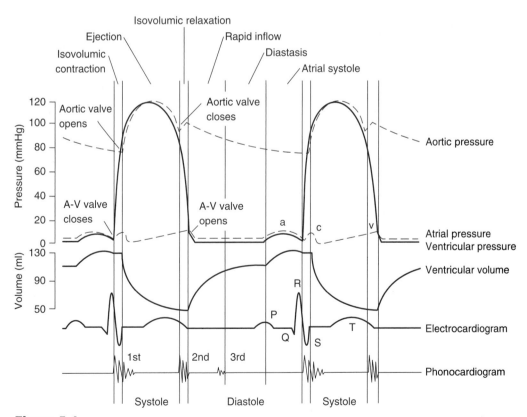

Figure 5-4.
Events of the cardiac cycle for left ventricular function, showing changes in aortic pressure, left atrial pressure, left ventricular pressure, ventricular volume, the ECG, and the phonocardiogram.

relaxation, where ventricular volume is constant because all valves are closed) and then opening of the A-V (mitral and tricuspid) valves. Because the aorta and large arteries are distensible, the blood pumped into the arterial system during systole is stored in these vessels and continues to flow through the systemic circulation even during diastole. At the end of diastole, the cycle begins again.

Note that the P wave of the ECG occurs slightly before atrial contraction; the QRS waves occur at the onset of ventricular contraction; and the T wave occurs at the time of ventricular relaxation.

Increased filling of the heart increases its overall pumping ability—the Frank-Starling mechanism. When increased amounts of blood flow into the heart from the veins and distend its chambers, the stretched cardiac muscle automatically contracts with increased force. This increased force, in turn, pumps the extra blood through the heart into the arterial system. This is called the **Frank-Starling law of the heart,** and it allows the heart to adjust its pumping capacity automatically to the amount of blood that needs to be pumped.

This mechanism also permits the heart to pump as much as two times the normal amount of blood even without increasing the heart rate, although there is a plateau in cardiac output at very high filling pressure. This plateau can be increased by sympathetic stimulation or decreased by heart failure. The **ejection fraction** is the ratio of stroke volume to end diastolic volume. Normally, this is about 0.6, but the ejection fraction can increase with sympathetic stimulation and decrease during heart failure.

Autonomic nervous control of the heart. There are two types of nerve fibers that supply the heart: **parasympathetic fibers,** which are carried in the vagus nerve, and **sympathetic fibers,** which are carried in the sympathetic nervous system. Stimulation of the sympathetic nerves releases **norepinephrine** at the nerve endings, which stimulates B_1 receptors and causes three main effects:

- Increased heart rate, called a **positive chronotropic effect**
- Increased conduction velocity, mainly in the A-V node, called a **positive dromotropic effect**
- Increased contractility of the heart muscle, called a **positive inotropic effect**

Thus, the net effect of sympathetic stimulation on the heart is increased heart rate and an increased force of contraction. Conversely, stimulation of the parasympathetic fibers carried to the heart by the **vagus nerves** releases **acetylcholine,** which stimulates **muscarinic receptors** and causes decreased heart rate **(negative chronotropic effect).**

In certain conditions, such as exercise, there is a simultaneous increase in sympathetic stimulation and an inhibition of parasympathetic stimulation, causing marked increases in heart rate and overall pumping ability. Conversely, during rest, the sympathetics are inhibited and the parasympathetics again transmit a moderate amount of impulses to the heart so that the degree of activity of the heart lessens.

DYNAMICS OF BLOOD FLOW THROUGH THE CIRCULATION

The circulation is a complete circuit. Contraction of the left heart propels blood into the systemic circulation through the aorta, which empties into smaller arteries, arterioles, and eventually the capillaries, where exchange of nutrients and waste products between the blood and the tissues occurs. Because blood vessels are distensible, each contraction of the heart distends the vessels; during relaxation the heart and vessels recoil, thereby continuing flow to the tissues even between heart beats.

Multiple regulatory mechanisms normally keep the average systemic arterial pressure within a range that is adequate for perfusion of the tissues. Pulsatile pressure in the **aorta** is dampened in the smaller, higher-resistance **arterioles** so that pressure and flows in the capillary feeding the tissues are relatively steady. The thin walls of the **capillaries** allow exchange of the nutrients and waste products between the blood and the tissues. Blood leaving the capillaries enters the **venules,** which coalesce into progressively larger **veins** that carry the blood to the right heart.

The right heart then pumps the blood through the pulmonary artery, smaller arteries, arterioles, and capillaries, where oxygen and carbon dioxide are exchanged between the blood and the tissues. From the pulmonary capillaries, blood flows into venules and large veins and empties into the left atrium and left ventricle before it is pumped again to the systemic circulation (Fig. 5-5).

Because blood flows around and around through the same vessels, *any change in flow in a single part of the circuit alters flow in other parts.* For example, strong constriction of the arteries in the systemic circulation can reduce the total cardiac output, in which case blood flow through the lungs will be decreased equally as much as flow through the systemic circulation. Another important aspect of the circulation is that sudden constriction of a blood vessel must always be accompanied by opposite dilation of another part of the circulation because blood volume cannot change rapidly and blood itself is incompressible. For instance, strong constriction of the veins in the systemic circulation displaces blood into the heart, dilating the heart and causing it to pump with increased force. This is one of the mechanisms by which cardiac output is regulated. With sustained constriction, changes in total blood volume can occur through exchange with the interstitial fluid or because of changes in fluid excretion by the kidneys.

Most of the blood volume is distributed in the veins of the systemic circulation. About 84% of the total blood volume is in the systemic circulation; 9% is in the pulmonary circulation; and 7% is in the heart. Within each of these circulations, about three-fourths of the blood is in the veins, about one-sixth in the arteries, and one-twelfth in the arterioles and capillaries. Thus, although the capillary blood exchanges nutrients and waste products in peripheral tissues

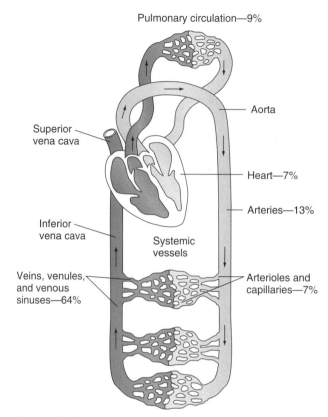

Figure 5-5.
Flow of blood and distribution of blood volume in the different portions of the circulatory system.

and gases in the lungs, only a small part of the total blood volume is in the capillaries at any given time.

Blood flow is determined by the pressure gradient and the resistance to blood flow. When a blood vessel has a high pressure at one end and a lower pressure at the other, the rate of blood flow through the vessel is directly proportional to the difference in pressure between the two ends of the vessel and inversely proportional to the resistance to blood flow along the vessel. This relationship can be expressed as:

Blood flow (Q) = pressure difference (ΔP) / resistance (R)

Blood pressure is usually expressed in millimeters of mercury (mmHg) and blood flow in milliliters per minute (ml/min). Therefore, vascular resistance is expressed as mmHg/ml/min. According to the **theory of Poiseuille,** vascular resistance is directly proportional to the viscosity of the blood and the length of the blood vessel, and inversely proportional to the radius of the vessel raised to the fourth power:

Resistance α (constant × viscosity × length) / radius^4

Thus, increased viscosity, determined principally by the concentration of red cells in the blood, raises vascular resistance. Likewise, decreased radius of the blood vessels increases vascular resistance. Because vascular resistance is inversely related to the fourth power of the radius, even a small change in radius can cause very large changes in resistance. For example, if the radius of a blood vessel increases from 1 to 2 (to 2 times normal), resistance will decrease to $1/16$ normal $(1/2^4)$ and flow will increase to 16 times normal, if the pressure gradient remains unchanged. Thus, small vessels in the circulation have the largest amount of resistance, while large vessels have very little resistance to blood flow.

For a parallel arrangement of blood vessels, as occurs in the systemic circulation in which each organ is supplied by an artery that branches into multiple vessels, the **total resistance** can be expressed as follows:

$$1/R_{total} = 1/R_1 + 1/R_2 + \cdots + 1/R_n$$

where R_1, R_2, R_n are the resistances of the various vascular beds in the circulation. The total resistance is less than the resistance of any of the individual vascular beds.

For a **series arrangement of blood vessels,** such as occurs within a tissue in which blood flows through arteries, arterioles, capillaries, and veins, the total resistance is the sum of the individual resistances:

$$R_{total} = R_1 + R_2 + \cdots + R_n$$

where R_1, R_2, R_n are the resistances of the different blood vessels in series within the tissue.

Vascular Compliance (Capacitance)

The total quantity of blood that can be stored in a given part of the circulation for each millimeter of mercury of pressure is called the **compliance** (also called **capacitance**) of the vascular bed:

$$\text{Vascular compliance} = \Delta \text{ volume} / \Delta \text{ pressure}$$

The greater the compliance of the vessel, the more easily it can be distended by pressure. Compliance is related to **distensibility** of the vessel as follows:

$$\text{Compliance} = \text{distensibility} \times \text{volume}$$

The compliance of a vein is about 24 times as great as its corresponding artery because it is about eight times as distensible and has a volume three times as great $(8 \times 3 = 24)$.

Systemic Arterial Pressure

The left ventricle normally pumps about 5 liters of blood into the aorta each minute. Each heart beat ejects approximately 70 ml of blood, called the **stroke volume output,** into the aorta. As a result, the arteries become greatly distended during cardiac systole, and during diastole the recoil of the arteries causes blood stored in the arterial tree to "run off" through the systemic vessels to the veins. Thus, the aortic pressure rises to its highest point, the **systolic pressure,** during systole and falls to its lowest point, the **diastolic pressure,** at the end of diastole. In the normal adult, the systolic pressure is approximately 120 mmHg, and diastolic pressure is 80 mmHg. This is usually written 120/80.

Pulse pressure is the difference between systolic and diastolic pressure (120 − 80 = 40 mmHg). The two most important factors that can increase pulse pressure are (1) increased **stroke volume** (the amount of blood pumped into the aorta with each heart beat), and (2) decreased **arterial compliance.** Decreased arterial compliance can result from "hardening" of the arteries that occurs with aging or arteriosclerosis.

Arterial pressure changes throughout the cardiac cycle. The **mean arterial pressure,** however, is not simply the value halfway between systolic and diastolic pressure because diastole usually lasts longer than systole. True mean arterial pressure can be measured with a catheter placed in the arteries, but a fairly accurate estimate of mean arterial pressure can also be obtained from the systolic and diastolic pressure as follows:

Mean arterial pressure

$$= 2/3 \text{ diastolic pressure} + 1/3 \text{ systolic pressure}$$

From the example above, mean arterial pressure is about $(2/3 \times 80) + (1/3 \times 120) = 93.3$ mmHg.

Because resistance to blood flow through the major arteries is so slight, the mean arterial pressure does not change markedly until the arteries become very small. In the arterioles, the major site of resistance to blood flow along the vasculature, there is a large drop in pressure to about 30 mmHg at the juncture of the arterioles and capillaries. The capillaries also contribute a moderate amount of resistance, causing the mean arterial pressure to fall to about 10 mmHg at the juncture of the capillaries and the veins. Resistance in the venous system is relatively slight, and pressure falls only 10 mmHg in the entire venous system. The pressure in the right atrium is approximately 0 mmHg.

Blood flow velocity is inversely related to total cross-sectional area of the blood vessels. The total cross-sectional area of the systemic circulation increases from about 2.5 cm^2 in the central aorta to 2500 cm^2 in the capillaries. The gathering of venules and veins decreases cross-sectional area to about 8 cm^2 at the level of the right atrium. The velocity of blood flow is inversely related to the cross-sectional area of the vessels:

$$\text{Flow velocity} = \text{blood flow} / \text{cross-sectional area}$$

Therefore, the blood flow velocity is greatest in the central aorta, reaches its lowest value in the capillary beds (about 1/100 as great as velocity in the aorta), then progressively increases in the venules and veins as the cross-sectional area decreases. The low velocity of blood flow in the capillaries is important in allowing efficient exchange of nutrients and waste products between the capillary blood and the tissues.

BLOOD FLOW REGULATION

The arterioles have a great capacity to change their resistance and to regulate blood flow. The arterioles have not only a high basal level of vascular resistance but also the capacity to change resistance by increasing or decreasing their diameters. Arteriolar walls are very muscular and respond to nervous, hormonal, and local control mechanisms that can constrict so intensely as to almost completely block blood flow, or dilate the vessels to increase blood flow as much as twenty times normal.

The local tissues autoregulate their blood flow. In most tissues, blood flow is **autoregulated,** which means that the tissue itself controls its own blood flow. This is beneficial to the tissue because it allows the rate of tissue delivery of oxygen and nutrients and removal of waste products to parallel the rate of tissue activity. Also, autoregulation permits blood flow to one tissue to be regulated independently of flow to another tissue.

The precise means by which tissues autoregulate their blood flow are still unknown. In many tissues, autoregulation appears to be linked to **oxygen delivery** or release from the tissues of **metabolic waste products,** such as adenosine and carbon dioxide, that cause vasodilation. For example, in metabolically active tissues, rapid utilization of oxygen tends to reduce oxygen tension in vascular smooth muscle. This, in turn, dilates the arterioles, increases blood flow, and causes more oxygen delivery to the active tissues. At the same time, a high rate of metabolism causes increased formation of vasodilatory metabolites that also increase blood flow. The higher rate of blood flow removes the waste products of metabolism, restoring their tissue levels toward normal. Some tissues, such as the kidneys, have special means of autoregulation that are not directly linked to metabolism.

The autoregulatory mechanisms also protect the tissues against changes in blood flow during variations in arterial pressure. For example, in most tissues, an acute decrease or increase in arterial pressure will cause an immediate decrease or increase in blood flow. However, within less than 1 minute, blood flow usually returns most of the way back toward normal (e.g., is autoregulated). In some tissues, such as skeletal muscle, this autoregulatory response is linked to maintenance of a constant supply of oxygen and other nutrients and a constant removal of metabolic waste products. For example, decreased arterial pressure tends to reduce blood flow to the tissues, which in turn causes a buildup of vasodilator metabolic waste products and decreased oxygen delivery, both of which tend to cause vasodilation. The vasodilation helps to restore tissue blood flow toward normal in the face of decreased arterial pressure. In many tissues, the autoregulatory mechanisms are capable of maintaining tissue blood flow within a few percent of normal despite variations in arterial pressure between 70 and 175 mmHg.

Endothelial cells release vasodilator (e.g., endothelium-derived relaxing factor) and vasoconstrictor (e.g., endothelin) substances. Endothelial cells of blood vessels, especially large blood vessels, release vasodilator and vasoconstrictor substances, which then diffuse to the adjacent vascular smooth muscle and induce either relaxation or vasoconstriction. One of these substances, **endothelium-derived relaxing factor** (EDRF) is believed to be **nitric oxide** and normally dilates only the local blood vessels near its release because it is rapidly destroyed in the blood. One stimulus for EDRF release is distortion of the endothelial cells caused by the shear stress of blood flowing along the vessel. With increased amounts of blood flowing into the vessel, increased shear stress on the endothelial cells releases EDRF, causing the vessel to dilate so that the increased quantity of blood can flow with greater ease at a lower velocity.

Endothelin is a powerful vasoconstrictor that is released by endothelial cells in response to damage to the endothelium. For example, after severe blood vessel damage caused by crushing of the tissues, endothelin release and subsequent vasoconstriction may help to prevent excessive bleeding from the vasculature.

Activation of the sympathetic nervous system reduces tissue blood flow. In almost all areas of the body, the arterioles are innervated by sympathetic nerve fibers that release **norepinephrine,** which then acts on α-**adrenergic receptors** in vascular smooth muscle to cause vasoconstriction and decreased blood flow. Conversely, decreased sympathetic stimulation relaxes the arterioles and increases tissue blood flow.

Activation of the sympathetic nervous system often constricts blood vessels in many parts of the body at the same time. For example, when a person stands up, many of the blood vessels of the body, especially the veins, are reflexly constricted. This offsets the tendency of blood to "pool" in the lower part of the body, allowing plenty of blood to flow back to the heart; rather than falling, the cardiac output remains near normal. The sympathetic nervous system also causes vasoconstriction in many regions of the circulation when there is a tendency for blood pressure or blood volume to decrease, as occurs with hemorrhage. This vasoconstriction helps to minimize the fall in blood pressure that would otherwise occur with hemorrhage.

Hormonal Regulation of the Circulation

Several hormones are released or secreted into the circulation and are then transported in the blood throughout the entire body. Some of the most important of these that can influence circulatory function include:

- **Norepinephrine and epinephrine** released by the adrenal medulla. These can act as vasoconstrictors in many tissues by stimulating α-adrenergic receptors. However, epinephrine is much less potent as a vasoconstrictor and may even cause mild vasodilation through stimulation of β-adrenergic receptors in some tissues, such as skeletal muscle.
- **Angiotensin II,** a powerful vasoconstrictor substance in most tissues. Angiotensin II is usually formed in response to volume depletion or decreased blood pressure.
- **Vasopressin,** which is released from the posterior pituitary. Vasopressin is one of the most powerful vasoconstrictors in the body. It is formed in the hypothalamus and transported to the posterior pituitary where it is released in response to decreased blood volume (as occurs in hemorrhage) or increased plasma osmolarity (as occurs in dehydration).
- **Prostaglandins.** These substances, formed in almost every tissue in the body, have important intracellular effects, but some of them are released into the circulation, especially prostacyclin and prostaglandins of the E-series, which are vasodilators. Some prostaglandins, such as thromboxane A_2 and prostaglandins of the F-series, are vasoconstrictors.
- **Bradykinin.** Formed in the blood and in tissue fluids, bradykinin is a powerful arteriolar vasodilator that also increases capillary permeability. For this reason, increased levels of bradykinin may cause marked edema as well as increased blood flow in some tissues.
- **Histamine,** which is a powerful vasodilator that is released into tissues when they become damaged or inflamed. Most of the histamine is derived from mast cells in damaged tissues or from basophils in the blood. Histamine, like bradykinin, increases capillary permeability and causes tissue edema as well as increased blood flow.

Long-term blood flow regulation occurs through changes in tissue vascularity and growth of new blood vessels. Most of the mechanisms previously discussed for controlling blood flow act within a few seconds or a few minutes. However, over a period of days and weeks, additional mechanisms are activated to maintain blood flow at the level needed to meet the needs of the tissues. For example, with chronic increases in blood pressure, the walls of the blood vessels become thicker and more muscular, leading to increases in vascular resistance. This increased resistance allows tissue blood flow to be maintained relatively normal in the face of high blood pressure.

Another mechanism for long-term blood flow regulation is the growth of new blood vessels—called **angiogenesis**—which occurs when the metabolic demands of the tissue change. If a tissue becomes chronically overactive and therefore requires an increased supply of nutrients and increased removal of waste products of metabolism, the blood supply usually increases within a few weeks to match the needs of the tissues. One mechanism by which this occurs is the growth of new, parallel blood vessels in the tissues, which decreases vascular resistance and increases blood flow.

The precise mechanisms that initiate angiogenesis are still being investigated, but changes in tissue levels of oxygen and waste products such as adenosine appear to be important. For example, decreased tissue levels of oxygen and increased levels of adenosine associated with increased metabolic activity of a tissue, both appear to initiate angiogenesis. Recently, several tissue growth factors have been discovered that appear to mediate the effects of changes in metabolic activity on blood vessel growth. Three of these that have been best characterized are **vascular endothelial growth factor** (VEGF), **fibroblast growth factor** (FGF), and **angiogenin,** each of which has been isolated either from rapidly growing tumors or from other tissues that have inadequate blood supply. Presumably, either a deficiency of tissue oxygen or increased production of metabolites, such as adenosine, leads to the formation of these angiogenic factors, which in turn promote new blood vessel growth.

SPECIAL AREAS OF THE SYSTEMIC CIRCULATION

Skeletal Muscle Circulation

Skeletal muscle blood flow is determined, to a large extent, by the metabolic activity of the tissue. During strenuous exercise, blood flow through skeletal muscle can increase as much as 25-fold. This increase is a result mainly of local regulatory mechanisms that cause vasodilation secondary to increased muscle metabolic activity. In most cases, skeletal muscle blood flow is proportional to the workload of the muscle, ranging from a low of 4 ml/min per 100 grams of tissue at rest to nearly 100 ml/min per 100 grams of tissue during strenuous exercise. Because a trained athlete's body often contains more than 20 kg of skeletal muscle, total muscle blood flow can be as high as 20 L/min in these individuals during strenuous exercise.

Skin Circulation

The skin plays a major role in regulating heat loss from the body. Skin blood flow, in turn, is controlled by the central nervous system via sympathetic nerves. When the body temperature rises above normal, sympathetic activity is reduced, causing the skin blood vessels to dilate and allowing rapid flow of warm blood into the skin and loss of heat. Conversely, when internal temperature of the body is too low, the skin vessels constrict, the skin becomes cold, and very little heat is lost.

Coronary Circulation

Blood flow through cardiac tissue is nearly proportional to myocardial metabolic activity, which in turn depends on the myocardial workload. Because the rate of coronary blood flow parallels very closely the rate of oxygen consumption of heart muscle, many physiologists believe that low **oxygen** concentration in the heart tissue is in some way involved in the autoregulatory process, either through the direct effect of low oxygen to cause arteriolar vasodilation or indirectly by causing the tissues to release **adenosine,** which also dilates the blood vessels.

   Coronary blood flow is normally about 4% of the resting cardiac output, but with increased myocardial workload, the coronary blood flow can increase as much as 3- to 5-fold.

Cerebral Circulation

Blood flow through the brain is normally almost constant due to autoregulation. When blood flow to the brain is threatened, as occurs with low blood pressure, this causes a buildup of carbon dioxide (CO_2) to the brain tissue, which in turn dilates cerebral arterioles and helps to restore blood flow toward normal. This interrelationship between brain tissue CO_2 and blood flow provides a stable environment for neuronal function that is highly dependent on changes in CO_2 and tissue pH.

   Decreased blood pH or decreased oxygen concentration will also increase brain blood flow when either of these effects is severe. However, the normal blood flow to the brain is relatively constant at about 700 to 800 ml/min, or about 15% of the cardiac output.

Splanchnic Circulation

The circulation of several abdominal organs, including the gastrointestinal tract, spleen, pancreas, and liver, is collectively referred to as the splanchnic circulation. Figure 5-6 shows the special arrangement of the venous circulation from the gastrointestinal tract, which is called the **portal circulation.** Note that blood flows from the gastrointestinal tract and from the spleen into the portal vein and then through the liver before emptying into the systemic veins. Normally the liver vessels offer significant amounts of resistance to blood flow and portal venous pressure is about 8 mmHg. However, in liver disease, such as **liver cirrhosis,** this resistance can increase so greatly that portal venous pressure may rise to as high as 20 mmHg or more.

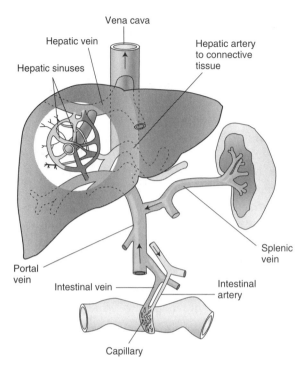

Figure 5-6.
The splanchnic circulation.

This causes the portal capillary pressure also to rise to a very high value, which in turn causes large amounts of fluid to leak out of the capillaries into the peritoneal cavity.

Together, the splanchnic organs receive about 25% of the cardiac output at rest. However, increased metabolic activity of the gastrointestinal organs, such as occurs after eating a large meal, can raise tissue blood flow as much as 50 to 100%. In emergency circumstances, such as acute hemorrhage, activation of sympathetic nerves can markedly reduce splanchnic blood flow.

REGULATION OF SYSTEMIC ARTERIAL PRESSURE

Because systemic arterial pressure is the driving force for blood flow through the tissues of the body, it is not surprising that it is carefully regulated. Under resting conditions, the mean arterial pressure is approximately 100 mmHg, but for short periods of time, such as during strenuous exercise, mean arterial pressure may rise to as high as 150 mmHg in the normal person. In chronic hypertension, the pressure remains elevated indefinitely, or until treatment is instituted to lower pressure.

Blood pressure regulation is so important for homeostasis that the body is endowed with multiple short-term and long-term control mechanisms that keep mean arterial pressure relatively constant. The short-term control mechanisms, especially the nervous reflexes, operate within seconds to minimize changes in arterial pressure during acute disturbances, such as during changes in body posture or sudden blood loss. Other mechanisms act very slowly, but powerfully, to keep mean arterial pressure relatively constant over a period of days, weeks, and months.

Short-Term Control of Systemic Arterial Pressure

The cardiovascular autonomic reflexes are powerful, rapid acting, short-term pressure control mechanisms. The **baroreceptor reflexes** are initiated by changes in mechanical stretch of receptors, called **baroreceptors,** located in the walls of the internal carotid arteries, the aorta, and in other regions of the circulation (Fig. 5-7). When the arterial pressure becomes excessively high in these vessels, these receptors are stimulated and impulses are transmitted to the brain to inhibit the sympathetic nervous system. As a result, the normal sympathetic impulses throughout the body are reduced, causing decreased heart rate, decreased strength of heart contraction, and decreased peripheral vascular resistance, which together help to reduce the blood pressure back toward normal. Conversely, a fall in blood pressure decreases the number of impulses transmitted by the baroreceptors. These impulses then no longer inhibit the

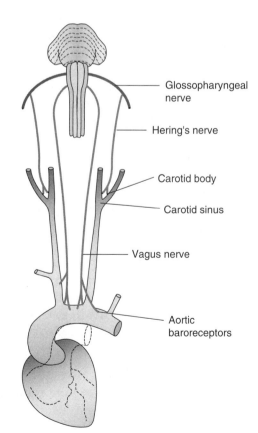

Figure 5-7.
The baroreceptor system including the aortic and carotid sinus baroreceptors, for controlling arterial pressure.

Glossopharyngeal nerve

Hering's nerve

Carotid body

Carotid sinus

Vagus nerve

Aortic baroreceptors

sympathetic nervous system so that it becomes very active, causing the blood pressure to increase back toward normal.

There are also stretch receptors located in other regions of the circulation, such as the atria, the ventricles, and the pulmonary artery. These receptors, called **cardiopulmonary baroreceptors,** also function in a manner similar to the arterial baroreceptors to keep the cardiovascular control centers informed about pressures in the venous side of the systemic circulation as well as in the pulmonary circulation. Increased pressure in these regions inhibits sympathetic activity, whereas decreased pressure stimulates sympathetic activity.

Chemoreceptors also exist in the brain and in the peripheral circulation. For example, an increase in CO_2 concentration excites the neurons of the **vasomotor centers of the brain stem,** resulting in strong sympathetic stimulation throughout the body and an increase in blood pressure. This mechanism helps to ensure adequate pressure during stressful conditions since physical stress to the body often increases the basal level of metabolism and the production of CO_2. There are also small structures known as **carotid and aortic bodies,** located in the arch of the aorta, that respond to changes in arterial blood oxygen. When blood oxygen tension decreases, these chemoreceptors cause reflex activation of the sympathetic nervous system, thereby raising blood pressure. The increased blood pressure, in turn, helps to maintain adequate delivery of oxygen to vital organs, especially the brain, in which sympathetic stimulation does not markedly increase vascular resistance. However, these receptors are much more important for control of respiration than for blood pressure regulation.

Cerebral ischemia, a lack of adequate blood flow to the brain, is also a potent stimulus for activation of the sympathetic nervous system and increased blood pressure. In brain ischemia, the vasomotor center of the brain automatically becomes highly excited, probably because of the failure of the blood to carry CO_2 out of the vasomotor center rapidly enough. As a result, **CNS ischemic reflexes** initiate strong sympathetic stimulation throughout the body, immediately elevating the arterial pressure. This, in turn, increases cerebral blood flow back toward normal and helps to relieve the effects of ischemia.

Hormones control arterial pressure. Several hormonal mechanisms also provide moderately rapid control of arterial blood pressure. Sympathetic stimulation to the adrenal medulla causes release of **norepinephrine** and **epinephrine,** both of which add to the vasoconstrictor effect of increased sympathetic stimulation.

A second hormonal system, involved in both short-term and long-term blood pressure regulation, is the **renin-angiotensin system.** This system acts in the following manner for acute blood pressure control (Fig. 5-8):

1. A decrease in blood pressure stimulates the juxtaglomerular cells of the kidney to secrete **renin** into the blood.
2. Renin catalyzes the conversion of **renin substrate (angiotensinogen)** into the peptide **angiotensin I.**

3. Angiotensin I is converted into **angiotensin II** by the action of **converting enzyme,** present in the lungs and many of the blood vessels. Angiotensin II, the primary active component of this system, is a potent vasoconstrictor and rapidly raises arterial pressure. As described in Chapter 2, angiotensin II also directly causes the kidneys to retain sodium and water, and stimulates **aldosterone** secretion, which also decreases renal sodium and water excretion and helps to expand blood volume—actions that are especially important in long-term blood pressure regulation.

Another hormone that is released when blood pressure falls too low is **vasopressin** (also called **antidiuretic hormone,** or ADH). Vasopressin has a direct vasoconstrictor effect on peripheral blood vessels and also decreases renal excretion of water, thereby increasing blood volume. These different hormonal mechanisms, together with the cardiovascular reflexes, provide a powerful means of resisting changes in blood pressure under acute conditions, such as when a person stands after having been in a lying position or when a person bleeds severely. However, the cardiovascular reflex mechanisms are probably not of great importance in long-term blood pressure

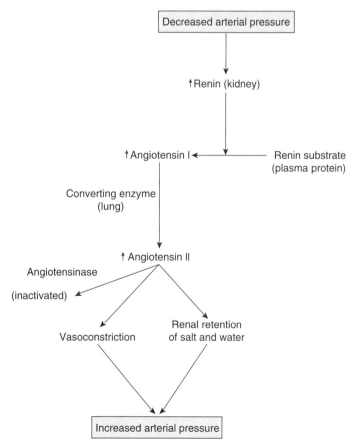

Figure 5-8.
The renin-angiotensin system and arterial pressure control.

regulation because most of them eventually adapt (or "reset") to the prevailing pressure level.

Long-Term Control of Systemic Arterial Pressure

The kidneys play a dominant role in long-term regulation of arterial pressure and circulatory volume. The most important mechanism for long-term control of arterial pressure is linked to control of circulatory volume by the kidneys, a mechanism known as the **renal-body fluid feedback mechanism** (Fig. 5-9). When arterial pressure rises too high, the kidneys excrete increased quantities of sodium and water because of **pressure natriuresis** as discussed in Chapter 2. As a result, the extracellular fluid volume and blood volume both decrease and continue to decrease until arterial pressure returns back to normal and the kidneys excrete normal amounts of sodium and water.

Conversely, when arterial pressure falls too low, the kidneys reduce their rate of sodium and water excretion and over a period of hours to days, if the person drinks enough water and eats enough salt to increase blood volume, arterial pressure will return to its previous level. This mechanism for blood pressure control is very slow to act, sometimes requiring several days or perhaps as long as 1 week or more to come to equilibrium. Therefore, it is not of major importance in acute control of arterial pressure. On the other hand, it is by far the most potent of all long-term arterial pressure controllers.

Hormonal factors increase the effectiveness of renal-body fluid feedback control. The basic mechanism for long-term control of blood volume and arterial pressure by the kidneys is enhanced by

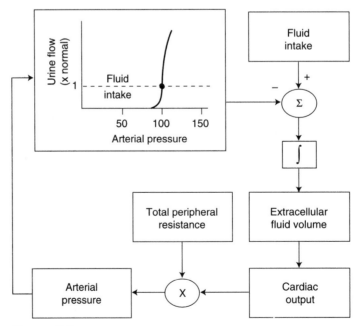

Figure 5-9.
Renal-body fluid feedback mechanism for long-term regulation of arterial pressure.

the hormonal and nervous mechanisms discussed above, especially the renin-angiotensin and aldosterone systems. For example, increasing intake of salt tends to raise blood volume and arterial pressure, which in turn increases renal salt and water excretion through pressure natriuresis, as previously discussed (see Fig. 5-9). The increased renal excretion eliminates the extra salt with relatively small changes in blood volume and arterial pressure as long as the renin-angiotensin and aldosterone systems are functioning normally.

Most persons can easily eliminate extra salt intake with minimal increases in arterial pressure and blood volume because increased salt intake also reduces the formation of angiotensin II and aldosterone, which helps to eliminate the additional sodium. As long as the renin-angiotensin-aldosterone systems are fully operative, salt intake can be as low as 1/10 normal or as high as 10 times normal with only a few mmHg change in blood pressure (Fig. 5-10). However, when the renin-angiotensin-aldosterone systems are excessively stimulated, the pressure natriuresis curve is not nearly as steep. Therefore, when sodium intake is raised, much greater increases in arterial pressure are needed to increase sodium excretion and to maintain sodium balance.

Drugs that block the renin-angiotensin system (e.g., angiotensin-converting enzyme [ACE] inhibitors that block the formation of angiotensin II) have proved to be clinically effective in improving the kidneys' ability to excrete salt and water. As shown in Figure 5-10, blockade of angiotensin II with a converting enzyme inhibitor shifts the renal-pressure natriuresis curve to lower blood pressures. This indicates enhanced ability of the kidneys to excrete sodium because normal sodium excretion can be maintained at reduced arterial pressures.

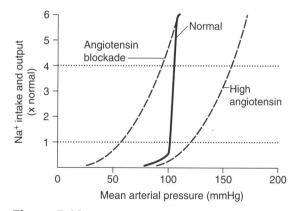

Figure 5-10.
Effect of excessive angiotensin II formation and the effect of blocking angiotensin II formation on pressure natriuresis. High levels of angiotensin II formation decrease the slope of pressure natriuresis, making blood pressure sensitive to changes in sodium intake. Blockade of angiotensin II formation shifts pressure natriuresis to lower blood pressures.

Hypertension

Hypertension is a disorder characterized by **elevated systemic arterial pressure.** A person is usually considered to be clinically hypertensive if the arterial pressure is greater than 140/90. Approximately 25 to 30% of the adult population in westernized societies is hypertensive, although the incidence of hypertension is much higher in elderly persons. Hypertension is one of the principal risk factors for the development of stroke, myocardial infarction, and kidney disease. Yet despite the great incidence of hypertension in the population and its important consequences, its precise cause in most people is still unknown. This type of hypertension is called **essential hypertension.** In the remaining cases, the cause is usually renal disease or nervous or hormonal disorders that affect the kidneys' pressure regulatory system.

Renal hypertension is caused by reduced ability of the kidney to excrete water and salt. Such conditions include pyelonephritis, glomerulonephritis, polycystic kidney disease, arteriosclerosis, renal artery stenosis, and many other types of renal disease.

One type of renal dysfunction that reduces water and salt excretion is renal vascular damage, such as stenosis of the arteries of the kidneys, constriction of the afferent arterioles, or increased resistance to fluid filtration through the glomerular membrane. Each of these factors reduces the ability of the kidneys to form glomerular filtrate, which in turn causes fluid and electrolyte retention, as well as increased blood volume and increased blood pressure. The rise in blood pressure then helps to return glomerular filtration rate toward normal and reduces tubular reabsorption, permitting renal excretory function to return to normal in the face of the vascular disorders.

The amount of long-term increase in blood volume required to cause hypertension is only a few percent and is much less than that required to cause acute increases in blood pressure. The reason for this is that acute increases in blood volume activate rapid blood pressure control mechanisms, especially the baroreceptor mechanisms, that oppose increased arterial blood pressure. Therefore, an increase in volume will not immediately raise blood pressure but instead will cause pressure to rise slowly over days or weeks. Even when hypertension is associated with marked increases in blood volume, the cardiac output normally rises significantly above normal only during the first 1 or 2 days of hypertension because the increased output also causes excess blood flow to the body's tissues, which automatically causes constriction of the arterioles of the tissues; this process, called **autoregulation,** helps to return tissue blood flow toward normal. Therefore, the cardiac output returns to normal while peripheral vascular resistance becomes elevated, although the initiating cause of the hypertension was increased blood volume and a rise in cardiac output.

Ischemia of the kidneys can also increase the arterial pressure by causing excess formation of renin and therefore **angiotensin II,** which constricts the arterioles throughout the body, thereby rapidly increasing arterial pressure. This effect occurs especially in malignant hypertension and in hypertension caused by unilateral renal artery disease when the renal ischemia fails to be relieved by the high blood pressure. However, increased angiotensin II levels are not elevated in most essential hypertensive patients.

Hypertension can be caused by excess vasoconstrictor or sodium-retaining hormones. Oversecretion of certain hormones, especially the adrenal medullary hormones and the adrenocortical hormones, can cause hypertension. For example, a tumor of the adrenal gland called **pheochromocytoma** occasionally secretes large amounts of norepinephrine and epinephrine. These two substances have almost exactly the same effect on the circulatory system as stimulation of the sympathetic nervous system, thereby elevating the mean arterial pressure.

Occasionally, a tumor of the adrenal cortex or hyperplastic adrenal cortices (called **primary aldosteronism**) secretes excessive quantities of adrenal cortical hormones, such as aldosterone or cortisone. These hormones can cause the kidneys to reabsorb large amounts of sodium and water from the tubules and this, in turn, leads to an elevated blood volume and high blood pressure.

Essential hypertension is caused by abnormal kidney function. Patients who develop essential hypertension slowly over many years almost always have significant changes in kidney function. Most important, the kidneys cannot excrete adequate quantities of water and salt at normal arterial pressures but instead require a high arterial pressure to maintain normal balance between intake and output of water and salt.

In theory, abnormal renal excretory capability could be caused either by renal vascular disorders that reduce glomerular filtration or tubular disorders that increase reabsorption of salt and water. Because patients with essential hypertension are very heterogeneous, with respect to the characteristics of their hypertension, it seems likely that both types of disorders contribute to increased blood pressure in hypertensive subjects.

Recent studies indicate that **excessive weight gain (obesity)** and **advanced age** contribute to increased blood pressure in a high percentage of essential hypertensive patients. The precise mechanisms by which aging and weight gain raise blood pressure are still not known, but some studies suggest that weight gain may activate the sympathetic nervous system and cause changes in the kidneys that lead to increased tubular reabsorption of salt and water. With advancing age, the development of atherosclerosis may also contribute to increased renal vascular resistance and further impairment of kidney function.

A widely held belief is that genetic abnormalities may contribute to increased blood pressure in 20% to 30% of hypertensive patients. For example, some investigators believe that hypertensive persons may have genetic abnormalities of vascular smooth muscle, perhaps of the smooth-muscle cell membranes, that cause excessive constriction of the small arterioles throughout the body, including the renal arterioles, which leads to hypertension.

CARDIAC OUTPUT REGULATION

Cardiac output equals venous return in the other steady state. Cardiac output is the amount of blood pumped by the left ventricle into the aorta and then to the circulation of the tissues each minute. **Venous return** is the amount of blood flowing from the tissues into the veins and then into the left or right atrium each minute. Although transient differences between cardiac output and venous return can occur, these two variables are equal in the steady-state.

The normal cardiac output (and venous return) in the adult is approximately 5 L/min but may increase to 4 to 5 times this value during strenuous exercise. Because the cardiac output is the sum of flow to all of the tissues of the body, it is regulated in proportion to the need for blood flow through the body tissues. That is, each tissue controls its own blood flow mainly by autoregulation, which matches the blood flows through the tissues to their metabolic needs, and the cardiac output is then regulated to supply the required blood flows. For example, cardiac output may reach over 25 L/min in a young athlete performing vigorous exercise, or it may be greatly reduced in a very inactive person with muscle atrophy. Loss of a limb by amputation or an organ because of injury or surgery likewise reduces the cardiac output by an amount equal to the flow to the lost limb or organ.

Changes in cardiac output can be effected in two different ways: (1) by **changing the pumping ability of the heart,** and (2) by **changing the venous return** of blood into the heart from the systemic vessels.

Pumping ability of the heart is regulated by ventricular preload and cardiac contractility. One of the two primary mechanisms that controls the pumping action of the heart is the **Frank-Starling mechanism,** which refers to the ability of the heart to modify its stroke volume in response to variations in the amount of diastolic filling. Within limits, the more the heart is filled during diastole, the greater the initial length of cardiac muscle fibers and the more forceful the ventricular contraction, causing a greater stroke volume. Figure 5-11 shows the cardiac output curve, in which cardiac output is plotted as a function of the cardiac filling pressure (right atrial pressure). The greater the atrial pressure, the greater the stretch of cardiac muscle fibers just before contraction begins. This is frequently referred to as **ventricular preload.** Within certain limits, the greater the ventricular preload, the greater the force of contraction of cardiac muscle. This mechanism has two important characteristics: (1) it occurs on a beat-by-beat basis to match cardiac pumping with venous return; and (2) it is a property of cardiac muscle fibers that is independent of innervation.

Contractility of cardiac muscle also influences cardiac pumping ability. This refers to the ability of cardiac muscles to contract independent of changes in the initial length of the muscle fibers.

Increased contractility shifts the relationship between left ventricular end diastolic pressure and cardiac output to higher levels. As discussed earlier in the chapter, increased sympathetic stimulation to the heart increases the cardiac contractility as well as the heart rate. Decreased cardiac contractility shifts the ventricular function curve downward. This can occur, for example, with myocardial infarction, which causes ischemia and impaired contraction of the cardiac muscle.

Regulation of venous return. Venous return is regulated by three primary factors: (1) the average pressure of the blood in the systemic circulation, called the **mean systemic filling pressure;** (2) the **right atrial pressure;** and (3) the **resistance to blood flow** between the peripheral vessels and the right atrium, often called **resistance to venous return.**

In the normal circulatory system, the right atrial pressure is not a primary factor in regulating venous return because the heart normally pumps all the blood that comes into it and therefore maintains the right atrial pressure near zero. For this reason, the two other factors, the mean systemic filling pressure and the resistance to blood flow through the vessels, are the major controllers of venous return.

The **mean systemic filling pressure** is the average of all the pressures in the systemic circulation, which is usually about 7 mmHg in the normal animal. It is this low because most of the blood is in the veins where the pressure is low rather than in the arteries where the pressure is high. Factors that can increase the mean systemic filling pressure and therefore return of blood to the heart include increased blood volume or constriction of the blood vessels. Sympathetic constriction of the veins is especially important because most of the blood is stored in the veins. When the veins constrict, large

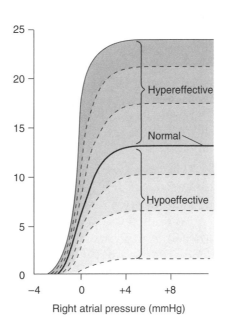

Figure 5-11.
Cardiac output curves for the normal heart and for hypoeffective and hypereffective hearts.
(Adapted from Guyton AC, Jones CE, and Coleman TG: *Circulatory Physiology: Cardiac Output and Its Regulation.* Philadelphia: WB Saunders, 1973.)

amounts of blood are forced into the heart, thus distending the heart chambers and transiently increasing the cardiac output.

Reducing the **resistance to blood flow** also increases flow from the systemic vessels toward the heart, thereby increasing venous return. This is a primary means by which venous return is controlled. When the tissues become metabolically active, their blood vessels dilate, causing blood to flow more rapidly from the arteries into the veins and increasing venous return and cardiac output. Thus, an adequate amount of blood flow is automatically made available to the active tissues while nonactive tissues maintain their same resting blood flow. Thus, increased cardiac output during strenuous exercise is caused mainly by increased flow to the actively exercising muscles.

Venous pressure is regulated by cardiac pumping ability and by venous return. The venous pressure is regulated mainly by the same two factors that regulate cardiac output—the ability of the heart to pump blood and the venous return to the heart. The normal heart is capable of pumping all of the blood that returns to it and, therefore, keeps the right atrial pressure at near 0 mmHg. However, if the heart is weakened by injury or disease, or if the venous return to the heart becomes so much greater than normal that the heart cannot pump it all, blood begins to dam up in the right atrium and the right atrial pressure rises. Thus, the balance between pumping of the heart and venous return determines the right atrial pressure. The right atrial pressure, in turn, is an important determinant of peripheral venous pressure. The greater the right atrial pressure, the higher the peripheral venous pressure must be to keep blood flowing toward the heart. In most cases, the right atrial pressure must rise about 5 to 6 mmHg before significant distention of the peripheral veins occurs.

The venous pump keeps blood flowing toward the heart. In the standing position, blood does not flow with ease uphill through the veins. However, the veins are provided with valves, and when the surrounding muscles intermittently contract and compress the veins, this acts as a pump to keep blood flowing toward the heart. If a person stands completely still and the veins are not intermittently compressed, or if the valves of the veins have been destroyed, as occurs in varicose veins, then the venous pump is no longer effective. Under these conditions, the weight of the blood in the veins makes the venous pressure in the foot of a standing person as high as 75 to 90 mmHg.

The relationship between cardiac output and venous return. As discussed previously, cardiac output and venous return must be equal under steady-state conditions. Figure 5-12 plots cardiac output and venous return as a function of right atrial pressure. The point where the two curves intersect represents the equilibrium point at which cardiac output equals venous return.

Cardiac output can be expressed as the product of stroke volume and heart rate. Although we have emphasized that cardiac output is determined by the venous return and pumping ability of the heart, the cardiac output can also be expressed as:

$$\text{Cardiac output} = \text{heart rate} \times \text{stroke volume}$$

Thus, the total amount of blood pumped by the heart each minute (cardiac output) is equal to the rate at which the heart beats multiplied by the volume of blood pumped with each heart beat (stroke volume).

Analysis of the quantitative importance of changes in heart rate on cardiac output is difficult because a change in heart rate also alters other factors that affect cardiac pumping. For example, an increase in heart rate decreases the duration of diastole and therefore reduces the time available for ventricular filling.

Experimental studies in animals and humans have shown that when heart rate is varied between about 100 to 200 beats/min, using a pacemaker, cardiac output is not markedly altered. However, at very slow heart rates (**bradycardia**) cardiac output may be reduced. This can occur, for example, in patients with complete **A-V block,** which may require the installation of a pacemaker to improve cardiac pumping.

At the other extreme, excessively high heart rates in patients with **ventricular tachycardia** or **supraventricular tachycardia** may require treatment because their ventricular filling time is so low at the high heart rates that cardiac output may not be sufficient to maintain the nutritional needs of the body. Sometimes the tachycardia can be converted to a more normal rhythm pharmacologically or by delivering a strong electrical current to the heart, a treatment called **cardioversion.**

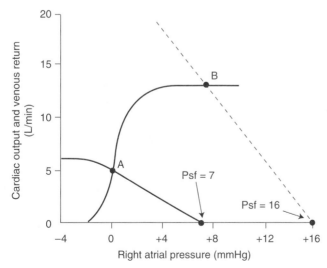

Figure 5-12.
The normal cardiac output and venous return curves.

PHYSIOLOGY OF HEART FAILURE

Coronary atherosclerosis and thrombosis are the most common causes of heart failure. One of the most important ailments requiring hospitalization is heart failure—the failure of the heart to pump enough blood to supply the needs of the body. The most frequent cause of heart failure is insufficient coronary blood flow to the heart muscle, which causes ischemia and weakness, and in some cases, destruction of the cardiac muscle.

The most common cause of insufficiency of coronary blood flow is **atherosclerosis,** fatty-fibrotic lesions of the coronary vessels that cause progressive, usually localized, blockage of blood flow. This occurs, to some extent, in most older individuals. In some persons, another condition, **coronary thrombosis,** occurs acutely as a result of a blood clot in the coronary artery, leading to the well-known "heart attack." Both coronary atherosclerosis and coronary thrombosis normally occur from **atheromata**—infiltration of the coronary wall with cholesterol and other fatty substances. The deposits of cholesterol attract the growth of fibrous tissue and thereby cause the sclerosis, or occasionally the cholesterol protrudes to the intima into the lumen of the vessel and causes a blood clot to form, thereby occluding the vessel rapidly and producing a heart attack.

Heart failure reduces cardiac output. One of the immediate effects of heart failure is decreased pumping ability of the heart, which, in turn, reduces cardiac output and causes blood to dam up in one or both atria. If cardiac output remains low and if the rise in atrial pressure is severe, edema can occur in peripheral tissues and in the lungs, which can rapidly cause death.

Heart failure activates the sympathetic nervous system. The fall in cardiac output that occurs immediately after an acute heart attack also decreases systemic arterial pressure and initiates intense cardiovascular reflexes that activate the sympathetic nervous system. The sympathetic nervous activation, in turn, helps to compensate for decreased pumping ability of the heart in two ways: Sympathetic stimulation increases the strength of contraction of the nondamaged portion of the heart, and sympathetic stimulation increases vasomotor tone throughout the systemic circulation, allowing increased venous return of blood to the heart and maintaining arterial pressure sufficiently high to perfuse vital organs, including the brain. In severe heart attacks, the nervous compensations are unable to return the cardiac output to normal, but nevertheless do help maintain perfusion of the brain and thereby help prevent death of the patient.

Heart failure causes renal fluid retention. With a reduction in blood pressure immediately after an acute heart attack, the urine output by the kidneys is also reduced for three reasons: (1) sympathetic reflexes cause intense afferent arteriolar constriction of the kidneys and reduced glomerular filtration rate as well as increased tubular reabsorption of sodium and water; (2) the low cardiac output and reduced arterial pressure also contribute to decreased glomerular filtration rate; and (3) the low cardiac output and reduced arterial pressure stimulate the kidneys to secrete large amounts of renin, causing increased angiotensin II formation. The increased angiotensin II levels directly stimulate sodium reabsorption by the kidneys and indirectly increase sodium reabsorption by stimulating aldosterone secretion. Consequently in severe, acute cardiac failure, a person may become completely **anuric** (complete cessation of urine formation), or in milder degrees of cardiac failure, the person may become **oliguric** (a state of reduced urine formation).

The fluid retention by the kidneys increases the extracellular fluid volume, but much of this leaks out of the capillaries into the interstitial spaces and causes edema. However, some of the volume remains in the blood and increases the blood volume. Moderate degrees of fluid retention are beneficial because the increased blood volume promotes return of blood to the heart, which allows increased cardiac pumping. Beyond a certain degree of venous return, the heart muscles can become overstretched, and further retention of fluid then becomes detrimental to heart function. Also, excessive fluid retention can promote pulmonary edema and death.

Effects of left heart failure compared with right heart failure. Since the heart is actually two separate pumps, one side of the heart can fail independently of the other. More often, the left heart fails because most coronary thromboses affect principally the left ventricle. However, right heart failure occurs in patients who have pulmonary hypertension or in patients with certain types of congenital heart defects.

Some of the differences between left and right heart failure are obvious. **Left heart failure** causes a shift of fluid into the lungs with resulting pulmonary edema, while right heart failure causes shift of fluid into the systemic circulation.

In **right heart failure,** only a small amount of venous congestion occurs immediately in the systemic circulation since only small amounts of extra blood are available in the lungs to shift into the systemic circulation. Therefore, peripheral edema does not occur to any significant extent immediately after acute right heart failure; instead, it must await retention of fluid by the kidneys.

On the other hand, in **acute left heart failure** a shift of blood into the lungs from the very voluminous systemic circulation can result in severe pulmonary edema and death within a matter of minutes. Sometimes pulmonary capillary pressure rises only a moderate amount immediately after the failure, not enough to cause significant edema. But during the next few days, as the kidneys retain fluid, pulmonary capillary pressure may rise still higher, causing more severe pulmonary edema and respiratory death.

PATHOPHYSIOLOGY OF CIRCULATORY SHOCK

Circulatory shock occurs when the cardiac output is so greatly reduced that tissues throughout the body begin to deteriorate for lack of adequate nutrition. Any circulatory abnormality that severely reduces cardiac output can cause circulatory shock. There are four major classifications of shock: cardiogenic, hypovolemic, septic, and neurogenic.

Cardiogenic shock. This occurs most frequently in acute heart failure in which the cardiac output falls because of impaired pumping ability of the heart itself.

Hypovolemic shock. This type of shock results from reduced blood volume caused by blood loss, plasma loss, and dehydration. Hypovolemic shock decreases the venous return to the heart thereby reducing cardiac output.

Septic shock. This condition, formally known as "blood poisoning," refers to widely disseminated infection to many areas of the body with the infection being borne through the blood from one tissue to another, causing extensive damage. Some of the typical causes of septic shock include:

1. **Peritonitis** caused by spread of infection in the gastrointestinal tract
2. **Generalized infection** resulting from spread of simple skin infection such as streptococcal infection
3. **Generalized gangrenous infection**
4. **Infection spreading into the blood** from the kidney or urinary tract
5. **Endotoxin shock,** which occurs when a large segment of the gut loses much of its blood supply and bacteria proliferate in the gut, releasing a toxin called **endotoxin.**

In septic shock, once the bacteria or endotoxin enter the bloodstream, severe dilation of the bloodvessels of the body occurs, resulting in reduced blood pressure. Also, further compounding circulatory depression is the direct effect of endotoxins on the heart to decrease myocardial contractility.

Neurogenic shock. This is a circulatory shock that results from sudden inhibition of the sympathetic nervous system throughout the body. This allows all the systemic vessels to dilate and the blood to "pool" in the lower part of the body rather than returning to the heart. If a person with loss of sympathetic tone is kept in the standing position, this can actually cause death, but if the person is placed in a horizontal or head-down position, sufficient blood will usually still flow back to the heart to allow survival.

Circulatory shock is associated with vicious cycles. One of the key features of circulatory shock is that it creates a vicious cycle that tends to make the shock itself worse. That is, the shock causes poor blood flow to the tissues of the body, including the tissues to the heart and the vascular system. This causes deterioration of the heart and blood vessels, causing still further decreases in cardiac output, and further tissue deterioration. The vicious cycle continues again and again unless the cardiovascular reflexes and other compensatory mechanisms overcome the progressive tendency of shock or, unless therapy is instituted, death occurs.

Circulatory shock can be irreversible. Another distinguishing characteristic of shock is that, beyond a certain stage of progressive deterioration, any amount of therapy becomes ineffective in preventing death of the patient. This is called the **irreversible stage of shock.** Often, different types of therapy such as blood transfusion will return the arterial pressure to normal and above normal. Yet, after a brief period, the blood pressure begins to fall again because the cardiovascular tissues have already been damaged too much and further therapy fails to keep this cycle from proceeding on to death.

Chapter 6

Respiratory System

The respiratory system supplies oxygen to the tissues and removes carbon dioxide (CO_2). Major functional events of respiration include ventilation, which is how air moves in and out of the alveoli; diffusion of oxygen (O_2) and CO_2 between the blood and alveoli; transport of O_2 and CO_2 to and from the peripheral tissues; and regulation of respiration.

VENTILATION OF THE LUNGS—MECHANICS

Breathing in humans requires the cyclical inflation and deflation of the lungs. Lung movement during inspiration results from the forces generated by the respiratory muscles.

Muscles of Respiration

The diaphragm is the most important muscle of inspiration. It is the only respiratory muscle used during rest. Contraction of the diaphragm elongates the thoracic cavity, causing the lungs to expand. During strenuous activity, muscles of inspiration also include the **external intercostals** and **accessory muscles** in the neck, which pull the rib cage upward and forward in a "bucket handle" motion, increasing the thickness of the chest cavity.

   Expiration occurs passively during quiet breathing by elastic recoil of the lungs and chest wall. The lungs and chest wall are elastic and tend to return to their resting positions following inspiration. Expiration becomes an active process during exercise and other strenuous activities in which breathing increases greatly. Expiration is also active when airway resistance is increased in diseases such as **asthma.** The major muscles of expiration are the abdominal muscles. Contraction of these forces the abdominal viscera upward against the bottom of the diaphragm. The **internal intercostals** help with expiration by pulling the chest cage downward, which decreases the thickness of the chest cavity.

The Breathing Cycle

Lung volume increases and decreases as the thoracic cavity expands and contracts. The lungs float freely in the thoracic cavity; whenever the length or thickness of the thoracic cavity increases or decreases, simultaneous changes in lung volume must also occur. The space between the visceral pleura of the lungs and the parietal pleura of the thoracic cage is called the **intrapleural space.** Continuous absorption of fluid by lymphatic channels keeps the space nearly empty except for a few milliliters of pleural fluid that provide lubrication for the moving lungs.

Changes in alveolar and pleural pressures during normal breathing. **Alveolar pressure** is the pressure of the air inside the lung alveoli. **Pleural pressure** is the pressure in the narrow space between the lung pleura and chest wall pleura. Pleural pressure can be estimated by a balloon catheter in the esophagus. Figure 6-1 shows the changes in lung volume, alveolar pressure, pleural pressure, and transpulmonary pressure during normal breathing:

1. **During rest before inspiration begins.** The pressures in all parts of the pulmonary tree are equal to the barometric pressure, which is considered to be 0 cm water. Thus, there is no pressure gradient for air to flow. Pleural pressure is subatmospheric because of the opposing tendencies of lungs to collapse and the chest to expand.

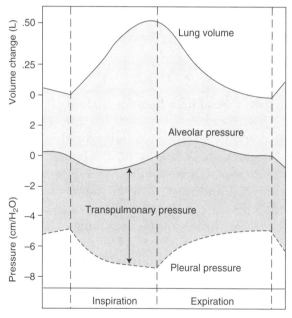

Figure 6-1.
Changes in lung volume, alveolar pressure, pleural pressure, and transpulmonary pressure during normal breathing. (Adapted from Guyton AC and Hall JE: *Textbook of Medical Physiology,* 9th ed. Philadelphia: WB Saunders, 1995, p 479.)

2. **During inspiration.** The thoracic cavity enlarges, creating an even greater suction force on the pleural fluid. Thus, the pleural pressure becomes more negative, and the transpulmonary pressure increases, causing the lungs to expand. Inward flow of air is caused by development of subatmospheric pressure (negative pressure) in the alveoli. When the lungs expand, the volume occupied by the gases in the alveoli is increased, causing alveolar pressure to decrease. A pressure gradient between the atmosphere and alveoli causes air to flow into the lungs. At the end of inspiration, alveolar pressure is again zero and air flow ceases.

3. **During expiration.** The respiratory muscles relax, and the elastic recoil of the lungs compresses the alveolar gases, increasing alveolar pressure to about +1 cm water. Since alveolar pressure is now greater than atmospheric pressure, air flows out of the lungs. Pleural pressure returns to its preinspiratory level. Thus, during a normal respiratory cycle, pleural pressure fluctuates solely in the negative range between about –5 and –8 cm water. At the end of expiration, alveolar pressure is once again equal to atmospheric pressure, causing air flow to stop.

Pulmonary Volumes and Capacities

Pulmonary volumes and capacities are measured using a spirometer. Figure 6-2 shows a recording for successive breath cycles at different depths of inspiration and expiration. The recording was made using an apparatus called a **spirometer**—a drum inverted in water with a tube extending from the air space in the drum to the mouth of the

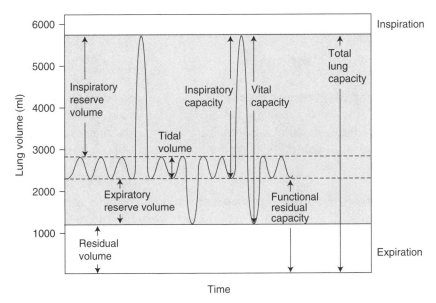

Figure 6-2.
Changes in lung volume during normal breathing and during maximal inspiration and maximal expiration. (Adapted from Guyton AC and Hall JE: *Textbook of Medical Physiology,* 9th ed. Philadelphia: WB Saunders, 1995, p 483.)

person being tested. As the person breathes in and out, the drum moves up and down, and a recording of the movement is made. Pulmonary volumes, except for residual volume, can be measured directly by spirometry. Pulmonary capacities consist or two or more pulmonary volumes. The four volumes and four capacities shown in Figure 6-2 are described below:

- **Tidal volume** (V_T) is the amount of air inspired and expired with each normal breath (500 ml).
- **Inspiratory reserve volume** (IRV) is the amount of air that can be inspired beyond the tidal volume (3000 ml). It is used during exercise and other strenuous activities.
- **Expiratory reserve volume** (ERV) is the amount of air that can be expired by forceful expiration at the end of a normal tidal expiration (1100 ml).
- **Residual volume** (RV) is the amount of air remaining in the lungs after a maximal expiration (1200 ml). It cannot be measured by simple spirometry.
- **Inspiratory capacity** ($IC = V_T + IRV$) is the maximum volume of air a person can inspire beginning from the end of a normal expiration (3500 ml).
- **Functional residual capacity** ($FRC = ERV + RV$) is the volume of air in the lungs after a normal expiration (2300 ml). It cannot be measured by spirometry because it includes the residual volume.
- **Vital capacity** ($VC = IRV + ERV + V_T$) is the range in lung volume from maximum inspiration to maximum expiration (4600 ml).
- **Total lung capacity** ($TLC = IRV + V_T + ERV + RV$) is the maximum volume of air that the lungs can hold after the greatest possible inspiration (5800 ml). It cannot be measured by spirometry because it includes the residual volume.

Ventilation Equations

Minute ventilation is the sum of all the air breathed during 1 minute. Minute ventilation, also called the total ventilation, is equal to the product of the normal respiratory rate (12 breaths per minute) and the normal tidal volume (500 ml) and is about 6000 ml/min:

$$\text{Minute ventilation} = \text{tidal} \times \text{volume respiratory rate}$$

Alveolar ventilation represents the amount of air available for gas exchange. Each time a person inspires, part of the new air must be used to fill the passageways between the nose and the alveoli. Since this portion of the inspired air does not reach the alveoli, it is called the **anatomic dead space** volume and is normally about 150 ml. Thus, with each normal tidal volume of air, 150 ml of the 500 ml fails to reach the alveoli and, therefore, is not available to aerate the blood. The portion of the air that does reach the alveoli is called the alveolar ventilatory air and normally amounts to about 350 ml with each breath. With a normal respiratory rate of 12 breaths per minute, this amounts to an alveolar ventilation of 4200 ml/min:

Alveolar ventilation

= (tidal volume – dead space volume) × respiratory rate

Note from the equation that the presence of dead space air dictates that a slow, deep pattern of breathing is more effective in ventilating the alveoli than is a rapid, shallow pattern. However, the pattern of breathing that a particular individual chooses to adopt is one that minimizes the expenditure of energy. For example, patients with **fibrotic lung disease** breathe rapidly and shallowly because expansion of the lungs is difficult.

There are three types of dead space air. **Anatomic dead space** is the air in the conducting airways that does not engage in gas exchange. **Alveolar dead space** is the air in the gas exchange portions of the lung that cannot engage in gas exchange. It is nearly zero in normal individuals. **Physiologic dead space** is the sum of the anatomic dead space and the alveolar dead space (i.e., the total dead space air). It is equal to the anatomic dead space in normal individuals.

Pulmonary Compliance

Lung compliance is the change in lung volume for each unit change in transpulmonary pressure (Δ volume / Δ pressure). Compliance is thus the slope of the pressure-volume curve shown in Figure 6-3. The curve is recorded by changing the pleural pressure in small steps and allowing the lung volume to come to a steady level at each pressure. The compliance of both human lungs together is normally

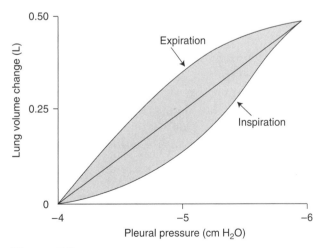

Figure 6-3.
Pressure-volume curve of the normal lung showing compliance of the lungs alone. (Adapted from Guyton AC and Hall JE: *Textbook of Medical Physiology*, 9th ed. Philadelphia: WB Saunders, 1995, p 479.)

about 200 ml/cm of water, which means that a 1-cm change in pleural pressure causes the lungs to change their volume by 200 ml. Note on the curve that the lung volume at any given pressure is greater during expiration than during inspiration. This behavior is know as **hysteresis.**

Compliance is determined by the elastic forces of the lung tissue and surface tension of the liquid film lining the alveoli. The elastic forces of the lung tissue are determined mainly by elastin and collagen fibers in the lung parenchyma. The elastic forces caused by surface tension at the air-liquid interface in the alveoli account for about two-thirds of the total elastic force of the lungs.

The compliance of a saline-filled lung is far greater than an air-filled lung. The volume of a saline-filled lung is greater at any given pleural pressure compared to the air-filled lung. Since there is no air-liquid interface in the saline-filled lungs, it is reasonable to conclude that the difference in compliance between the two curves is due to surface tension effects, and that the compliance of the saline-filled lungs is determined entirely by the elastic forces of the lung tissue. Also, hysteresis is nearly absent in the saline-filled lungs.

Patients with emphysema have a higher than normal FRC because the tendency of the lungs to collapse is decreased. In patients with emphysema, many of the alveoli coalesce, with dissolution of many alveolar walls. This loss of lung tissue decreases the elasticity of the lungs; thus, the tendency of the lungs to collapse is decreased. Therefore, the lung and chest wall seek a new equilibrium volume at FRC, which is much higher than normal, explaining why the patient develops a **barrel-shaped chest.**

Pneumothorax causes the lungs to collapse. When a hole is made in the chest wall, the elastic lungs immediately collapse and the chest wall springs outward, sucking air into the chest. This is called a "pneumothorax." When the person tries to breathe, air moves in and out of the hole in the chest and death can occur by suffocation.

Attraction between water molecules in the liquid layer lining the alveoli attempts to contract the alveoli. Water molecules on the surface of water are attracted to each other so that the surface is always attempting to contract. This is the force that holds droplets of water together. This attraction between water molecules attempts to collapse the alveoli and thus accounts for the surface-tension elastic forces of the lung.

Surfactant reduces the work of breathing (increases compliance) by decreasing surface tension in the alveoli. If the alveoli were lined with pure water, the surface tension would be so great that the alveoli would likely remain collapsed all the time. Fortunately, surfactant is secreted into the alveoli by type II epithelial cells that line the alveoli. This substance, which consists primarily of **dipalmitoyl phosphatidyl choline,** acts to decrease the surface tension of the fluid lining the alveoli, allowing normal expansion of the lungs. Surfactant is formed relatively late in fetal life, and some newborns without adequate quantities may develop **neonatal respiratory distress syndrome,** which is characterized by hemorrhagic edema, patchy atelectasis (collapsed alveoli), and profound hypoxemia.

The pressure created by a smaller bubble is greater than that of a larger bubble. This phenomenon can be explained by the **law of Laplace,** which states that

$$\text{Pressure} = (2 \times \text{surface tension}) / \text{radius}$$

Note from the equation that at any given surface tension, a smaller bubble will generate a greater pressure than a larger bubble. This is why the air from a smaller bubble flows into a larger bubble when they are connected together by a tube.

Factors that stabilize the sizes of the alveoli and thus help prevent alveolar collapse (i.e., atelectasis). Alveoli are surrounded by other alveoli and fibrous tissue so that if a group of alveoli begin to collapse, they are pulled open by elastic forces that have developed in the surrounding tissues. The phenomenon is called **interdependence. Surfactant** reduces surface tension, allowing interdependence to overcome surface-tension effects. Surfactant decreases surface tension to a greater extent as an alveolus becomes smaller because the molecules of surfactant become squeezed together.

Airway Resistance

Airway resistance is the most important determinant of air flow to the lungs. Air flow to the lungs is directly proportional to the difference in pressure between the atmosphere and the alveoli and inversely proportional to airway resistance:

$$\text{Flow} = \Delta \text{ pressure} / \text{resistance}$$

Recall that resistance is inversely proportional to the fourth power of the radius ($R \propto 1/r^4$) so that a 2-fold increase in radius causes a 16-fold decrease in resistance.

The medium-sized airways are the major site of resistance to air flow. The very small bronchioles, less than 2 mm in diameter, account for less than 20% of the total airway resistance because of their large number and parallel arrangement. The bronchi are tethered to the lung tissue, causing their diameter to increase when the lungs expand. This is one of the reasons that patients with constricted airways (e.g., **asthma**) breathe at higher than normal lung volumes. **Bronchoconstriction** narrows the airways and thus increases the resistance to air flow. **Parasympathetic stimulation** constricts the airways as does direct application of **acetylcholine.** Constriction is also caused by **chronic bronchitis, asthma,** and irritants such as cigarette smoke. **Bronchodilation** opens the airways and thus decreases the resistance to air flow. **Sympathetic stimulation** dilates the airways by way of β_2-receptors, as do sympathetic agonists such as **isoproterenol.**

PRINCIPLES OF GAS EXCHANGE AND DIFFUSION

Gas Partial Pressures

Respiratory gases move from areas of high partial pressure to areas of low partial pressure. In essentially all respiratory studies, gas concentration is expressed not in terms of percentage but in **partial pressures.** The partial pressure of a gas is the amount of pressure exerted by that gas alone. Partial pressures are used to express the concentrations of gases because it is pressure that causes the gases to move by diffusion from one part of the body to another.

The partial pressure of a gas is calculated by multiplying its fractional concentration by the total pressure. For example, dry atmospheric air is 20.93% oxygen. The partial pressure of oxygen (P_{O_2}) at sea level is equal to 760 mmHg (atmospheric pressure) $\times$ 20.93/100 = 159.1 mmHg. Air becomes humidified in the warm, moist respiratory airways, and its vapor pressure is 47 mmHg. When water vapor pressure is subtracted from the total pressure (760 − 47), the total pressure is now 713 mmHg. Therefore, the O_2 tension of moist tracheal air is 713 $\times$ 20.93/100 = 149.2 mmHg.

The partial pressures of O_2 and CO_2 in the pulmonary capillary blood become equal to the alveolar values. O_2 diffuses from the alveoli into the mixed venous blood entering the pulmonary capillaries along a partial pressure gradient of 104 − 60 or 64 mmHg, as shown in Figure 6-4. The curve below the capillary shows that the blood P_{O_2} increases progressively as the blood passes through the capillary, reaching equilibrium with the alveolar gas at 104 mmHg within the first third of the distance along the vessel. CO_2 diffuses in

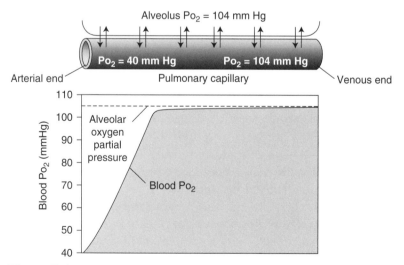

Figure 6-4.
Diffusion of oxygen from the alveolus into the pulmonary capillary.
(Adapted from Guyton AC and Hall JE: *Textbook of Medical Physiology,* 9th ed. Philadelphia: WB Saunders, 1995, p 514.)

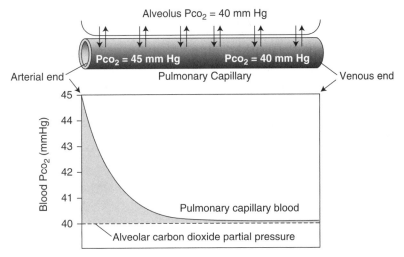

Figure 6-5.
Diffusion of carbon dioxide from the pulmonary capillary into the alveolus. (Adapted from Guyton AC and Hall JE: *Textbook of Medical Physiology,* 9th ed. Philadelphia: WB Saunders, 1995, p 515.)

the opposite direction along a gradient of only 5 mmHg (Figure 6-5). The reason that CO_2 can equilibrate fully with only a 5 mmHg pressure gradient is because it diffuses 20 times as rapidly as O_2.

Diffusion through Pulmonary Membrane

The arterial P_{O_2} is lower than the alveolar and pulmonary capillary P_{O_2}. The P_{O_2} of pulmonary capillary blood is 104 mmHg, but this has decreased to about 100 mmHg or less by the time the blood arrives in the left atrium. One of the causes of this decrease in P_{O_2} is the addition of "shunt" flow from the bronchial circulation that has not passed through the gas exchange areas of the lung and is, therefore, venous blood. This so-called **venous admixture** of blood

TABLE 6-1.

Values of Total and Partial Pressure for Respiratory Gases in mmHg

	Dry Air	Moist Tracheal Air	Alveolar Gas	Arterial Blood	Mixed Venous Blood
P_{O_2}	159.1	149.2	104	100	40
P_{CO_2}	0.3	0.3	40	40	46
P_{H_2O}	0.0	47.0	47	47	47
P_{N_2}	600.6	563.5	569	573	573
P total	760.0	760.0	760	760	706

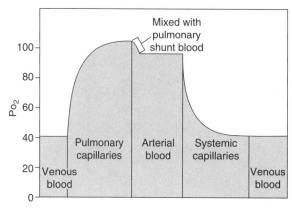

Figure 6-6.
Changes in P_{O_2} in the blood showing the effect of venous admixture. (Adapted from Guyton AC and Hall JE: *Textbook of Medical Physiology,* 9th ed. Philadelphia: WB Saunders, 1995, p 514.)

causes the arterial P_{O_2} to decrease by several mmHg, as shown in Figure 6-6.

The rate of gas diffusion across the pulmonary membrane is proportional to the membrane area and partial pressure gradient and inversely proportional to membrane thickness. The total surface area of the pulmonary membrane is 50 to 100 m², and its thickness averages about 0.5 µm. The rate of diffusion is also proportional to the diffusion constant for the gas, which is proportional to the solubility of the gas and inversely proportional to the molecular weight of the gas. Since CO_2 has a much greater solubility (20 times greater) compared to O_2 but a similar molecular weight, CO_2 diffuses through the pulmonary membrane far more easily than does O_2.

The diffusing capacity of the lungs for CO_2 is 20 times greater than for O_2. The rate at which a gas will diffuse from the alveoli into the blood for each mmHg pressure difference is called the **diffusing capacity** of the lungs for that particular gas. The diffusing capacity of the lungs for O_2 when a person is at rest is approximately 22 ml/min/mmHg. The diffusing capacity for CO_2 is about 20 times this value, or approximately 440 ml/min/mmHg.

Oxygen Transport

The oxygen-hemoglobin dissociation curve shows the percent saturation of hemoglobin plotted as a function of P_{O_2}. Note the following features shown in Figure 6-7. *Point A is arterial blood.* When the P_{O_2} is 100 mmHg, hemoglobin is almost 100% saturated with O_2. The O_2 content is about 20 ml/100 ml blood. An average of four molecules of O_2 are bound to each hemoglobin molecule. *Point B is mixed venous blood.* When the P_{O_2} is 40 mmHg, hemoglobin is 75% saturated with O_2. The O_2 content is about 15 ml/100 ml blood. An

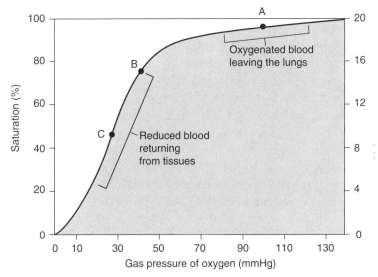

Figure 6-7.
Oxygen-hemoglobin dissociation curve for pH 7.4, Pco_2 40 mmHg, and
37° C. (Adapted from Guyton AC and Hall JE: *Textbook of Medical
Physiology,* 9th ed. Philadelphia: WB Saunders, 1995, p 516.)

average of three molecules of O_2 are bound to each hemoglobin
molecule. *Point C is mixed venous blood during exercise.* When the Po_2
is 25 mmHg, hemoglobin is 50% saturated with O_2. (The Po_2 at 50%
saturation is the **P_{50}.**) The O_2 content is about 10 ml/100 ml blood,
and an average of two molecules of oxygen are bound to each hemoglobin molecule.

**About 97% of the oxygen is carried to the tissues in chemical
combination with hemoglobin.** The remaining 3% is carried to the
tissues in the dissolved state in the water of the plasma and cells.
Hemoglobin combines with large quantities of oxygen when the Po_2
is high and then releases the oxygen when the Po_2 falls. Therefore,
when blood passes through the lungs, where the blood Po_2 rises to
100 mmHg, hemoglobin picks up large quantities of O_2. Then as it
passes through the tissue capillaries, where the Po_2 falls to about 40
mmHg, large quantities of O_2 are released from the hemoglobin.
The free O_2 then diffuses to the tissue cells.

**Hemoglobin releases O_2 to the tissues at a fairly constant Po_2
between 20 and 50 mmHg, regardless of very wide fluctuations in
the Po_2 in the air.** Hemoglobin is normally about 97% saturated
with O_2 at a normal alveolar Po_2 of 104 mmHg. If the O_2 in the alveolar air rises even as high as 1000 mmHg, the hemoglobin can
become saturated with O_2 only 100%; thus, even with large increases
in atmospheric O_2, essentially the same amount of O_2 is carried by
the hemoglobin to the tissues. As the blood passes through the capillaries, its Po_2 normally falls to about 40 mmHg before it enters the
veins. During high metabolic activity of the tissues, this value may
fall to as low as 15 to 20 mmHg.

The sigmoid shape of the oxygen-hemoglobin dissociation curve results from stronger binding of O_2 to hemoglobin as more molecules of O_2 become bound. Each molecule of hemoglobin can bind four molecules of O_2. After one molecule of O_2 has bound, the affinity of hemoglobin for the second molecule is increased, and so forth. The affinity for the fourth molecule of O_2 is the greatest. Therefore, the affinity for O_2 is high in the lungs where the P_{O_2} is about 100 mmHg (Figure 6-7, *flat portion of curve*) and low in the peripheral tissues where the P_{O_2} is about 40 mmHg (Figure 6-7, *steep portion of curve*).

The sigmoid shape of the oxygen-hemoglobin dissociation curve has important consequences for loading and unloading of oxygen. The flat portion of curve, where O_2 affinity is high, facilitates loading of hemoglobin with O_2 in the lungs. The atmospheric P_{O_2} can vary widely without having much effect on the amount of O_2 carried by hemoglobin. The steep portion of curve, where O_2 affinity is low, facilitates unloading of O_2 in the peripheral tissues where the P_{O_2} is low.

The maximum amount of O_2 transported by hemoglobin is about 20 ml of O_2 per 100 ml of blood. In a normal person, each 100 ml of blood contains about 15 grams of hemoglobin, and each gram of hemoglobin can bind with about 1.34 ml of O_2 when it is 100% saturated ($15 \times 1.34 = 20$ ml $O_2/100$ ml blood). The hemoglobin in venous blood leaving the peripheral tissues is about 75% saturated with O_2 so that the amount of O_2 transported by hemoglobin in venous blood is about 15 ml $O_2/100$ ml blood. Therefore, about 5 ml of O_2 is normally transported to the tissues in each 100 ml of blood.

The oxygen-hemoglobin dissociation curve shifts to the right in metabolically active tissues where temperature, P_{CO_2}, and hydrogen ion concentration are increased. The curve shifts to the right when the affinity for O_2 is low, and thus facilitates the unloading of O_2. Note in Figure 6-8 that for any given value of P_{O_2}, the percent saturation with O_2 is low when the curve is shifted to the right. The shift to the right is thus beneficial in exercising muscles and in other instances where O_2 delivery is poor or metabolic activity is increased. The oxygen-hemoglobin dissociation curve is also shifted to the right as an adaptation to chronic hypoxemia associated with life at high altitudes. **Chronic hypoxemia** increases the synthesis of **2,3-diphosphoglycerate** (2,3-DPG), a factor that binds to hemoglobin, decreasing the affinity for O_2. Again, when the curve is shifted to the right, the percent saturation of hemoglobin is reduced at each value of P_{O_2} (i.e., the P_{50} is increased).

Carbon monoxide (CO) interferes with oxygen transport because it has about 250 times the affinity of O_2 for hemoglobin. For this reason, relatively small amounts of CO can tie up a large portion of the hemoglobin, making it unavailable for O_2 transport. A patient with severe CO poisoning can be helped by administration of pure O_2, because O_2 at high alveolar pressures displaces CO from its combination with hemoglobin more effectively than O_2 at low atmospheric pressures.

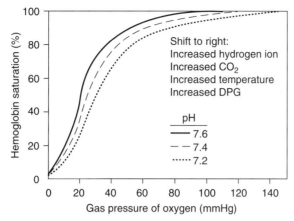

Figure 6-8.
Effects of pH, P_{CO_2}, temperature, and DPG on the oxygen-hemoglobin dissociation curve. (Adapted from Guyton AC and Hall JE: *Textbook of Medical Physiology,* 9th ed. Philadelphia: WB Saunders, 1995, p 518.)

Carbon Dioxide Transport

Under resting conditions, about 4 ml of CO_2 are transported from the tissues to the lungs in each 100 ml of blood. Figure 6-9 shows the ways in which CO_2 is transported in the blood. About 70% of

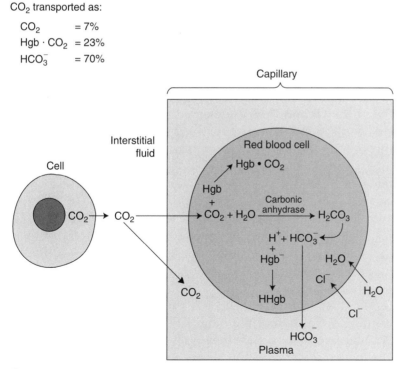

Figure 6-9.
Transport of carbon dioxide in the blood.

the CO_2 is transported to the lungs in the form of **bicarbonate ions.** Dissolved CO_2 reacts with water inside red blood cells to form **carbonic acid.** This reaction is catalyzed by an enzyme in the red cells called **carbonic anhydrase.** Most of the carbonic acid immediately dissociates into bicarbonate ions and hydrogen ions; the hydrogen ions in turn combine with hemoglobin. Approximately 23% of the CO_2 produced in the tissues combines directly with hemoglobin to form **carbaminohemoglobin,** and an additional 7% is transported in the dissolved state in the water of the plasma and cells. When the blood arrives in the lungs, the CO_2 diffuses from the blood into the alveoli, causing rapid reversal of these chemical reactions.

Binding of O_2 with hemoglobin tends to displace CO_2 from the blood. This so-called Haldane effect results from the fact that hemoglobin becomes a stronger acid when it combines with O_2. The increase in acidity of hemoglobin displaces CO_2 for two reasons: (1) The excess hydrogen ions released from hemoglobin bind with bicarbonate to form carbonic acid. The carbonic acid then dissociates into water and CO_2, and the CO_2 diffuses from the blood into the alveoli. (2) Highly acidic hemoglobin has less tendency to bind with CO_2 to form carbaminohemoglobin. The CO_2 is thus displaced from its hemoglobin-binding sites, allowing it to diffuse into the alveoli. The Haldane effect nearly doubles the amount of CO_2 released from the blood in the lungs where the P_{O_2} is high, and nearly doubles the amount of CO_2 picked up by the blood in the tissues where the P_{O_2} is low.

PULMONARY CIRCULATION

Flow, Pressure, and Resistance

Pulmonary blood flow is equal to cardiac output. Because the circulatory system is a circuit, the same amount of blood flowing through the entire systemic circulation must also flow through the lungs. However, resistance to blood flow in the lungs is only about one-eighth of that in the systemic circulation, and the pressures are correspondingly smaller. The pulmonary arterial systolic pressure averages 25 mmHg, and diastolic pressure averages 8 mmHg. The mean pulmonary artery pressure is about 15 mmHg, and left atrial pressure is 2 mmHg, giving a total pressure drop through the pulmonary circulation of only 13 mmHg. The pulmonary capillary pressure is about 7 mmHg, which is only a few mmHg greater than the left atrial pressure. This low capillary pressure is important in keeping the alveoli of the lungs dry.

Increased blood flow decreases pulmonary vascular resistance. During strenuous exercise and in other states of physiologic stress, cardiac output sometimes increases as much as four- to fivefold, and blood flow through the lungs also increases by the same amount. However, pulmonary arterial pressure rises only a moderate amount for two reasons: (1) As flow increases, many pulmonary capillaries that are normally closed open up, and those that are already open dilate even more; and (2) the pulmonary vessels are highly distensible and therefore stretch as blood flow increases, decreasing pulmonary vascular resistance. As a result, the pulmonary vascular resistance decreases markedly as cardiac output increases, and mean pulmonary pressures usually rise only a few mmHg.

Blood shifts between the systemic and pulmonary circulation during heart failure. Normally, only about 10% of the blood in the circulatory system is in the lungs and about 80% is in the systemic circulation, the remainder being in the heart. However, the amounts of blood in the two circulations can change when one side of the heart fails. If the right heart fails, the pressure in the right atrium increases and some of the blood normally pumped by the right heart builds up in the systemic circulation. Conversely, if the left heart fails, a portion of the systemic blood fails to be pumped into the systemic circulation and is displaced into the lungs, sometimes increasing pulmonary blood volume as much as two times normal. This can lead to pulmonary edema, as discussed below.

Decreased alveolar oxygen concentration causes constriction of local pulmonary blood vessels. This effect of **hypoxia** to constrict the pulmonary vasculature serves the important function of distributing blood flow to better ventilated areas of the lung. In other words, the pulmonary blood flow is shunted to areas of the lung where oxygenation of blood can occur most effectively. It should be pointed out that hypoxia has the opposite effect on the systemic vasculature, causing vasodilation and thus increasing blood flow to poorly perfused tissues.

Blood Flow Zones

Hydrostatic pressure gradients in the lung create three zones of pulmonary blood flow. In the normal adult, the distance between the apex and base of the lungs is about 30 cm, which creates a 23-mmHg difference in blood pressure. This pressure gradient has a large effect on blood flow in different regions of the lung. The three zones of blood flow are shown in Figure 6-10.

Zone 1—No Blood Flow (Top of Lung)

The capillaries are pressed flat because the arterial and venous pressures are both greater than the alveolar pressure. Zone 1 does not occur during normal conditions, but can occur when pulmonary artery pressure is decreased (e.g., following hemorrhage) and when alveolar pressure is increased (e.g., during positive-pressure ventilation).

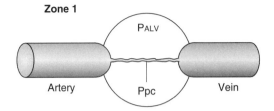

Zone 1

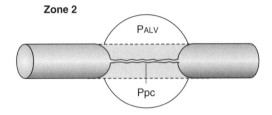

Zone 2

Figure 6-10.
Blood flow in three different
blood flow zones of the lungs.
(Adapted from Guyton AC
and Hall JE: *Textbook of
Medical Physiology,* 9th ed.
Philadelphia: WB Saunders,
1995, p 494.)

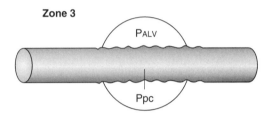

Zone 3

Zone 2—Intermediate Blood Flow (Middle of Lung)

Blood flow is determined by the difference between arterial and
alveolar pressures because alveolar pressure is greater than venous
pressure. Blood flow may be intermittent at times, occurring only
during systole when the arterial pressure is high.

Zone 3—High Blood Flow (Bottom of Lung)

Blood flow is determined by the difference between arterial and
venous pressures.

Pulmonary Edema

Low pulmonary capillary pressure helps to keep the alveoli dry.
The normal pulmonary capillary pressure is only about +7 mmHg,
while the interstitial fluid pressure is –8, the plasma colloid osmotic
pressure is +28 mmHg, and the interstitial fluid colloid osmotic pressure is +14 mmHg. Thus, the net driving pressure causing fluid to
move out of the capillary equals +1 mmHg. If the pulmonary capillary pressure were as high as the system capillary pressure (+17
mmHg), the net driving pressure would be +11 mmHg, and significant extra amounts of fluid would then filter through the capillary
walls.

   **Pulmonary edema is the accumulation of fluid in the pulmonary
interstitium.** Whenever excess blood shifts into the lungs, as a result

of failure of the left heart to pump adequately, the pressures throughout the lungs increase, including the capillary pressure. As long as the capillary pressure remains less than about 30 mmHg, the alveoli of the lungs will remain dry, but as soon as capillary pressure rises above the plasma colloid osmotic pressure, large amounts of fluid immediately begin to filter out of the capillaries into the interstitial fluid, and usually also through the alveolar membranes into the alveoli. This condition is called **pulmonary edema,** and it greatly impairs gas exchange between the alveoli and the blood capillaries.

Normally, there are safety factors that prevent pulmonary edema with moderate increases in pulmonary capillary pressure. If pulmonary capillary pressure rises acutely to about 50 mmHg, however, sufficient pulmonary edema can develop in 20 minutes to cause death; if pulmonary capillary pressure rises acutely to only 30 mmHg, sufficient pulmonary edema can still develop in 3 to 6 hours to cause death. Yet in chronic conditions such as mitral valve stenosis, pulmonary capillary pressure can remain as high as 40 mmHg for long periods of time without causing severe pulmonary edema, probably because very large lymphatic vessels develop in the lungs and provide extra drainage of fluid from the tissues.

VENTILATION-PERFUSION MISMATCH

The ventilation/perfusion ratio is the ratio of alveolar ventilation to pulmonary blood flow. In the ideal lung, all of the alveoli are ventilated equally and the blood flow to each alveolus is identical. In reality, some areas of the lung are well ventilated but have only limited blood flow, while other areas have excellent blood flow but are poorly ventilated. In other words, ventilation-perfusion mismatch occurs in the normal lung. Also, an individual can have severe respiratory distress even though the total pulmonary blood flow and total alveolar ventilation are both normal. The concept of **ventilation/perfusion ratio** ($\dot{V}_A\dot{Q}$) was developed to help us understand **gas exchange abnormalities** when there is imbalance between the ventilation of a given alveolus or lung unit ($\dot{V}_A$) and the blood flow to the same lung unit ($\dot{Q}$).

The ventilation/perfusion ratio can range from zero to infinity. The effects of changing the ventilation/perfusion ratio ($\dot{V}_A/\dot{Q}$) on the P_{O_2} and P_{CO_2} are shown in Figure 6-11. When $\dot{V}_A/\dot{Q}$ is normal, alveolar ventilation and alveolar blood flow are both normal and the exchange of O_2 and CO_2 is nearly optimal. When $\dot{V}_A/\dot{Q}$ equals zero, the lung unit has blood flow but no ventilation. Air trapped in the alveolus comes to equilibrium with the O_2 and CO_2 in the alveolar blood because the gases diffuse through the alveolar membrane. The P_{O_2} and P_{CO_2} of alveolar air equal that of **mixed venous blood.** When $\dot{V}_A/\dot{Q}$ equals infinity, the lung unit has ventilation but no

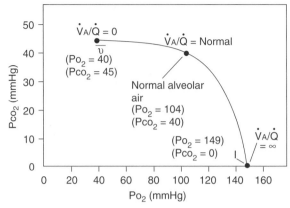

Figure 6-11.
The normal P_{O_2}-P_{CO_2} ventilation/perfusion ($\dot{V}_A/\dot{Q}$) diagram. (Adapted from Guyton AC and Hall JE: *Textbook of Medical Physiology,* 9th ed. Philadelphia: WB Saunders, 1995, p 510.)

blood flow. The inspired air loses no O_2 and gains no CO_2 because there is no capillary blood flow. The P_{O_2} and P_{CO_2} of alveolar air equal that of inspired air.

The ventilation/perfusion ratio is high at the top of the lung and low at the bottom. Both blood flow and ventilation increase from the top to the bottom of the lung, but blood flow increases to a greater extent. The ventilation-perfusion ratio, therefore, is higher at the top of the lung compared with the ratio at the bottom. These differences in ventilation and perfusion at the top and bottom of the lung and their effect on the regional P_{O_2} and P_{CO_2} are summarized in Table 6-2.

REGULATION OF RESPIRATION

Central Respiratory Centers

The respiratory rhythm is caused by intermittent nerve impulses originating in the respiratory center. The respiratory center is located in the brain stem in the reticular substance of the

TABLE 6-2.

$\dot{V}_A/\dot{Q}$ Characteristics at the Top and Bottom of the Upright Lung

Area of Lung	Ventilation	Perfusion (Blood Flow)	$\dot{V}_A/\dot{Q}$	Local Alveolar P_{O_2}	Local Alveolar P_{CO_2}
Top	Low	Lower	Highest	Highest	Lowest
Bottom	High	Higher	Lowest	Lowest	Highest

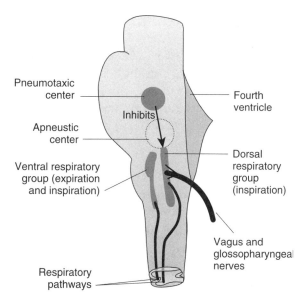

Figure 6-12.
The respiratory center. (Adapted from Guyton AC and Hall JE: *Textbook of Medical Physiology,* 9th ed. Philadelphia: WB Saunders, 1995, p 526.)

medulla and the pons. This center has a basic oscillating mechanism that causes it to emit rhythmic impulses to the respiratory muscles. However, the intensity of this rhythmic excitation of respiration can be increased or decreased by changes in the chemical composition of the blood and also by sensory signals from the lungs.

The respiratory centers are comprised of three main groups of neurons. The respiratory centers are shown in Figure 6-12. The **dorsal respiratory group** generates inspiratory action potentials in a steadily increasing ramp fashion, and is thus responsible for the basic rhythm of respiration. It receives input from peripheral chemoreceptors and other types of receptors by way of the vagus and glossopharyngeal nerves. The **pneumotaxic center** transmits inhibitory signals to the dorsal respiratory group and thus controls the filling phase of the respiratory cycle. It limits inspiration and therefore has a secondary effect of increasing respiratory rate. The **ventral respiratory group** is inactive during normal quiet breathing. It is important for stimulating abdominal expiratory muscles when high levels of respiration are required.

The Hering-Breuer reflex prevents overinflation of the lungs. This reflex is initiated by nerve receptors located in the bronchi and bronchioles that detect the degree of stretch of the lungs. When the lungs become overly inflated, the receptors send signals through the vagi into the dorsal respiratory group that affect inspiration much the same as inhibitory signals from the pneumotaxic center. That is, inspiration is switched off and the rate of respiration is increased.

Chemical Control of Respiration

Chemical control of respiration serves to maintain proper concentrations of O_2, CO_2, and H^+ in the tissues. Excess amounts of H^+ and CO_2 stimulate mainly the respiratory center, increasing the rate and depth of breathing. O_2 has little direct effect on the respiratory center and instead acts mainly on peripheral chemoreceptors, which in turn send signals to the respiratory center for control of respiration.

The central chemoreceptor area in the medulla is excited greatly by direct exposure to H^+. However, it is stimulated only slightly by increased H^+ concentration in the blood because the blood-brain barrier is relatively impermeable to H^+. It is excited greatly when CO_2 levels increase in the arterial blood. The CO_2 diffuses through the blood-brain barrier and combines with water to form H^+. The H^+ then stimulate the chemoreceptor area. The central chemoreceptor area is not stimulated directly by low levels of O_2.

Peripheral chemoreceptors in the carotid and aortic bodies are stimulated by decreased O_2 in the arterial blood. However, the arterial PO_2 must decrease to below 60 mmHg for stimulation to occur. The peripheral chemoreceptors are affected to a small extent by H^+ concentration and CO_2 in the arterial blood. These **chemoreceptors** are special nerve receptors located in minute **carotid** and **aortic bodies,** which lie, respectively, in the carotid bifurcations and along the aorta. Each of these bodies has a special artery that supplies abundant amounts of arterial blood to the chemoreceptors. When the arterial oxygen concentration falls, signals from the chemoreceptors are transmitted to the respiratory center where they cause an increase in alveolar ventilation.

CO_2 is the most powerful blood stimulus for increasing the rate and depth of alveolar ventilation. When increased quantities of CO_2 are formed in the body cells and carried in the blood to the respiratory center in the brain stem, the ventilation sometimes increases to as high as ten times normal. This, in turn, "blows off" the extra quantity of CO_2 from the lungs.

Increased blood H^+ concentration (i.e., decreased pH) increases alveolar ventilation. This increase in alveolar ventilation can be as much as four times normal. The effects of blood H^+ concentration on ventilation are thought to be mediated by way of peripheral chemoreceptors in addition to direct effects on the respiratory center.

The effects of CO_2 on alveolar ventilation are mediated almost entirely by H^+. CO_2 itself has little direct effect in stimulating the respiratory center. However, CO_2 reacts with water to form carbonic acid, which in turn dissociates into hydrogen and bicarbonate ions. The increase in ventilation causes increased quantities of CO_2 to be blown off from the blood, which in turn decreases the amount of blood carbonic acid. Since carbonic acid is in constant equilibrium with H^+ in the blood, the H^+ concentration also decreases back toward normal.

Why is it that blood CO_2 has a more potent effect in stimulating the respiratory center than do blood H^+? The blood-brain barrier is almost totally impermeable to H^+ so that increases in the blood H^+ concentration have relatively little effect on the H^+ concentration in the vicinity of the respiratory center. CO_2, on the other hand, permeates the blood-brain barrier with ease and immediately reacts with water to form H^+. Thus, more H^+ are released in the respiratory center when the blood CO_2 concentration increases than when the blood H^+ concentration increases.

Stimulation of alveolar ventilation by excess CO_2 and low pH is great compared to the O_2-lack stimulus. A maximal increase in CO_2 can increase alveolar ventilation about 10-fold; maximal increase in H^+ concentration can increase it about 4-fold, but maximal O_2 lack (under acute conditions) can increase alveolar ventilation only by about one and two-thirds.

Why is oxygen lack a relatively poor stimulus for alveolar ventilation? One often wonders why the evolutionary processes have made O_2 lack such a poor stimulus of respiration in comparison with CO_2 and H^+. However, O_2 concentration in the tissues is regulated principally by the manner in which it is released from hemoglobin, as discussed above, while CO_2 and H^+ concentration is regulated almost entirely by alveolar ventilation. Therefore, there usually is less need for O_2 to control respiration than for CO_2 and H^+ concentration to control it.

Cheyne-Stokes breathing is a respiratory disorder most commonly characterized by periodic breathing. The patient breathes deeply for a short time and then breathes slightly or not at all for a short time, with the cycle repeating itself again and again. The period of overbreathing decreases the CO_2 level in the pulmonary blood (and increases the O_2 level), but it takes several seconds for the pulmonary blood to reach the respiratory center so that the person has continued to overventilate for a few extra seconds. When the changed pulmonary blood does finally reach the respiratory center, the center becomes depressed. The cycle then repeats itself over and over again, as shown in Figure 6-13. The two causes of Cheyne-Stokes breathing are as follows:

1. **Delayed transport of blood from the lungs to the brain** often occurs in patients with severe heart failure because the left heart is large and blood flow is slow.
2. **Increased negative feedback gain of the respiratory center** causes a given decrease in the level of CO_2 in the blood to result in a greater than normal depression in respiration.

Exercise can cause alveolar ventilation to increase as much as 30-fold. This increase in alveolar ventilation is even greater than the increase that occurs as a result of maximal CO_2 or maximal H^+ stimulation. The precise cause of the greatly increased respiration during exercise has not been determined, but it is believed to result from nerve signals transmitted during exercise from other centers

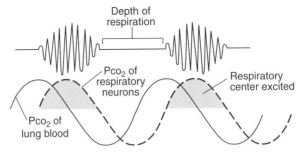

Figure 6-13.
Cheyne-Stokes breathing showing changes in pulmonary blood P_{CO_2} and delayed changes in P_{CO_2} at the respiratory center. (Adapted from Guyton AC and Hall JE: *Textbook of Medical Physiology,* 9th ed. Philadelphia: WB Saunders, 1995, p 535.)

of the brain that provide the nervous drive for the exercise itself, and possibly from sensory signals originating in the active muscles.

RESPIRATORY INSUFFICIENCY AND ITS DIAGNOSIS

Obstructive and Restrictive Diseases

Obstructive lung disease is characterized by increased resistance to air flow and high lung volumes. Patients with obstructive lung disease find it easier to breathe at high lung volumes because this tends to increase the caliber of the airways (increased radial traction), thus decreasing the resistance to air flow. The airway lumen may be partially obstructed by **excessive secretions** (chronic bronchitis), **edema fluid,** or by **aspiration of food or fluids.** The airway wall may be thickened because of inflammation and edema (asthma, bronchitis) or smooth muscle may be contracted (asthma). Outside the airway, the destruction of lung parenchyma may decrease radial traction, causing the airways to be narrowed (emphysema). In addition, swollen lymph nodes or neoplasm may compress the airway.

    Restrictive lung disease is characterized by low lung volumes. Patients with restrictive lung disease find it easier to breathe at low lung volumes because it is difficult to expand the lungs. Expansion of the lung may be restricted because of the following:

1. **Abnormal lung parenchyma** in which excessive fibrosis increases lung elasticity (pulmonary fibrosis, silicosis, asbestosis, tuberculosis)
2. **Problems with the pleura,** such as pneumothorax and pleural effusion
3. **Neuromuscular problems,** such as polio and myasthenia gravis

Pulmonary Function Tests

The Maximum Expiratory Flow-Volume Curve

Figure 6-14 shows the instantaneous relationship between pressure and flow when the patient expires with as much force as he can after having inspired as much air as possible. Thus, expiration begins at total lung capacity and ends at the residual volume. The curve shows the maximum expiratory flow at all lung volumes. Note that the expiratory flow reaches a maximum value of over 400 L/min at a lung volume of 5 L and then decreases progressively as lung volume decreases. An important aspect of the curve is that the expiratory flow reaches a maximum value beyond which the flow cannot be increased further with additional effort. In other words, the descending portion of the curve representing the maximum expiratory flow is **effort independent.**

The maximum expiratory flow is limited by dynamic compression of the airways. During a forced expiration, both the pleural and alveolar pressures increase, but the airway pressure increases to a lesser extent because of the resistance to flow in the airways. The airways collapse when the pressure pushing against the airways from the outside becomes greater than the pressure inside the airways.

A

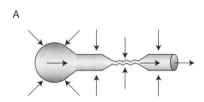

B

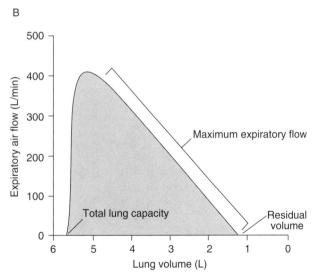

Figure 6-14.
Flow-volume curve. (Adapted from Guyton AC and Hall JE: *Textbook of Medical Physiology,* 9th ed. Philadelphia: WB Saunders, 1995, p 538.)

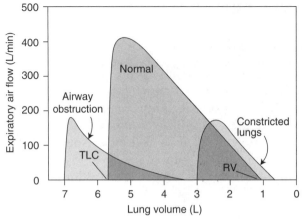

Figure 6-15.
Effect of obstructive lung disease and restrictive lung disease on the maximum flow-volume curve. (Adapted from Guyton AC and Hall JE: *Textbook of Medical Physiology,* 9th ed. Philadelphia: WB Saunders, 1995, p 538.)

The maximum expiratory flow is determined by the difference between the alveolar pressure and the pressure outside of the airways at the point of collapse (**Starling resistor effect**). This difference in pressure (normally about +8 cm water) is the static recoil pressure of the lungs and is independent of respiratory effort.

The maximum expiratory flow-volume curve is useful in determining whether an obstructive or a restrictive lung disease is present. Figure 6-15 shows a normal maximum flow-volume curve along with curves generated from patients with obstructive lung disease and restrictive lung disease. The flow-volume curve in a **restrictive lung disease** (e.g., interstitial fibrosis) is characterized by low lung volumes and slightly higher than normal expiratory flow rates at each lung volume. The flow-volume curve in **obstructive lung diseases** (e.g., chronic bronchitis, emphysema, asthma) is characterized by high lung volumes and lower than normal expiratory flow rates. The curve may also have a scooped-out appearance.

Hypoxia

Hypoxia signifies insufficient availability of oxygen to support normal tissue metabolism. If the hypoxia is severe enough, it can depress mental activity, sometimes causing coma, and reduce the work capacity of the muscles. The possible **causes of hypoxia** are as follows:

1. Inadequate oxygenation of normal lung
   Deficiency of oxygen in atmosphere
   Hypoventilation (neuromuscular disorders)
2. Pulmonary disease
   Hypoventilation (airway obstruction or decreased pulmonary compliance)

Uneven alveolar ventilation/pulmonary capillary blood flow
Decreased respiratory membrane diffusion
3. Venous-to-arterial shunts ("right-to-left" cardiac shunts)
4. Inadequate transport of oxygen by blood to tissues
   Anemia or abnormal hemoglobin
   General or local circulatory deficiency
   Tissue edema
5. Inadequate tissue capacity to use oxygen
   Poisoning of cellular enzymes
   Diminished cellular metabolic capacity caused by factors such as
     toxicity and vitamin deficiency

Oxygen therapy can relieve certain types of hypoxia. This is particularly true of atmospheric hypoxia, hypoventilation hypoxia, and hypoxia caused by impaired alveolar membrane diffusion. In each of these instances, an increase in the oxygen concentration increases the PO_2 in the alveoli and thereby promotes increased oxygen diffusion into the blood. In other types of hypoxia, the problem is mainly diminished transport of O_2 to the tissues or diminished use of O_2 by the tissues. In these types of hypoxia, oxygen therapy may be of some benefit but not nearly so much as in the types mentioned above.

AVIATION AND HIGH-ALTITUDE PHYSIOLOGY

Hypoxia at high altitudes can impair mental acuity and end in coma. A major problem in aviation physiology is the progressive decrease in PO_2 at increasingly higher altitudes. A normal person often becomes lethargic and loses much mental alertness at about 12,000 to 15,000 feet. At 18,000 feet, a person can become so disoriented that judgment is lost; pilots may actually fly still higher rather than returning to a lower level to correct the hypoxic condition. And at about 23,000 feet, an unacclimatized aviator will become comatose in 20 to 30 minutes.

Breathing pure O_2 can improve tolerance to high altitude. If pure O_2 is breathed rather than normal air, a pilot can ascend to an altitude of about 45,000 feet before becoming hypoxic because the O_2 replaces the nitrogen that normally fills the major amount of space in the alveoli.

Acclimatization to hypoxia can begin within a few days and continue on for weeks to months. Although an aviator almost never remains at a high altitude long enough to become adjusted to the altitude, mountain climbers often become acclimatized sufficiently that they can live and work at altitudes many thousand feet higher than normal persons. Acclimatization results from three major physiologic changes:

1. **Increased pulmonary ventilation.** The oxygen-lack mechanism for control of pulmonary ventilation normally increases ventilation only about 65%, but after a person remains at high altitudes for several days, this mechanism becomes progressively more effective and increases ventilation about 400% instead of the normal 65%, thus providing much greater amounts of oxygen for the alveoli.

2. **Increase in red blood cells and hemoglobin.** When a person stays at a high altitude for several weeks, the hypoxia causes greatly increased production of red blood cells, sometimes increasing the total red cell mass to as much as 80% above normal and the hematocrit to 50% above normal. This obviously increases the ability of the blood to transport oxygen to the tissues.

3. **Increased capillarity in the tissues.** Associated with the increased blood cell mass is a slight increase in the number of blood vessels in the tissues or in their sizes so that increased quantities of blood can flow through the tissues, thus again increasing the available oxygen in the tissues.

Chapter **7**

Nervous System

SECTION 1
CENTRAL NERVOUS SYSTEM

BASIC ORGANIZATION

The nervous and endocrine systems provide most of the control systems of the body. The nervous system generally controls rapid functions of the body, such as muscle contractions, and the endocrine system mainly controls metabolic functions of the body. Three main parts of the nervous system important for controlling bodily functions include the sensory system, the motor system, and the integrative system.

Sensory receptors initiate most activities of the nervous system. The sensory receptors include any type of nerve ending in the body that can be stimulated by some physical or chemical stimulus originating either outside or within the body. Examples include visual receptors, auditory receptors, tactile receptors, taste receptors, and so forth. The sensory experience can produce an immediate reaction or information can be stored for many weeks, months, or years before a final reaction takes place.

The motor system performs functions dictated by the nervous system. Some of the most important motor functions include contraction of skeletal muscle, contraction of smooth muscle in the internal organs, and secretion by endocrine or exocrine glands. The muscles and glands that perform the **motor functions** are called **effectors** because they perform the motor functions dictated by the nerve signals.

The reflex arc is a basic means by which the nervous system controls functions in the body. A reflex arc is a complete neuronal network extending from the peripheral receptor through the central nervous system (CNS) and then to the peripheral effector. Reflex arcs can be as simple as withdrawing the hand from a hot object or blinking when the cornea is touched, or they can involve more complicated actions such as coughing, sucking, sneezing, and protecting the body from the environment. Other reflexes include circulatory reflexes, digestive reflexes, sexual reflexes, and respiratory reflexes.

The integrative centers of the nervous system process information. Those parts of the nervous system that put many different types of sensory signals together before causing a reaction or that first store the information and later cause a reaction are called the **integrative centers** of the nervous system. The brain stem is the integrative center for most respiratory control, for most nervous control of arterial pressure, and for control of swallowing. The motor area of the cerebral cortex, the cerebellum, the basal ganglia, and large parts of the reticular substance of the brain stem are major parts of the integrative centers for control of muscular movement.

SYNAPTIC TRANSMISSION

The Synapse

Nerve signals are transmitted through the synapse. The neurons of the nervous system are arranged so that each neuron stimulates other neurons, and these in turn stimulate still others until the functions of the nervous system are performed. The point of contact between successive neurons, illustrated in Figure 7-1, is called a **synapse,** and the terminal endings of the nerve filaments that synapse with the next neuron are called **presynaptic terminals, synaptic knobs, boutons,** or simply **end feet.** Usually, there are many thousand presynaptic terminals on each neuron, these having originated from preceding neurons. Each terminal secretes a particular transmitter substance that may either excite the next neuron or inhibit it. These substances are called **excitatory** or **inhibitory transmitters**.

Synaptic Transmitters

Transmitter substances provide a means of communication between neurons. More than 40 different types of chemical substances have been postulated to act as synaptic transmitters. Each presynaptic terminal generally secretes one characteristic transmitter substance, but often more than one. Each transmitter is synthesized within the terminal and stored in thousands of small vesicles. When an action potential spreads over the end of the nerve fiber, the depolarization of the terminal causes migration of a few of the vesicles to the membrane surface of the terminal, and these vesicles then extrude their contents of transmitter substance into the synaptic cleft between the

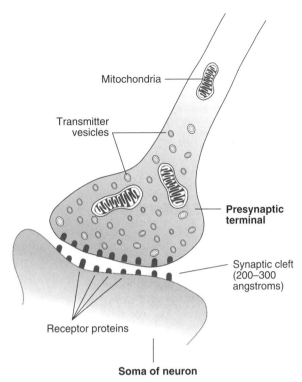

Mitochondria

Transmitter vesicles

Presynaptic terminal

Synaptic cleft (200–300 angstroms)

Receptor proteins

Soma of neuron

Figure 7-1.
Physiologic anatomy of the synapse.

terminal and the membrane of the succeeding neuron. The transmitter then combines with a receptor (a protein molecule) that is an integral part of the subsequent neuronal membrane. This opens a channel through the receptor in the membrane and allows ions to move through the channel.

Acetylcholine is a classic excitatory transmitter. Acetylcholine is released by a large number of presynaptic terminals in the CNS. Acetylcholine stimulates the successive neuron in exactly the same way that it stimulates a muscle fiber at the neuromuscular junction—that is, by increasing the permeability of the neuronal membrane to sodium. Sodium leaks rapidly to the interior of the cell, causing a sudden loss of electrical potential across the membrane (i.e., causing depolarization). Other transmitter substances that often function as excitatory transmitters include **norepinephrine, epinephrine, glutamic acid, enkephalin,** and **substance P.** However, some of these also function as inhibitory transmitters in the presence of inhibitory receptor types.

γ-Aminobutyric acid (GABA) and glycine are classic inhibitory transmitters. Other transmitters that sometimes, but not always, serve as inhibitory transmitters (in the presence of inhibitory receptors) include **norepinephrine, epinephrine, serotonin,** and **dopamine.** Presynaptic terminals secrete inhibitory transmitters rather than excitatory transmitters. In fact, there are many more inhibitory synapses in the CNS than one might imagine, for the function of large parts of the brain, including the cerebral cortex, the basal ganglia, the thalamus, and the cerebellum, depends almost as much on inhibition of neurons as upon excitation. It is probable that as many as a third or even more of the synapses are of the inhibitory type rather than of the excitatory type.

Whether a given transmitter substance will be excitatory or inhibitory depends on both the transmitter and the nature of the receptor. Some transmitters can be either excitatory or inhibitory, depending on the type of receptor with which it binds. However, other transmitters are almost always either inhibitory or excitatory.

Postsynaptic Potentials

A given neuron can be in a resting state, an excited state, or an inhibited state. The three states of a neuron are shown in Figure 7-2. Whether a neuron becomes excited or inhibited depends upon the nature of the membrane receptor as well as the type of transmitter substance. A given receptor may be an excitatory receptor or an inhibitory receptor.

If the receptor is excitatory, it opens sodium channels, allowing sodium ions to move selectively to the inside of the membrane. This

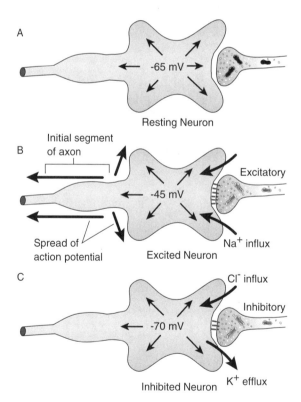

Figure 7-2.
Three states of a neuron. *A:* Resting neuron; *B:* excited neuron; *C:* inhibited neuron. (Adapted from Guyton AC and Hall JE: *Textbook of Medical Physiology,* 9th ed. Philadelphia: WB Saunders, 1995, p 576.)

action partially **depolarizes** the neuron (makes the inside of the membrane less negative), and therefore stimulates it. The increase in voltage above the normal resting membrane potential (to a less negative value) is called the **excitatory postsynaptic potential** (or EPSP).

If the receptor is inhibitory, the channels become permeable to chloride ions or potassium ions or both. Movement of these ions through the membrane **hyperpolarizes** the neuron (makes the inside of the membrane more negative), inhibiting the neuron rather than exciting it. Note in Figure 7-2 that chloride ions and potassium ions move in opposite directions through the cell membrane. The decrease in voltage below the normal resting membrane potential to a more negative value is called the **inhibitory postsynaptic potential** (or IPSP).

Summation at Synapses

Summation of nerve impulses is a means for transmitting signals of different strengths. Since nerve impulses are an all-or-nothing function, the means by which the nervous system transmits weak or strong signals is to "summate" many numbers of nerve impulses, that is, the greater the number of impulses transmitted, the greater the effect at the other end of the nerve.

Spatial summation occurs when excitatory impulses from two or more presynaptic neurons arrive simultaneously at a postsynaptic neuron. The amount of depolarization caused by each excitatory input is summated, and if enough synapses fire at the same time, the threshold potential is reached and an action potential is elicited, as shown in Figure 7-3.

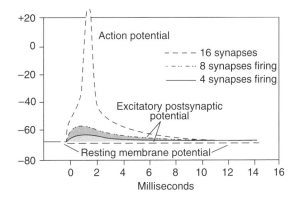

Figure 7-3.
Spatial summation. An action potential can be elicited when several excitatory synapses fire at the same time and the summated potential reaches the threshold potential. (Adapted from Guyton AC and Hall JE: *Textbook of Medical Physiology,* 9th ed. Philadelphia: WB Saunders, 1995, p 578.)

Temporal summation occurs when two or more impulses from the same excitatory synapse occur in rapid succession. The postsynaptic depolarizations overlap in time and add together in stepwise fashion.

In either instance, the effect at the opposite end of the nerve is very much the same because the degree of reaction is dependent on the number of impulses arriving at the end of the nerve in a given period.

NEURONAL CIRCUITRY

Each part of the brain usually contains large numbers of similar types of neurons that lie close to each other and are interconnected by means of many fine nerve filaments. Each such group of neurons is called a **neuronal pool.** Different patterns of nerve filament interconnections exist in different pools, and the type of pattern in turn determines the manner in which the pool operates in the overall function of the brain. In general, three basic types of circuits occur in neuronal pools: (1) the **diverging circuit;** (2) the **converging circuit;** and (3) the **repetitive firing circuit.**

Each type of circuit has special characteristics that allow it to emit a certain pattern of output impulses in response to incoming signals. By combining the functional characteristics of the many different pools in the nervous system, almost any type of integrative function can be achieved in one portion of the nervous system or another.

Diverging Circuits

In a diverging circuit, the signals entering a neuronal pool excite many more nerve fibers leaving the pool. This circuit, shown in Figure 7-4, is the simplest of all that occur in the neuronal pools. The nerve fibers entering the pool divide many times so that a few nerve fibers entering the pool branch many times and excite many different new neurons, causing a large number of impulses to leave the pool.

The diverging circuit is typified by the nervous control of muscular activity. Stimulation of a single large neuron in the motor cortex can stimulate many interneurons in the spinal cord, and these in turn might then stimulate as many as 50 to 100 anterior motor neurons that, in turn, stimulate thousands of muscle fibers.

Converging Circuits

In a converging circuit, the signals from multiple inputs converge to excite a signal neuron. A converging circuit is one that, after receiv-

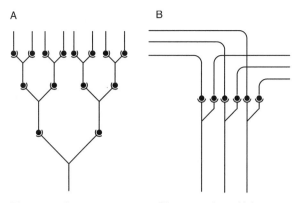

Divergence in same tract   Divergence in multiple tracts

Figure 7-4.
Diverging circuits. Circuit A causes amplification of the signal. Circuit B shows divergence into multiple tracts, which send the signals to separate areas. (Adapted from Guyton AC and Hall JE: *Textbook of Medical Physiology,* 9th ed. Philadelphia: WB Saunders, 1995, p 590.)

ing incoming signals from several sources, determines the level of reaction that will occur. That is, impulses "converge" into the pool, some from inhibitory nerves, some from excitatory nerves, some from peripheral nerves, and some from parts of the brain. Figure 7-5 shows two types of converging circuits.

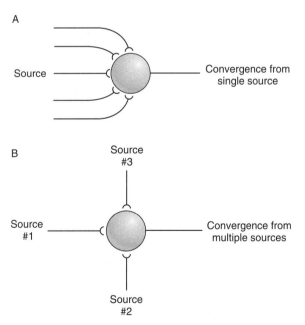

Figure 7-5.
Converging circuits. Circuit A shows convergence of multiple inputs from a single source. Circuit B shows convergence of single inputs from multiple neurons. (Adapted from Guyton AC and Hall JE: *Textbook of Medical Physiology,* 9th ed. Philadelphia: WB Saunders, 1995, p 590.)

The overall response of a converging circuit depends on multiple factors. These include:

- Basic excitability of the neurons in the pool
- Number of excitatory impulses entering the pool
- Number of inhibitory impulses entering the pool
- Whether or not there might be some diverging circuits also in the pool
- Distribution of excitatory and inhibitory impulses to the different neurons, and so forth

From this list of possible factors that can affect the output from the neuronal pool, one readily understands that basic differences in the anatomic organization of different neuronal pools can give thousands of different responses to incoming signals.

Repetitive Firing Circuits

In a repetitive firing circuit the neuronal pool emits a series of impulses lasting long after the incoming signal is over. This is among the most important types of circuits in the nervous system. Three types of circuits can cause repetitive firing: (1) A **pool of neurons** consisting of very excitable neurons, each one of which has a natural tendency to fire repetitively. (2) A **long chain of neurons** arranged one after another so that an incoming stimulus activates each one in succession. From each neuron of the chain a nerve fiber extends to some outlying neuron. Thus, this outlying neuron receives repetitive impulses from the successive neurons of the chain, but after all these have fired, the repetitive firing from the output neuron ceases. (3) The **reverberating circuit** is discussed below.

In a reverberating circuit, an incoming impulse is passed along a succession of neurons until one of the neurons restimulates an earlier neuron in the succession. This is probably the most important type of repetitive firing circuit. The impulse goes around the reverberating circuit again and again. Every time around the circuit, collateral impulses are emitted into outgoing nerve fibers that spread to other parts of the nervous system. (Figure 7-6 shows reverberating circuits of increasing complexity.) Theoretically, this type of circuit might continue to oscillate indefinitely, but more usually the oscillation ceases when some of the neurons in the circuit become too fatigued to continue. The continual respiratory rhythm represents a continually reverberating circuit, while the thought processes of the cerebral cortex probably represent circuits that reverberate for short periods until neurons in the circuit fatigue or are inhibited so that the thought ceases.

Neuronal pools can have high or low thresholds. A pool may have a high threshold into which many excitatory impulses must arrive before an effect will occur, or it might be a low-threshold pool into which only a few impulses must arrive before an effect will occur. The low-threshold circuit is typified by the neuronal response

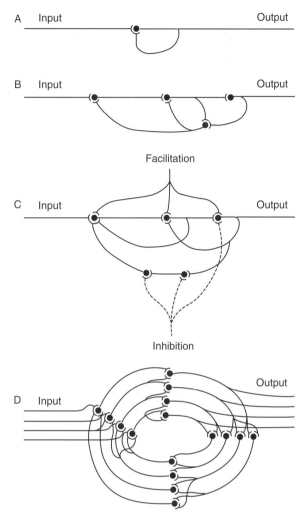

Figure 7-6.
Reverberating circuits. (Adapted from Guyton AC and
Hall JE: *Textbook of Medical Physiology,* 9th ed.
Philadelphia: WB Saunders, 1995, p 592.)

that causes withdrawal of a limb when only a few pain receptors are
stimulated, while the high-threshold circuit is typified by withdrawal
of a hand only when tremendous numbers of touch receptors are
stimulated.

CEREBRATION

The bases of cerebration are the individual thoughts. Many
thoughts occur directly as a result of incoming sensory impulses. For
instance, when a person is looking at a beautiful scene, the impulses
from the eyes certainly generate a number of different thoughts.

The precise mechanisms of thoughts in the brain are not understood, but one of the suggestions is that a thought represents a pattern of impulses passing through particular neurons in multiple simultaneous areas of the conscious brain.

Memory

Memory is believed to result from permanent facilitation of synapses. This means simply that excitation of a synapse repetitively over time will cause that synapse to become more and more effective in stimulating the neuron. In other words, the fact that an impulse passes through a synapse once makes it easier for successive impulses to pass through the same synapse. Therefore, if a thought pattern is evoked over and over by incoming sensory stimuli, eventually the pathway for transmission of impulses through that particular thought channel becomes facilitated so that even the slightest stimulus entering this pathway at a later time can elicit the entire thought. For instance, such a facilitated thought pathway might develop in response to seeing the beautiful scene referred to above. Then 1 year later, some stray impulse from another part of the brain might enter this particular thought pathway and enable the person to see the scene again in his or her mind. This is believed to be the basis of memory.

The cerebral cortex is the portion of the brain most concerned with memory. All through the cerebral cortex are located neuronal pools that can be facilitated by sensory impulses so that subsequent signals entering these pools will evoke specific reactions. The storage of information in the brain is mostly lost when the cortex is gone.

Experiments in lower animals have demonstrated one possible type of memory circuit. When repetitive signals are passed through the synapses of the brain's memory system, the surface areas of the activated presynaptic terminals grow larger. Also, increased numbers of synaptic vesicles appear adjacent to the new presynaptic membrane area. Therefore, at any later time when still newer signals enter the same neuronal pathway, far greater quantities of synaptic transmitter are secreted into the synaptic cleft. Obviously, this enhances the sensitivity of the memory pathway and allows one to reactivate the memory circuit with ease.

Programming

Thoughts must be programmed for cerebration to occur. Everyone is familiar with the fact that different thoughts usually occur in rapid succession, and that each succeeding thought usually has some association with the preceding thought. Many sequences of thoughts are initiated by **incoming sensory signals,** whether these originate from the skin, from the eyes, from the ears, and so forth. For instance, a sudden knife cut on the leg would elicit first a thought of pain, then another thought that localizes the cut on the body, this followed by integrative processes that make the person turn the eyes and head to look at the pained area, followed by visual input impulses that combine with the previous thoughts to determine the nature of the stimulus causing the pain, and, finally, a series of integrations that cause motor movements to remove the painful object from the body. In this **sequence of cerebration,** the person must call forth memories from past experiences in order to understand why and how the leg is being pained, for, if he or she has never seen a knife before and is not familiar with its cutting capabilities, simply looking at the leg and seeing a knife against the skin will not explain the cause of the pain. In short, for cerebration to occur, the thoughts must be programmed.

Some part of the brain must determine where the attention of the mind will be directed. For example, the mind may be directed to the incoming sensory signals from the leg, to the movement of the head and the eyes, or to one of the memory circuits to call forth information. The nature of this programming system of the brain is still unclear. However, the anatomic locations of the thalamus and the reticular substance of the mesencephalon have made many neurophysiologists point to these two areas as possible programming centers. Also, stimulation of specific points in these two areas causes highly specific patterns of reaction to occur in other parts of the brain and cord.

SECTION 2
SOMATIC SENSORY SYSTEM

The general plan for transmission of sensory signals from all parts of the body into the CNS is illustrated in Figure 7-7. The somatic sensory portion of the nervous system is often divided into three different systems: the **exteroceptive system,** which transmits impulses from the skin; the **proprioceptive system,** which transmits impulses mainly from the muscles and joints relating to the momentary physical condition of the body; and the **visceral sensory system,** which transmits impulses from the viscera.

Modalities of Sensation

It is common knowledge that many different types of sensations can be perceived from the skin, including light touch, tickle, pressure, pain, cold, and warmth. These are called **modalities of sensation.** Proprioceptive modalities of sensation include sense of position of the limbs, degree of tension in the muscle tendons, degree of stretch of the muscle fibers, and deep pressure on different parts of the body. In addition to these, other modalities that are not trans-

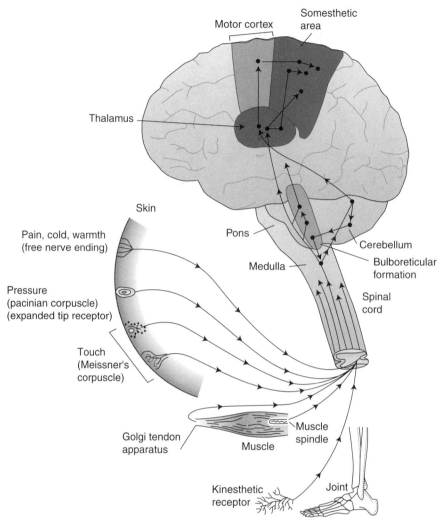

Figure 7-7.
Somatic sensory system. (Adapted from Guyton AC and Hall JE: *Textbook of Medical Physiology,* 9th ed. Philadelphia: WB Saunders, 1995, p 566.)

mitted by the somatic sensory system are vision, hearing, equilibrium, smell, and taste.

Sensory Receptors

The five basic types of sensory receptors are listed below:

- **Mechanoreceptors** detect various mechanical stimuli with some receptors responding to high-frequency vibrations and others to constant pressure.
- **Thermoreceptors** can be subdivided into warm receptors and cold receptors that respond to a respective rise or fall in temperature.
- **Nociceptors (pain receptors)** detect physical or chemical damage occurring in the tissues.

- **Electromagnetic receptors** detect light on the retina of the eye.
- **Chemoreceptors** detect taste, smell, oxygen level, carbon dioxide level, osmolarity in body fluids, and other chemical substances.

The most common type of sensory receptor is the free nerve ending. The free nerve ending, illustrated in Figure 7-7, is nothing more than a filamentous end of a nerve usually interwoven with other filamentous nerve endings. Different types of free nerve endings can transmit relatively crude sensations such as pain, crude touch, tickle, heavy pressure, and temperature. In addition to the free nerve endings, the skin contains a number of specialized endings that are adapted to respond to some specific type of physical stimulus. For instance, one of these endings, called a **Meissner's corpuscle,** is most numerous in the skin over the fingertips and responds specifically to light touch.

Proprioceptive sensations are detected by specialized sensory receptors. These include **joint receptors,** which detect the degree of angulation of the joints; **pacinian corpuscles,** which detect high-frequency vibration and very rapid changes in pressure; **Golgi tendon apparatuses,** which detect the degree of tension in the tendons; and **muscle spindles**, which detect the degree of elongation of the muscle fibers. In general, the proprioceptive impulses are transmitted by **large type A nerve fibers** that can transmit at velocities as high as 100 m/sec. This rapid velocity is especially important when a person is moving quickly because the nervous system needs to know during each split second the positions of all parts of the body.

The Labeled-Line Law

Each type of sensory nerve fiber transmits only one modality of sensation—the labeled-line law. If a sensory nerve fiber is stimulated by an electrical stimulus, a person will perceive only one particular modality of sensation. For instance, excitation of a pain fiber will cause pain, and excitation of a warmth fiber will cause the sensation of warmth. Thus, each type of sensory nerve fiber transmits only one modality of sensation; this is called the **labeled-line law.**

Modalities of sensation are determined by the brain, not the receptor. It is not the type of receptor that determines the modality of sensation transmitted, but, instead, it is the point in the CNS to which the fiber from the receptor connects that determines the modality. For instance, pain fibers end in a slightly different point in the thalamus from the warmth fibers and in a different point from the cold fibers.

Sensory centers in the brain stem and thalamus can determine some sensory modalities, but localization is a function of the somatic sensory cortex. Since some modalities of sensation, such as pain and temperature sensations, can still be perceived even when the sensory portions of the cerebral cortex are removed, it is believed that sensory centers in the brain stem and the thalamus can determine at least some of the different sensory modalities. On the other hand, a person cannot localize sensations in different parts of his or her body accurately when the sensory portions of the cerebral cortex have been destroyed. Therefore, discrete localization is principally a function of the somatic sensory cortex, although the thalamus by itself is capable of crude localization to general areas of the body.

Pathways for Transmission of Somatic Sensations into the CNS

Somatic sensory signals may be transmitted up the spinal cord in one of two different pathways. The impulses generated in the sensory receptors are transmitted first into the spinal nerves and then through the dorsal roots of the spinal nerves into the spinal cord. The nervous signals are then carried up the spinal cord by way of the **dorsal column-medial lemniscal system** or the **anterolateral system,** depending on the origin of the sensory signal.

The **dorsal column-medial lemniscal system allows rapid transmission of information with high spatial orientation.** The dorsal column-medial lemniscal system is composed of large, myelinated nerve fibers that transmit signals to the brain at velocities ranging from 30 to 110 m/sec (Table 7-1). The anterolateral system is composed of much smaller myelinated fibers and can only transmit signals at lower velocities. The dorsal column-medial lemniscal system also has a much higher degree of spatial orientation with respect to the origin of the nerve signals on the surface of the body. A special feature of the anterolateral system is the capability to transmit a broad spectrum of sensory modalities, such as pain, warmth, cold, crude tactile sensations, tickle and itch, and sexual sensations (Table 7-2). The dorsal column-medial lemniscal system transmits position sensations and more refined pressure and touch sensations.

TABLE 7-1.

Modalities of Sensation Conducted by Way of the Dorsal Column-Medial Lemniscal System

Touch sensations requiring a high degree of spatial orientation
Touch sensations requiring transmission of fine gradations of intensity
Phasic sensations (e.g., vibratory sensations)
Sensations that signal movement against the skin
Pressure sensations that require a high degree of intensity judgment
Position sensations

TABLE 7-2.
Modalities of Sensation Conducted by Way of the Anterolateral System

Pain

Thermal sensations (warm and cold sensations)

Crude touch and pressure sensations

Tickle and itch sensations

Sexual sensations

Types of Nerve Fibers

It is important for some signals to be transmitted to or from the CNS very rapidly for information to be useful. Such is the case with information that apprises the brain of the position of the legs when an individual is running. Other types of sensory information, such as prolonged aching pain, can be transmitted to the CNS very slowly along slowly conducting nerve fibers.

Figure 7-8 shows physiologic classifications and functions of various types of nerve fibers. Two systems of classification are indicated in the figure: a general classification system that includes sensory and motor fibers as well as the autonomic nervous system, and a sensory nerve classification system for sensory nerve fibers only.

In the **general classification scheme,** the large type A fibers are further subdivided into α, β, γ, and δ fibers. A type Aα myelinated fiber has a diameter ranging from 10 to 20 μm and a conduction velocity ranging from 60 to 120 m/sec. They innervate skeletal muscle and Golgi tendon organs, and comprise the annulospiral endings of muscle spindles. Under the **sensory nerve classification** system, Golgi tendon organs are innervated by type Ib fibers, and type Ia fibers comprise the annulospiral endings of muscle spindles, as shown in Figure 7-8.

PAIN

Pain Stimulus

Pain fibers are stimulated any time a tissue is being damaged or overstressed. Once damage to tissue is complete, the pain sensation generally disappears in a few minutes or sometimes even in a few seconds. Pain nerve endings can be stimulated by mechanical

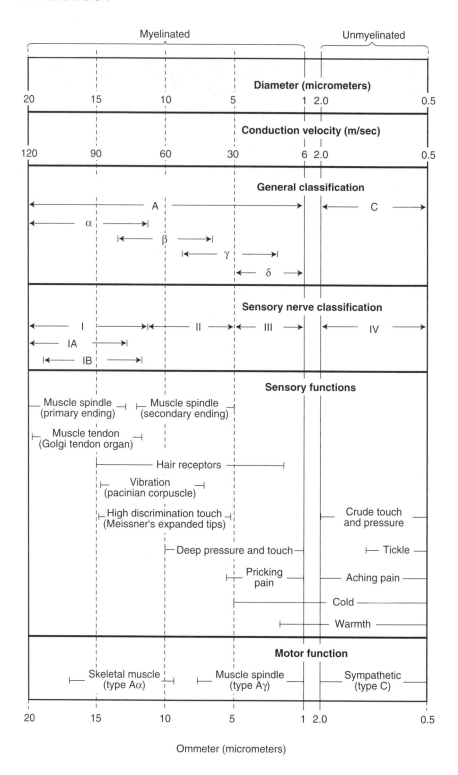

Figure 7-8.
Physiologic classification and functions of nerve fibers. (Adapted from Guyton AC and Hall JE: *Textbook of Medical Physiology,* 9th ed. Philadelphia: WB Saunders, 1995, p 588.)

trauma to the tissues; excess heat; excess cold; chemical damage; certain types of radiation damage, such as the pain associated with sunburn; and even lack of adequate blood flow to a tissue area, which causes ischemic pain.

Pain Threshold

It is often said that some people perceive pain far more easily than others. However, this usually is not true. Actual research on various degrees of injury in relation to the perception of pain shows that all persons begin to perceive pain at almost exactly the same degree of injury. Yet, once the pain is perceived, the degree of transmission of pain signals in the nervous system, as well as the reactivity of different persons to the pain, varies tremendously. The reactivity to pain is partly an inherited characteristic of certain people, but it is also greatly influenced by previous training.

Control of Pain by the Nervous System

The brain has the capability to control the sensitivity of pain pathways. Sensitivity is controlled by sending centrifugal inhibitory signals to the brain stem and spinal cord to control pain signal transmission. The analgesia system has three main components: Neurons from the **periaqueductal gray area** surrounding the aqueduct of Sylvius in the mesencephalon and upper pons send signals to the **raphe magnus nucleus** located in the lower pons and upper medulla. The signals are then transmitted to a **pain inhibitory complex** located in the dorsal horns of the spinal cord where the analgesia signals can block pain before it is relayed to the brain.

Enkephalin and serotonin are transmitter substances involved in the analgesia system. Nerve fibers derived from the periaqueductal gray area secrete enkephalin at their terminations in the raphe magnus nucleus. Fibers originating in the raphe magnus nucleus secrete serotonin at their endings in the dorsal horns of the spinal cord, which in turn excite other dorsal horn neurons to secrete enkephalin. The enkephalin then acts on the pain-conducting neurons to block the pain signals.

Visceral Sensation

Visceral pain usually occurs only on stimulation of pain endings over a wide area. Internal organs of the body have a sparse supply of pain endings compared to the skin and are relatively insensitive to a sharp knife cut since there are insufficient pain endings in any minute area to cause pain. Therefore, pain from the viscera is far more likely to be a generalized aching or burning rather than a sharp pain, and the pain usually occurs only on stimulation of pain endings over a wide area. However, this is not true in the periosteum of the bones, in the walls of the arteries, in the parietal pleura, and in the parietal peritoneum. These areas are almost equally as susceptible to pain as the skin.

Visceral pain can be caused by many types of stimuli. The different types of stimuli that are particularly prone to cause visceral pain are as follows:

- **Overdistention** of a hollow organ
- **Spasm** of the smooth muscle of an organ
- **Inadequate blood flow** to an organ
- **Chemical damage,** such as that produced by spillage of acid gastric juice into the peritoneal cavity through a ruptured peptic ulcer

Referred Pain

Referred pain is felt in a part of the body that is distant to the tissues actually causing the pain. Pain in a visceral organ is not always felt directly over the organ itself but may be referred to a distant area of the body. For instance, pain originating in the heart is often felt mainly in the left arm or shoulder. This is called **referred pain.** Referred pain usually results from collateral neuronal connections between the visceral pain fibers and the somatic pain pathways in the cord, the visceral impulses exciting the somatic pathways, and the person localizing the pain in some nonvisceral part of the body.

CORD REFLEXES

Many CNS functions occur locally in the spinal cord without the aid of the brain. The cord especially integrates many specific reflexes that help to control muscle movements.

The Stretch Reflex

The stretch reflex utilizes the muscle spindle to prevent the length of a muscle from changing rapidly. It is elicited by stretching the muscle. In its simplest form, the stretch reflex involves only two neurons, the **sensory neuron** from the muscle spindle to the anterior motor neuron, and the **anterior motor neuron** back to the muscle. Stretch of the muscle spindle increases the number of impulses transmitted by the spindle, and this increases the number of impulses transmitted by the anterior motor neuron back to the muscle. Therefore, muscle stretch enhances the contractile tension in the muscle. This tension in turn tends to shorten the muscle back to its initial length. Thus, the stretch reflex opposes changes in muscle length.

The stretch reflex has both a dynamic and a static component. These components are called, respectively, the **dynamic stretch reflex** and the **static stretch reflex.** The dynamic effect occurs only when the muscle is stretched rapidly because the spindles are very strongly stimulated during the actual instant of stretching. The strong signal from the spindle causes extreme feedback contraction of the muscle to oppose the sudden stretch. The static stretch reflex is much weaker than the dynamic reflex, but it maintains muscle contraction for minutes or hours when the muscle remains stretched beyond its normal length.

Signals from the brain stem can alter the overall reactivity of the muscles. The muscle spindles themselves are provided with excitatory nerve fibers from the spinal cord called **gamma efferent fibers,** and these in turn are controlled by signals from the reticular formation of the brain stem. Impulses transmitted through the gamma fibers can increase the degree of activity of the muscle spindle and therefore can also increase the intensity of either the dynamic stretch reflex or the static reflex.

Muscle spindles help to maintain a certain amount of basal tone in the muscle. Muscle spindles continually send impulses into the spinal cord to excite the anterior motor neurons. These, in turn, transmit impulses back to the respective muscle. This continual flow of impulses helps to maintain muscle tone.

The knee-jerk reflex is used to test the sensitivity of the stretch reflex. Tapping on the patellar tendon stretches the quadriceps and initiates a dynamic stretch reflex that causes the lower leg to jerk forward. Similar reflexes can be obtained from other muscles by striking the tendon or the belly of the muscle.

Withdrawal Reflex

Withdrawal reflexes function to move any pained part of the body away from the object causing the pain. For instance, if the hand is placed on a hot stove, impulses are transmitted from the pain receptors to the cord and immediately back to the flexor muscles of the arm to withdraw the hand. Because the flexor muscles are involved in this instance, this particular withdrawal reflex is called a **flexor reflex.** Part of the withdrawal response involves impulses transmitted to the opposite side of the body to extend the opposite limb, thereby pushing the whole body away from the vicinity of the painful object. This extensor effect is called the **crossed-extensor reflex.**

Positive Supportive Reflex

The positive supportive reflex helps to support the weight of the body against gravity. Pressure on the bottoms of the feet causes the extensor muscles of the legs to tighten, helping the legs to support the weight of the body against gravity. This reflex is integrated entirely in the few segments of the spinal cord that control the activity of each respective limb.

Walking Reflexes

Walking movements can be performed in the limbs of lower animals with the spinal cord transected in the neck. In an opossum with a transection in the thorax, the hind limbs can "walk" but without coordination with the movements of the forelimbs. If the cord is transected in the neck, rhythmic to-and-fro coordinated walking movements among all four limbs can occur. Occasionally, trotting movements also occur and, very rarely, galloping movements. However, with a neck transection, equilibrium cannot be maintained, so the animal cannot actually make forward progression.

The basic patterns for walking and other movements of locomotion are integrated in the spinal cord. The nerve fiber tracts that coordinate the functions of the superior and the inferior segments of the cord are the **propriospinal fiber pathways** that lie near the cord gray matter and account for approximately one-half of all the fiber tracts in the cord.

FUNCTIONS OF THE BRAIN STEM

Support of the Body against Gravity

The human body cannot walk or even stand without the aid of CNS centers higher than the spinal cord. Although the spinal cord is capable of providing both the positive supportive reflex that helps to support the body against gravity and the walking reflexes, the human body still cannot stand and certainly cannot walk without the aid of higher CNS centers. With progression from lower phylogenetic types to the higher types of animals, more and more of the control systems have gradually shifted from the cord toward the brain. As stated above, a lower animal, such as an opossum, can still walk quite well with its hind limbs even when its spinal cord is transected in the thorax. In the dog, basic walking movements can occur in the hind limbs with the thoracic cord transected, but these cannot be coordinated sufficiently to provide functional walking. In the human, even these walking reflexes are crude when the cord is cut.

Maintenance of Equilibrium

The vestibular and reticular nuclei of the brain stem are important for controlling whole body movement and equilibrium. The vestibular and reticular nuclei transmit impulses especially to the extensor muscles, tightening the muscles of the trunk, the buttocks, the thighs, and the lower legs to allow the body to stand in an upright position. Therefore, it is frequently said that the brain stem supplies the nervous energy required for supporting the body against gravity.

The vestibular apparatus is the organ that detects sensations of equilibrium. Closely associated with the support of the body against gravity is the maintenance of equilibrium. The vestibular and reticular nuclei of the brain stem can vary the degree of tension in the different extensor muscles in proportion to the need for maintenance of equilibrium. To do this, these nuclei in turn are controlled by the vestibular apparatuses located on the two sides of the head in close association with the ears.

The maculae of the utricle and saccule form one of the receptor systems of the vestibular apparatus. They contain large numbers of small calcified crystals called **otoliths** that lie on "hairs" projecting from sensory receptor cells called **hair cells.** Leaning the head to one side or forward or backward causes these otoliths to fall toward the direction of leaning, thus bending the hairs. Because the different hair cells are oriented in all the different directions, this bending of the hairs causes signals to be transmitted into the brain informing the brain of the position of the head in relation to the direction of gravitational pull.

Another receptor system of the vestibular apparatus is the semicircular ducts. This system consists of three circular ducts on each side of the head. Each of the ducts is oriented in one of the three planes of space. The ducts are filled with fluid so that any time the head rotates in any plane, inertia of the fluid causes it to move through one or more of the ducts and thereby stimulate hair cells located in the **ampullae** of the semicircular ducts. Thus, rotating movements of the head are also made known to the nervous system.

INTEGRATION OF SENSORY AND MOTOR FUNCTIONS

Primary Sensory Areas of the Cortex

Signals from each type of sensory receptor are transmitted to a specific area of the cerebral cortex:

* **Somatic sensations** are relayed by the thalamus directly to the somatic sensory cortex located anteriorly in the parietal lobes.

- **Visual sensations** are relayed from the optic tract by the lateral geniculate bodies of the thalamus directly to the visual cortex in the calcarine fissure area of the occipital lobes.
- **Auditory impulses** from the auditory nerves are relayed by the medial geniculate bodies of the thalamus to the auditory cortex in the central portion of the superior temporal gyri.
- **Taste impulses** are relayed through the nuclei of the tractus solitarius and the thalamus to a small area of the cerebral cortex deep in the fissure of Sylvius.
- **Olfactory sensations** are relayed to the amygdala (a subcortical mass of neurons in the anterior temporal lobe) and the pyriform area of the cortex.

Spatial orientation of sensory information is maintained in the cerebral cortex. In the visual system, each minute area of the retina is connected directly to a minute area of the visual cortex. In the somatic sensory system, each point on the surface of the body connects with a specific point in the somatic sensory cortex, as illustrated in Figure 7-9, so that stimulation of a finger, for instance, will excite only a minute area of the cortex. In the auditory cortex, certain sound frequencies stimulate one portion of the auditory cortex, while others stimulate other portions.

In general, therefore, the types of information transmitted into the cerebral cortex by the different sensory systems are (1) the locations in the body from which the signals are arriving; (2) the types of receptors detecting the sensations; and (3) the intensities of the sensory stimuli. Once this information has entered the primary sensory areas, signals are relayed to other portions of the

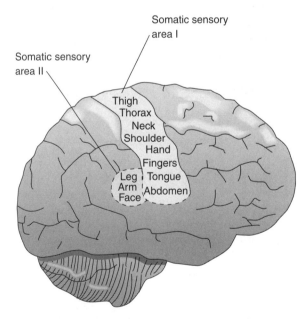

Figure 7-9.
Localization of sensory perception in the cerebral cortex.

brain where the different types of information begin to be assembled into usable thoughts.

Sensory Association Areas

Memories of past sensory associations are stored in the sensory association areas. Located immediately adjacent to the primary sensory areas are the sensory association areas, which receive direct communications from the primary sensory areas. In the sensory association areas, many memories of past sensory associations are stored, and here the new information arriving from the primary sensory areas is compared with information that has been stored from the past. In this way, the significance of the new sensory signals is determined. For instance, when a person hears a word, he or she will not know that it is a word unless memory of that word has been stored in the auditory association areas. Likewise, when a person sees an airplane, the primary visual cortex is unable to determine the nature of the object, but, on transmission of appropriate information into the visual association areas, the person becomes aware that he is seeing an object that he has seen before and classifies it as an airplane. Similar functions are performed by somatic, taste, and smell association areas.

Gnostic Function of the Brain

Wernicke's area of the brain provides a gnostic function. Brain surgeons have found that destruction of the posterior part of the superior gyrus of the temporal lobe, an area called **Wernicke's area,** in the left hemisphere of the right-handed person will destroy the ability to put together information from the different sensory association areas and thereby determine the overall meaning. For this reason, this region of the brain has been called the **gnostic center,** which means simply the "knowing center." This area is well located for this purpose because it lies at the juncture of the temporal, the parietal, and the occipital lobes in very close association with most of the sensory association areas of the cortex.

Ideomotor Function of the Brain

Wernicke's area is necessary for ideomotor function of the brain. Once all the information from the different sensory association areas has been integrated into a distinct conscious meaning, the brain then decides what type of physical reaction should occur, from no reaction at all to very violent reaction. This is called the **ideomotor function** of the brain. Again, in neurosurgical patients it has been found that damage to Wernicke's area will cause a person to lose ideomotor ability. After all sensory information is put together, appropriate signals are then sent to the motor portion of the brain, which in turn causes muscular movements.

MOTOR PATHWAYS

Primary Motor Cortex

The motor axis of the nervous system for controlling skeletal muscle contraction is shown in Figure 7-10. Muscle contraction can be controlled at many different levels in the CNS. These include the spinal cord, the reticular substance of the brain stem, the basal ganglia, the cerebellum, and the motor cortex.

The motor cortex is somatotopically organized. In other words, there is point-to-point communication between the primary motor cortex and specific muscles everywhere in the body. The primary motor cortex is a strip of the cortex averaging about 2 cm wide and lying horizontally all the way across the cortex, located immediately

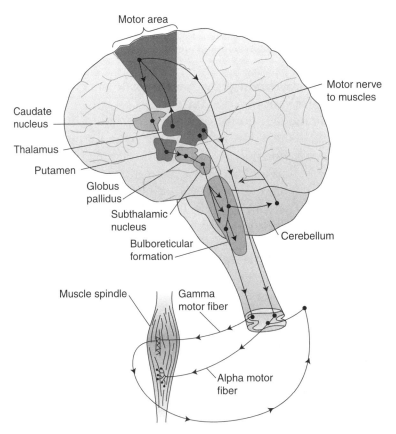

Figure 7-10.
Motor nervous system. (Adapted from Guyton AC and Hall JE: *Textbook of Medical Physiology,* 9th ed. Philadelphia: WB Saunders, 1995, p 567.)

in front of the central sulcus of the brain. Stimulation of discrete points in the primary motor cortex will cause contraction of discrete muscles in the body. For instance, stimulation of the primary motor cortex at a point on top of the brain where it dips into the longitudinal fissure will cause contraction of a leg muscle on the opposite side of the body, while stimulation of the primary motor cortex where it begins to dip into the fissure of Sylvius will contract a muscle somewhere on the opposite side of the face.

The thumb, finger, mouth, and throat muscles are represented to the greatest extent in the motor cortex. Stimulation of a small area of the motor cortex might cause a large group of muscles in the trunk to contract; stimulation of the same amount of cortical tissue in the mouth area might cause only one small muscle to contract. The degree of representation of the mouth and throat and also of the thumb and fingers is as much as 100 times that of the trunk muscles. This high level of representation allows the cerebral cortex to control with extreme fidelity the movement of the hands, and the special representation of the mouth and throat accounts for the ability of the human being to talk.

Pyramidal System and Pyramidal Tracts

The pyramidal tract (corticospinal tract) is the most important output pathway from the motor cortex. In the primary motor cortex of each hemisphere there are some 30,000 large neuronal cells called **pyramidal** or **Betz cells.** Fibers from these cells pass downward through the **pyramidal tracts** all the way into the spinal cord. This means that in humans some of the axons must be over 1 m long. The majority of pyramidal axons cross in the brain stem to the contralateral side and end on **interneurons** located in the posterolateral gray matter of the cord.

The interneurons also receive many nerve endings from the following: (1) sensory nerve fibers entering the spinal cord through the spinal nerves; (2) propriospinal fibers originating in the other segments of the cord; and (3) other nerve fibers from the brain. The interneurons, after integrating the signals from all these sources, in turn transmit impulses to the **anterior motor neurons** located in the anterior gray matter of the cord. These neurons receive additional impulses directly from (1) proprioceptive nerve fibers from the spinal nerves, (2) a few nuclei in the brain stem, and (3) other segments of the spinal cord. After integrating these signals, the anterior motor neurons in turn send impulses through the peripheral nerves to all the skeletal muscles of the body.

Premotor Cortex

Different patterns of movements can be learned and stored in the premotor cortex. Located anterior to the motor cortex is still another strip, averaging about 2.5 cm wide, called the **premotor cortex.** Stimulation of a discrete point in the premotor cortex usually does not cause contraction of a discrete muscle but, instead, causes a "pattern" of muscle contraction. That is, it might cause the whole arm to rise upward, or it might cause the whole hand to flex forward, or stimulation of still another point might cause the thumb and the forefinger to move toward each other as if cutting with scissors.

Perhaps not more than a few hundred different patterns of movement are stored in the normal premotor cortex. However, considering the thousands to millions of different combinations into which these patterns of movement can be organized, even this small number of movements could allow almost any type of activity. It is believed that the ideomotor function of the cortex from Wernicke's area controls the sequence of patterns of movement.

Extrapyramidal Pathways

Unlike the pyramidal pathways, the extrapyramidal pathways do not cross to the contralateral side and have one or more synapses along their route to the spinal cord. Another important difference is that the extrapyramidal motor pathways originate not only in the motor cortex but also in other brain structures such as the cerebellum and the vestibular nuclei. Extrapyramidal pathways are thought to transmit many of the stereotyped and subconscious movements of the body.

One major extrapyramidal pathway is from the motor and premotor cortex to the reticular formation of the brain stem and from there through the reticulospinal tracts to the spinal cord. Another extrapyramidal pathway is from the motor and premotor cortex to the basal ganglia, then to the brain stem, and finally to the spinal cord.

THE BASAL GANGLIA

The basal ganglia form an accessory motor system that functions in close association with the cerebral cortex and pyramidal system. Major portions of the basal ganglia include the caudate nucleus, the **putamen,** and the **globus pallidus.** Physiologically related structures include the **subthalamus** and **substantia nigra.** The basal ganglia have very extensive neuronal connections with the premotor and primary motor portions of the cortex, with the somatic sensory cortex, with the thalamus, and with some nuclei of the brain stem.

The basal ganglia in association with the premotor and the motor cortex operate to help control most of the patterns of movements. Damage to certain areas of the basal ganglia will cause abnormal and often continuous movements such as choreiform movements, writing movements, and so forth.

The basal ganglia help to control the basal degree of activity of the entire motor system. Damage to certain areas of the basal ganglia can cause portions of the motor system to become greatly overexcitable, resulting in intense tonic contraction of either localized portions of the body or of the whole body. This results in a state of rigidity.

The basal ganglia operate in conjunction with the nuclei of the brain stem to damp the antagonistic movements of the postural muscles. For instance, if an extensor muscle should attempt to extend a limb, this would immediately elicit certain proprioceptive reflexes that would make flexor muscles tend to contract. This, in turn, would tend to make the extensor muscles contract again, and, as a result, a continuous state of oscillation would develop. However, this effect is normally damped by some of the lower basal ganglia so that antagonistic movements throughout the body are normally very smooth rather than tremorous. But in patients who have **Parkinson's disease,** which results from damage to the **substantia nigra,** one of the brain stem nuclei connected with the basal ganglia, a continuous tremor exists between the antagonistic pairs of muscles either in the entire body or in certain affected areas.

The basal ganglia become activated before the primary motor cortex when a person performs voluntary muscle activity. Other studies have shown that, before the onset of muscle activity, portions of the sensory cortex also become activated even before the basal ganglia. Therefore, a suggested scheme to explain voluntary motor activity is as follows: First, the nature of the motor act is probably conceived in the sensory cortex. Then, signals are sent to the middle regions of the brain's motor system such as the basal ganglia, the reticular formation in the brain stem, and even the cerebellum to initiate the more gross aspects of the motor act. Finally, the primary motor cortex is called into play to control the more discrete actions of the peripheral parts of the body such as the hands, fingers, and feet.

THE CEREBELLUM

The cerebellum helps the motor system to stop when the mission has been accomplished. The cerebellum receives collateral signals from the pyramidal and extrapyramidal fibers whenever they are stimulated by the primary motor cortex, by the premotor cortex, and by the basal ganglia, and it also receives impulses from proprioceptor nerves originating in all peripheral parts of the body. Thus, each time a motor movement is instituted by the brain, the cerebellum receives direct information of the projected movement from the cerebrum as well as information from the peripheral parts of the body telling it whether the movement has been accomplished and how much so. Putting these different types of information together, the cerebellum helps the motor system to stop the movements when the mission has been accomplished. To do this, the cerebellum performs two basic functions:

- **Predictive function.** From the proprioceptor impulses, the cerebellum can tell how rapidly a part of the body is moving and from this can predict when the part will get to a desired position. As it approaches the appropriate point, impulses are transmitted from the cerebellum through the thalamus to the motor cortex and basal ganglia, there initiating the motor signals that stop the movement.
- **Damping function.** The damping function of the cerebellum is closely associated with the predictive function. Impulses from the cerebellum, for example, act to delimit the movement of a limb before it gets to the desired point, preventing the momentum from carrying the limb beyond its intended position. However, if the cerebellum has been destroyed, the momentum will carry the limb beyond the position. Then the other areas of the brain attempt to bring it back again to the desired position, but again the limb overshoots; this continues several times until the intended movement is finally accomplished.

Cerebellar damage can cause tremors very similar to those resulting from basal ganglia damage. However, there is one particular difference: The basal ganglia tremor is almost continuous when the person is awake, while the cerebellar tremor occurs only during movements associated with specific voluntary motor acts such as intentional movement of the hand from one point to another. In other words, cerebellar damage produces an **intention tremor.**

The cerebellum is a system for helping to control rapid motor movements. Failure of the predictive and the damping functions of the cerebellum causes a person to walk with ataxic movements, causes rapid hand movements to be jerky, and even causes speech to become dysarthric—meaning that some sounds are overemphasized or held too long while other sounds are underemphasized to such an extent that the words are frequently unintelligible. It should be emphasized, though, that a person without a cerebellum can still perform most functions, even with precision, if he or she performs them very, very slowly. Therefore, the cerebellum is a system for helping to control rapid motor movements while they are actually occurring, keeping them precise despite rapidity of movement.

AUTONOMIC NERVOUS SYSTEM

The motor impulses from the CNS to the visceral portions of the body are transmitted differently than those to the skeletal muscles.

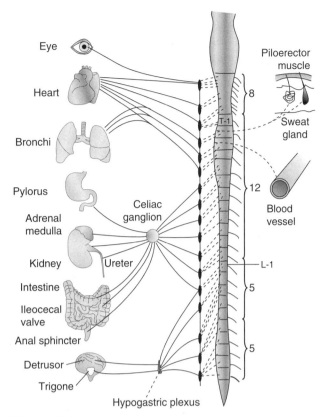

Figure 7-11.
Sympathetic nervous system. (Adapted from Guyton AC and Hall JE: *Textbook of Medical Physiology,* 9th ed. Philadelphia: WB Saunders, 1995, p 770.)

The former pass through two different divisions of the autonomic nervous system called the **sympathetic** and the **parasympathetic systems,** illustrated in Figures 7-11 and 7-12.

Sympathetic Nervous System

The sympathetic nervous system originates in neurons located in the lateral horns of the gray matter in the spinal cord between the first thoracic cord segment and the second lumbar segment. Nerve fibers pass by way of the anterior spinal roots first into the spinal nerves and then branch immediately into the sympathetic chain. From here, fiber pathways are transmitted to all portions of the body, especially to the different visceral organs and to the blood vessels.

Most sympathetic nerve endings secrete norepinephrine. Norepinephrine excites most of the visceral structures but inhibits a few. In general, it excites the heart and most of the blood vessels of the body, causing increased force of cardiac contraction and increased total peripheral resistance, with a resultant rise in arterial pressure. It inhibits the activity of the gastrointestinal tract,

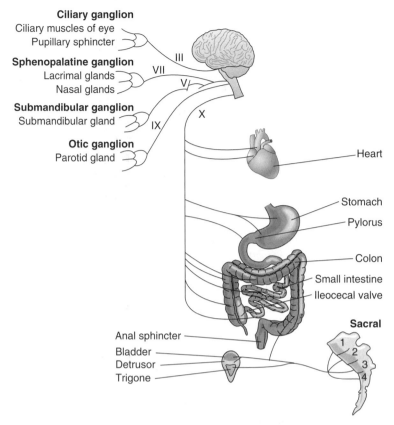

Figure 7-12.
Parasympathetic nervous system. (Adapted from Guyton AC and Hall JE: *Textbook of Medical Physiology,* 9th ed. Philadelphia: WB Saunders, 1995, p 771.)

thereby slowing peristalsis, and it inhibits the urinary bladder, dilates the pupil of the eye, excites the liver to cause release of glucose, and increases the rate of metabolism of essentially all cells of the body.

Epinephrine and norepinephrine are secreted by the adrenal medullae. Sympathetic nerves also control the rate of secretion of both epinephrine and norepinephrine by the adrenal medullae, the central portions of the two adrenal glands. These hormones are carried by the blood and cause essentially the same effects in most parts of the body as those caused by direct sympathetic stimulation in each respective part. Furthermore, these hormones · reach some cells that have no sympathetic nerve supply. They especially increase the rate of metabolism in all cells of the body, an effect that is much more potent for epinephrine than norepinephrine.

The adrenal medullae represent a second means by which the CNS can cause sympathetic effects throughout the body. When sympathetic nerves to some organs have been destroyed, the sympathetic hormones can still elicit the usual sympathetic functions when the overall sympathetic nervous system is excited.

TABLE 7-3.
Adrenergic Receptors and Their Functions

Alpha Receptor	Beta Receptor
Vasoconstriction	Vasodilation (β_2)
Iris dilation	Cardioacceleration (β_2)
Pilomotor contraction	Increased myocardial contractility (β_2)
Intestinal relaxation	Intestinal relaxation (β_2)
Intestinal sphincter contraction	Uterus relaxation (β_2)
Bladder sphincter contraction	Bronchodilation (β_2)
	Calorigenesis (β_2)
	Glycogenolysis (β_2
	Lipolysis (β_2)
	Bladder wall relaxation (β_2)

Adrenergic Receptors

Table 7-3 shows the effects of stimulating alpha and beta receptors in some of the organs and systems controlled by the sympathetic nervous system.

Parasympathetic Nervous System

Parasympathetic fibers pass mainly through the vagus nerves. A few fibers pass through several of the other cranial nerves and through the anterior roots of the sacral segments of the spinal cord. Parasympathetic fibers do not spread as extensively through the body as do sympathetic fibers, but they do innervate some of the thoracic and abdominal organs, as well as the pupillary sphincter and ciliary muscles of the eye and the salivary glands.

Parasympathetic nerve endings secrete acetylcholine. Like norepinephrine, acetylcholine stimulates some organs and inhibits others. In general, it inhibits the heart and those very few blood vessels that have parasympathetic innervation, but it excites the ciliary and the pupillary sphincter muscles of the eye, the glandular and motor functions of the gastrointestinal tract, the urinary bladder, and the gallbladder.

Autonomic Effects on Various Organs

Table 7-4 lists the effects of sympathetic stimulation and parasympathetic stimulation on visceral functions. Note in the table that sympathetic stimulation causes excitatory effects in some organs and

TABLE 7-4.

Autonomic Effects on Various Organs of the Body

Organ	Effect of Sympathetic Stimulation	Effect of Parasympathetic Stimulation
Eye		
Pupil	Dilated	Constricted
Ciliary muscle	Slight relaxation (far vision)	Constricted (near vision)
Glands		
Nasal, lacrimal, parotid, submandibular, gastric, pancreatic	Vasoconstriction and slight secretion	Stimulation of copious secretion (containing many enzymes for enzyme-secreting glands)
Sweat glands	Copious sweating (cholinergic)	Sweating on palms of hands
Apocrine glands	Thick, odoriferous secretion	None
Blood vessels	Most often constricted	Most often little or no effect
Heart		
Muscle	Increased heart rate	Decreased heart rate
	Increased contractility	Decreased contractility (mainly in the atria)
Coronaries	Dilated (β_2); constricted (α)	Dilated
Lungs		
Bronchi	Dilated	Constricted
Blood vessels	Mildly constricted	Dilated (?)
Gut		
Lumen	Decreased peristalsis and tone	Increased peristalsis and tone
Sphincter	Increased tone (most times)	Relaxed (most times)
Liver	Glucose released	Slight glycogen synthesis
Gallbladder and bile ducts	Relaxed	Contracted
Kidney	Decreased output and renin secretion	None
Bladder		
Detrusor	Relaxed (slight)	Contracted
Trigone	Contracted	Relaxed
Penis	Ejaculation	Erection
Systemic arterioles		
Abdominal viscera	Constricted	None
Muscle	Constricted (adrenergic α)	None
	Dilated (adrenergic β_2)	
	Dilated (cholinergic)	
Skin	Constricted	None
Blood		
Coagulation	Increased	None
Glucose	Increased	None
Lipids	Increased	None
Basal metabolism	Increased up to 100%	None
Adrenal medullary secretion	Increased	None

TABLE 7-4.

Autonomic Effects on Various Organs of the Body *(continued)*

Organ	Effect of Sympathetic Stimulation	Effect of Parasympathetic Stimulation
Mental activity	Increased	None
Piloerector muscles	Contracted	None
Skeletal muscle	Increased glycogenolysis	None
	Increased strength	
Fat cells	Lipolysis	None

inhibitory effects in others. The parasympathetic system can also cause excitatory and inhibitory effects.

Control of Autonomic Nervous System

Autonomic nerves are controlled by the CNS. The activities of the sympathetic and the parasympathetic nerves are controlled in four different levels in the CNS:

- **Spinal cord.** Autonomic cord reflexes have to do principally with local reactions in discrete parts of the body. For instance, excess filling of the rectum causes a parasympathetic reflex from the sacral cord that promotes emptying of the rectum. Visceral pain from the small intestine causes reflex sympathetic inhibition of the gastrointestinal tract, and excess heat to a skin area causes reflex sympathetic vasodilation and sweating, which help to reduce the local skin temperature.
- **Brain stem.** The brain stem controls such factors as blood pressure, swallowing, vomiting, salivary secretion, stomach and pancreatic secretion, and, to a certain extent, emptying of the urinary bladder.
- **Hypothalamus.** The autonomic centers of the hypothalamus control such functions as body temperature, degree of overall excitability of the body, and various responses of the viscera to emotions. To perform these functions the hypothalamus transmits signals into the lower brain stem and thence either into the vagus nerves or down into the spinal cord to stimulate the spinal autonomic centers.
- **Cerebral cortex.** Centers in the cerebral cortex can elicit almost any type of autonomic response. These responses are often of an emotional nature, such as fainting caused by widespread vasodilation through the body. Also, some are associated with muscular exercise, such as a rise in blood pressure and vasodilation in the muscles. The responses caused by the cerebral cortex are

transmitted mainly through the autonomic centers in the hypothalamus and the lower brain stem.

SECTION 3
SPECIAL SENSES

VISION

The Eye as a Camera

The lens system of the eye functions as a camera. The eye is constructed very much like a camera, as illustrated in Figure 7-13. The **retina** is analogous to the film in a camera; the **cornea** and the **lens** of the eye are analogous to the lens system of a camera; and the **pupil** is analogous to the diaphragm of a camera.

The retinal image is upside down and reversed. Because light rays travel at different velocities in the eye fluids, the cornea, and the lens, the rays are refracted; that is, they are bent. Refraction occurs at four different corneal and lens surfaces: (1) the anterior surface of the cornea, (2) the posterior surface of the cornea, (3) the anterior surface of the lens, and (4) the posterior surface of the lens. This bending of the light rays allows an image of the scene in front of the eyes to be focused on the retina in exactly the same way that an image is focused by the lens system of a camera on the film. The image is upside down and reversed to the opposite side from the orientation of the object in front of the eyes.

Total refractive power = 59 diopters

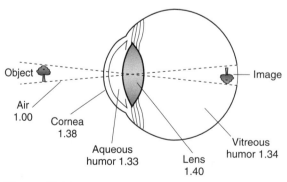

Figure 7-13.
The eye as a camera. Numbers are the refractive indices, which are the reciprocals of the light velocities.

Mechanism of Accommodation (Focusing)

The image is focused on the retina by adjusting the curvature of the lens. For a clear image to be formed on the retina, the surfaces of the cornea and of the lens of the eye must have the appropriate curvatures in relation to the distance of the retina behind the lens system. That is, the image must be focused on the retina. The eye can change the curvature of the lens in the following way: Attached around the periphery of the lens are approximately 70 ligaments that pull continually to the side, keeping the lens normally in a flattened, ovoid shape. The lens itself is a very elastic structure so that when these ligaments are loosened, it assumes a round, globular shape. When an object comes close to the eye, the more rounded shape of the lens is required to focus a clear image on the retina.

The shape of the lens is controlled by the ciliary muscle. This muscle is a circular sphincter extending all the way around the peripheral attachments of the ligaments. On contraction the circle of the sphincter becomes smaller so that the ligaments are loosened. This automatically allows the lens to change from its normal ovoid shape to a more rounded shape, enhancing the curvature and allowing adequate focusing of the images of nearby objects.

The ciliary muscle is controlled by the cerebral cortex. The tension on the lens ligaments must be very exact or the lens might become too round for adequate focusing. This tension is controlled by the cerebral cortex. If the image is in poor focus, the visual image in the brain is indistinct, and appropriate impulses are transmitted back through the visceral nucleus of the third nerve and finally through the third cranial nerve to the ciliary muscle to adjust the degree of contraction.

Pupillary Diameter

The pupil controls the amount of light that reaches the retina. The diameter of the pupil of the eye is controlled by a nervous reflex originating in the retina called the **light reflex.** Signals caused by strong light on the retina are transmitted along the optic nerve and optic tract into the pretectal nuclei of the midbrain, from there to the visceral nucleus of the third nerve, and then back to the pupillary constrictor to decrease the pupillary aperture, thus decreasing light intensity on the retina. Conversely, in darkness, lack of light signals from the retina reverses the reflex and causes the diameter of the pupil to increase.

The pupil alters the depth of focus of the eye. When the image on the retina is not in exact focus, the light rays passing through the peripheral edges of the lens will be much more out of focus than those passing through the very center of the lens. However, as the pupillary diameter becomes smaller, the light rays entering the peripheral edges of the eye are blocked and do not reach the retina. Therefore, a person whose lens is not in exact focus will still have fairly clear vision when the size of the pupil is constricted, as occurs in bright light; that is, the person has increased "depth of focus."

Errors of Refraction

Hyperopia (farsightedness) and myopia (nearsightedness) are errors of refraction. The eye is considered to be normal or **emmetropic** when parallel light rays from distant objects are sharply focused on the retina when the ciliary muscle is completely relaxed. **Hyperopia** occurs when an eyeball is too short or when a lens system is too weak. In both cases, the parallel light rays are not bent enough by the relaxed lens system to focus the image on the retina, as shown in Figure 7-14. Contraction of the ciliary muscle increases the strength of the lens, and thereby overcomes this abnormality. **Myopia** usually occurs because an eyeball is too long, but it can result from too much refractive power in the lens system. In each case, the parallel light rays are bent too much by the relaxed lens system, and the light is focused in front of the retina. Because the eye cannot decrease the strength of its lens when the ciliary muscle is fully relaxed, the myopic person lacks the mechanism to focus distant objects sharply on the retina. Figure 7-15 shows how myopia can be corrected with a concave lens and how hyperopia can be corrected with a convex lens.

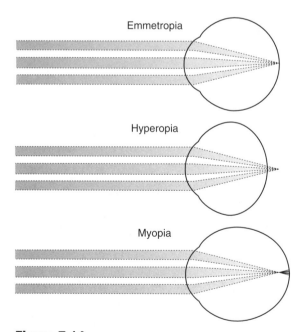

Figure 7-14.
Parallel light rays focus on the retina in emmetropia, behind the retina in hyperopia, and in front of the retina in myopia.

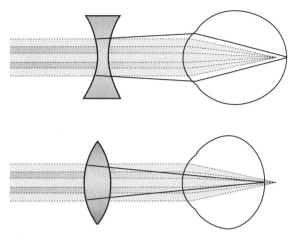

Figure 7-15.
Correction of myopia with a concave lens, and correction of hyperopia with a convex lens.

The Retina

Rods and cones are the photoreceptor cells of the retina. The rods outnumber the cones in a ratio of about 125 million to 4 million. The rods distinguish only the white and the black aspects of an image while the cones are capable of distinguishing its colors as well. In general, one or more cones plus 50 to 400 rods are connected to a single optic nerve fiber. However, in the very central portion of the retina a single cone is connected to a single optic nerve fiber. As a result, minute points of light on the retina can be localized to very discrete positions by the cones but can be localized far less acutely by the rods. Thus, very acute and clear vision of objects is mediated by the cones, while only a more diffuse type of vision is mediated by the rods.

The foveal region of the retina is capable of very sharp vision. A person normally has very acute vision only in the central portion of the visual field. The reason is that a small area in the center of the retina having a diameter of only 0.4 mm is especially capable of detailed vision. This special area, called the **fovea,** has only cones; the cones are smaller in diameter; and the ratio of cones to optic nerve fibers is close to one. Also, the blood vessels and nerves are pulled to one side, so that light can pass with ease directly to the deep layers of the retina where the cones are located. The peripheral areas, which contain progressively more rods, have progressively more diffuse vision.

Photochemistry of Vision

Rods utilize rhodopsin to convert light energy into nerve impulses. For a person to see an image, the light energy entering the eye must

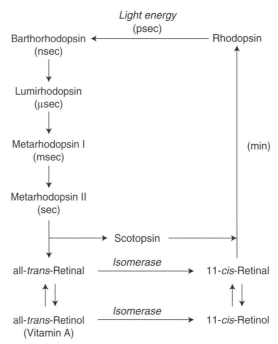

Figure 7-16.
Rhodopsin-retinal visual cycle in the rod.

be changed into nerve impulses. In the rods, this is accomplished by means of a chemical system called the rhodopsin-retinal cycle. Large quantities of the light-sensitive substance **rhodopsin,** also known as **visual purple,** are present in the rods. When light impinges on the rods, a series of chemical reactions occurs in which a number of unstable compounds are formed, each lasting for a fraction of a second, as shown in Figure 7-16. But, during the split second while the rhodopsin is being degraded, the rod becomes excited, sending nerve impulses from the retina into the optic nerve. The degeneration products of the reactions are two substances called **retinal** and **scotopsin** (a protein). The retinal and scotopsin are gradually recombined by the metabolic processes of the rod to reform rhodopsin, thereby continually supplying the rod with new rhodopsin.

Night blindness occurs in severe vitamin A deficiency. Retinal is derived from vitamin A. Therefore, when a person has a very serious deficiency of vitamin A in the diet, the retina is not able to form adequate quantities of retinal and thus becomes relatively insensitive to light.

Light and Dark Adaptation

Light adaptation is caused by decreasing concentrations of rhodopsin, and dark adaptation is caused by increasing concentrations of rhodopsin. The retina is capable of adapting its sensitivity so that the eye can see almost equally as well in both very bright light and in dim light. This adaptation is a far more powerful mechanism than the pupillary adaptation discussed above, although it requires several minutes to several hours to develop fully each time the person changes to a new level of light intensity. The mechanism of dark and light adaptation is the following:

- **Light adaptation**. When a person remains in very bright light for a long time, extremely large quantities of rhodopsin are split into retinal and scotopsin. This reduces the quantity of rhodopsin in the rods and therefore causes them to become insensitive to light.
- **Dark adaptation.** When a person spends a long time in darkness, only very small amounts of rhodopsin are split while the metabolic systems of the rods are continually building more and more rhodopsin. Consequently, rhodopsin collects in very high concentration after a while and greatly increases the sensitivity of the retina.

Color Vision

Cones utilize color-sensitive photopsins to convert light energy into nerve impulses. The cones of the eye function in very much the same way as the rods except that the light-sensitive chemicals are slightly different from rhodopsin. These chemicals still utilize retinal as the basis for light sensitivity, but the retinal combines with a different **photopsin** for each of the three primary colors rather than with scotopsin. Each photopsin, like scotopsin, is a protein, but slightly different from other photopsins. The nature of this protein determines the color sensitivity of the light-sensitive chemical. There are three major groups of cones that respond especially intensely to certain colors of light. These cones are classified as **blue** cones, **green** cones, and **red** cones.

The eye determines the color of an object by the relative intensities of stimulation of the different types of cones. For instance, yellow is a color with a wavelength midway between green and red. Therefore, it stimulates the green and the red cones about equally, which gives one the sensation of seeing the color yellow. Orange has a wavelength somewhat closer to that of red light than of green light. Therefore, it stimulates the red cones about twice as much as it does the green cones, giving the sensation of orange. Finally, pure red light stimulates the red cones very strongly while stimulating the green and blue cones only weakly. This gives the sensation of red. The same principles hold true for the different shades of color among green, blue, and yellow.

Neurophysiology of Vision

Visual images undergo analysis even before they leave the retina. In the retina are several other types of neuronal cells in addition to the rods and cones. These are the bipolar cells, the horizontal cells, the amacrine cells, and the ganglion cells. Some of these are inhibitory while others are excitatory. When excitatory and inhibitory signals from the excited rods and cones combine, the signals that finally reach the ganglion cells of the retina are initiated almost entirely by three types of visual effects: (1) **spots or borders** in the retinal image where there is a sudden change from light to dark or dark to light, that is, sudden change in contrast in the image; (2) **sudden increases or decreases in light intensity** from one instant to another; and (3) **changes in color** from one area to another. In other words, those portions of the visual image that do not have any contrast in intensity or color in them do not stimulate the ganglion cells to a great extent. It is mainly where contrast borders occur that the ganglion cells are strongly stimulated. Also, some ganglion cells are stimulated strongly when there is a sudden change in light intensity. Thus, these contrast borders and the changes in light intensity send most of the signals to the primary visual cortex.

Nervous signals are transmitted from the retina to the cerebral cortex. Each point of the retina connects with a discrete point in the **visual cortex** of the brain located in the calcarine fissure area of the occipital cortex. Therefore, every time a single point on the retina is stimulated, a corresponding point is stimulated in the visual cortex. From here other signals pass to the visual association areas and then to Wernicke's area for analysis of the visual images, as described earlier in the chapter.

AUDITION

Tympanic Membrane and Ossicular Lever System

Sound is transmitted from the tympanic membrane to the cochlea by way of the ossicular lever system. Figure 7-17 illustrates the functional parts of the ear. Sound is caused by compression waves that travel through the air at a velocity of about 0.2 mile/sec. As each compression wave strikes the **tympanic membrane** (or "eardrum"), the membrane is forced inward; between compressions it moves outward. The center of the tympanic membrane is connected to the **ossicular system,** which consists of three bony levers (the malleus, incus, and stapes) that transmit the sound vibrations into the cochlea at the oval window.

The tympanic membrane and the ossicular system function as a sound "transformer." The tympanic membrane has a surface area some 20 times the surface area of the oval window. Therefore, the force of **sound vibrations** reaching the oval window has been increased 20-fold.

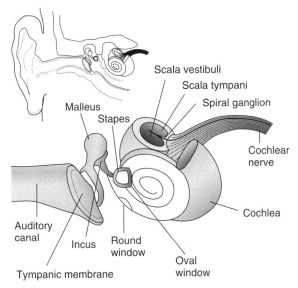

Figure 7-17.
The tympanic membrane, the ocular system of the middle ear, and the inner ear.

The Cochlea

When the stapes moves inward against the oval window, the round window bulges outward into the middle ear. Figure 7-18 illustrates the fluid system in the cochlea. The cochlea is composed of two major fluid-filled tubes, the **scala vestibuli** and the **scala tympani,** which lie side by side in a coil and are separated by the basilar membrane. Inward movement of the stapes against the oval window, which is at the end of the scala vestibuli, pushes the fluid deeper into the scala vestibuli, and this in turn causes the basilar membrane to bulge and push the fluid in the scala tympani. Finally, the fluid in the scala tympani pushes outward against the round window, which is at the end of this scala. Thus, every time the stapes moves inward, the round window bulges outward into the middle ear.

Frequency of sound (pitch) is determined by resonance in the cochlea. Low-frequency sound causes the stapes to move back and forth very slowly, allowing the pressure waves to travel far up into the scala vestibuli before maximum bulging of the basilar membrane into the scala tympani occurs. On the other hand, high-frequency sound waves cause very rapid vibration of the stapes, so the pressure waves have enough time to travel only a short distance into the scala vestibuli before maximum bulging occurs. In this way, a form of resonance occurs in the cochlea, with low-frequency waves causing maximum back-and-forth vibration of the basilar membrane near the far tip of the cochlea and high-frequency sound causing vibration of the basilar membrane near the base of the cochlea, that is, close to the

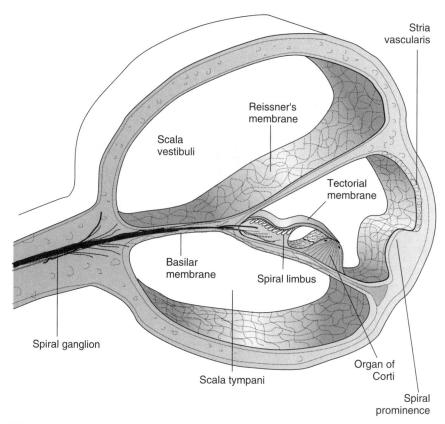

Figure 7-18.
Vibration of the basilar membrane in response to sound. (Adapted from Guyton AC and Hall JE: *Textbook of Medical Physiology,* 9th ed. Philadelphia: WB Saunders, 1995, p 665.)

oval and round windows. The brain determines the pitch of the sound mainly from the portion of the basilar membrane that vibrates: high pitch near the base of the cochlea and low pitch near the apex.

Organ of Corti

The organ of Corti is the receptor organ that generates nerve impulses in response to vibration of the basilar membrane. The actual sensory receptors of the organ of Corti are hair cells that synapse with a network of cochlear nerve endings. Bending of the hair cells in one direction depolarizes the hair cells, and bending in the opposite direction hyperpolarizes them. The more forceful the vibration and subsequent bending of the hair cells, the greater the rate of nerve impulses. The loudness of the sound is determined by the rate of impulse transmission from the hair cells into the brain by way of the cochlear nerve.

Central Auditory Mechanisms

Spatial orientation of auditory impulses is maintained throughout the brain and the meanings of the signals are interpreted in the auditory association areas. Approximately 25,000 nerve fibers are attached to the hair cells of the cochlear apparatus. The auditory signals in the auditory nerves go first to the cochlear nuclei located in the brain stem. From here they pass upward to the inferior colliculus, then to the medial geniculate body, and finally to the **primary auditory cortex** in the middle of the superior temporal gyrus of the temporal lobe. The spatial orientation of the nerve fibers is maintained all the way from the basilar membrane to the auditory cortex so that one sound frequency excites specific areas of the auditory cortex while another sound frequency excites other areas. The meanings of the auditory signals are then interpreted in the auditory association areas immediately adjacent to the primary auditory cortex.

CHEMICAL SENSES

Sense of Taste

The sensory receptor for taste is the taste bud. Located on the surfaces of the tongue, especially on the papillae and most importantly on the **circumvallate papillae,** which lie in a V line on the posterior part of the tongue, and also in small numbers on the lateral walls of the pharynx, are many small taste receptor organs called taste buds. One of these is illustrated in Figure 7-19. Each taste bud has a hollow cavity that communicates through a small **taste pore** with the mouth. Lining the cavity are sensory taste receptor cells, and cilia called taste "hairs" protrude from the ends of these cells into the pore. Certain types of chemicals diffuse into the taste pores and excite the hairs of the taste cells.

The four types of taste buds respond to saltiness, sweetness, sourness, and bitterness. The first three of these taste sensations help a person select the quality of food that he eats and in some instances even makes him desire certain substances such as salt that may be deficient in his body. The last of the taste sensations, bitterness, is principally for protection, because most of the naturally occurring poisons among plant foods elicit a bitter taste that normally will cause an animal to reject the food.

Taste signals are transmitted to the primary taste cortex. Most of the taste signals are transmitted by way of the **chorda tympani** into the **seventh nerve** and then into the brain stem; the remainder are transmitted through the ninth and tenth nerves into the brain stem. The

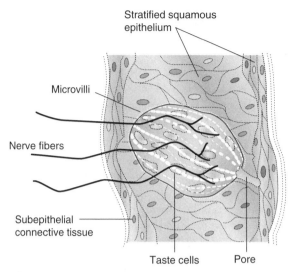

Figure 7-19.
The taste bud. (Adapted from Guyton AC and Hall JE:
Textbook of Medical Physiology, 9th ed. Philadelphia:
WB Saunders, 1995, p 676.)

signals pass first to the nucleus of the **tractus solitarius,** then to the **thalamus,** and finally to the **primary taste cortex,** which lies far laterally in the parietal cortex immediately posterior to the central sulcus of the brain and deep in the fissure of Sylvius (also called the lateral fissure).

Sense of Smell

The sense of smell is vested in the olfactory epithelium. Located in the superiormost part of each nostril is a small area, having a surface of about 2.5 cm², called the olfactory epithelium. This contains large numbers of nervous receptors called **olfactory cells** that send long cilia ("olfactory hairs") into the mucus on the surface of the epithelium. Almost any chemical substance that can diffuse through the mucus and then into these cilia will stimulate one or more of the olfactory cells. Since the cilia themselves have lipid membranes, lipid-soluble substances stimulate the cells much more readily than nonlipid-soluble substances. The precise types of chemicals that stimulate different types of olfactory cells are not known, for it has been very difficult to study by either subjective or objective means the olfactory stimulus from a single olfactory cell. It is believed that there might be 7 to 50 primary sensations of smell, and that the many thousands of different smells to which we are accustomed are actually combinations of these primary sensations.

From the olfactory cells, signals are transmitted first to the olfactory bulb of the first cranial nerve and then through the olfactory tract into several midline nuclei of the brain that lie superior and anterior to the hypothalamus, and also into the pyriform cortex, the amygdala, and the thalamus. The midline nuclei are associated with the crude functions of smell, such as eliciting salivation or licking the lips. The pyriform cortex, amygdala, and thalamus deal with olfactory conditioned reflexes that determine appetite and the social responses to food.

Chapter **8**

Gastrointestinal System

The alimentary tract provides water, electrolytes, and nutrients for the body. This requires propulsion and mixing of gastrointestinal contents; secretion of digestive juices; digestion of food; and absorption of digestion products, water, and electrolytes. The entire alimentary tract is shown in Figure 8-1.

HORMONES—PARACRINES—NEUROCRINES

Hormones

Four gastrointestinal peptides are considered to be official hormones. The gastrointestinal hormones are released into the portal circulation and have physiologic actions on target cells with specific receptors for the hormone; the effects of the hormones persist even after all nervous connections between the site of release and the site of action have been severed. The four gastrointestinal hormones outlined in Table 8-1 are **gastrin, cholecystokinin** (CCK), **secretin,** and **gastric inhibitory peptide** (GIP). There are also a number of gastrointestinal peptides, called **candidate hormones,** that for one reason or another have failed to satisfy all the criteria required to become an official hormone.

Paracrines

Somatostatin and histamine are considered to be gastrointestinal paracrines. Paracrines are released from endocrine cells in the same manner as hormones, but the amounts that enter the general circulation are insufficient to produce physiologic actions. The paracrines thus act locally, reaching target cells by simple diffusion or by moving short distances through the microcirculation. **Somatostatin** is secreted throughout the gastrointestinal tract in response to luminal acid and it inhibits the release of all gastrointestinal hormones. Vagal stimulation inhibits the release of somatostatin. **Hista-**

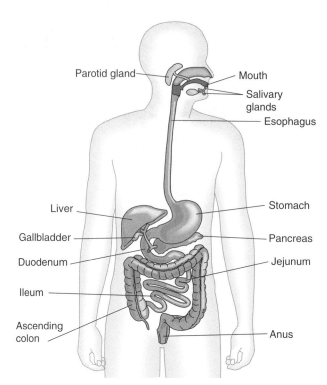

Figure 8-1.
The alimentary tract.

mine is a non-proteinaceous paracrine agent that is secreted by mast cells in the gastric mucosa. It is not released by specific stimuli in the classic manner, but background levels of histamine potentiate the effects of gastrin and acetylcholine to stimulate gastric acid secretion.

Neurocrines

Gastrointestinal neurocrines are released from nerves in the mucosa and smooth muscle of the gut. The three neurocrines of the gut are **vasoactive intestinal peptide** (VIP), **gastrin-releasing peptide** (GRP, also called bombesin), and **enkephalins** (met-enkephalin and leu-enkephalin). Table 8-2 lists their physiologic actions and sites of release.

MOTILITY

The intestinal wall has an outer longitudinal muscle layer and an inner circular muscle layer required for the two basic types of movements in the gastrointestinal tract: **propulsive movements** that pro-

TABLE 8-1.

Gastrointestinal Hormones

Hormone	Actions	Stimulus for Secretion	Site of Secretion
Gastrin	Stimulates:	Small peptides	G cells of the antrum
	Gastric acid secretion	Amino acids	
	Growth of gastric mucosa	Gastric distention	
		Vagal stimulation	
CCK	Stimulates:	Small peptides	I cells of the duodenum and jejunum
	Pancreatic enzyme secretion	Amino acids	
	Pancreatic bicarbonate secretion	Fatty acids	
	Gallbladder contraction		
	Growth of exocrine pancreas		
	Inhibits:		
	Gastric emptying		
Secretin	Stimulates:	Acid	S cells of the duodenum
	Pepsin secretion	Fatty acids	
	Pancreatic bicarbonate secretion		
	Biliary bicarbonate secretion		
	Growth of exocrine pancreas		
	Inhibits:		
	Gastric acid secretion		
	Effect of gastrin on growth of gastric mucosa		
GIP	Stimulates:	Fatty acids	Duodenum and jejunum
	Insulin release	Amino acids	
	Inhibits:	Oral glucose	
	Gastric acid secretion		

pel food forward along the tract and **mixing movements** that mix food with the gastrointestinal secretions.

Single-unit (unitary) smooth muscle predominates in the intestine, allowing action potentials initiated in one cell to spread to neighboring cells, thus facilitating coordinated muscular activity in the gut. Contraction of circular smooth muscle leads to a decrease in diameter, whereas contraction of longitudinal smooth muscle produces a decrease in the length of the gut segment.

Slow Waves

Most gastrointestinal contractions are regulated by the frequency of slow waves in the smooth muscle membrane potential. Slow waves

TABLE 8-2.

Gastrointestinal Neurocrines

Peptide	Actions	Site of Secretion
VIP	Relaxes gastrointestinal smooth muscle	Nerves in mucosa and smooth muscle of gastrointestinal tract
	Stimulates intestinal secretion	
	Stimulates pancreatic secretion	
	Inhibits gastric acid secretion	
GRP or bombesin	Stimulates gastrin release	Nerves in gastric mucosa
Enkephalins	Contracts gastrointestinal smooth muscle	Nerves in mucosa and smooth muscle of gastrointestinal tract
	Inhibits intestinal secretion of fluid and electrolytes	

are oscillating changes in the resting membrane potential that determine the frequency of action potentials, although the slow waves themselves are not action potentials. The slow waves control the frequency of spike potentials that occur intermittently when the resting membrane potential becomes more positive than about –40 mV, as shown in Figure 8-2. The spike potentials are true action potentials that initiate contraction of the gut smooth muscle. The mechanism of slow-wave generation is poorly understood, but it is thought to result from cyclic activity of the cell membrane sodium-potassium pump.

The frequency of slow waves varies along the gastrointestinal tract, although each segment has a characteristic frequency. Slow-wave frequency is ~3/min in the stomach, ~12/min in the duodenum, and ~8 or 9/min in the terminal ileum. The generation of slow waves is not dependent upon neural or hormonal input; however, this input does affect the frequency and amplitude of slow waves and thus the frequency of spike potentials and subsequent contractions.

Types of Gastrointestinal Movements

Peristalsis is a wave of contraction passing along the gastrointestinal tract. Peristaltic contractions serve to propel the contents of the gut in a caudal direction. The bolus moves along the gut as contraction occurs behind the bolus and relaxation occurs in front of the bolus. The intensity and frequency of peristalsis vary greatly from one part of the tract to another. In the stomach, the intensity of peristalsis is usually sufficient to cause movement of the food through the stomach within 1 to 3 hours after a meal. The peristaltic waves are less intense in the small intestine, and they spread along only 10 cm to

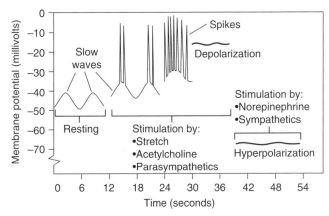

Figure 8-2.
Membrane potentials in intestinal smooth muscle, showing slow waves and intermittent spike potentials. (Adapted from Guyton AC and Hall JE: *Textbook of Medical Physiology,* 9th ed. Philadelphia: WB Saunders, 1995, p 794.)

15 cm at a time so that about 3 to 10 hours are required to move the food through the small intestine. In the large intestine, the propulsive movements are a modified type of peristalsis, called **mass movements;** these are often very strong but last only a fraction of an hour out of each day.

Distention is the usual stimulus for peristalsis. If food collects at any point in the gut and stretches the gut wall, a constrictive ring moves forward along the gut, pushing any material in the gut ahead of the constriction. This is known as the **law of the intestines.** Peristalsis is caused by nerve impulses that move along a nerve plexus, called the **myenteric plexus,** in the wall of the gut. Stimulation at any point of the plexus causes nerve impulses to travel around the gut and also in both directions along the gut. Impulses traveling around the gut cause it to constrict and impulses traveling lengthwise along the gut cause the constriction to move in the analward direction.

Peristalsis is regulated by the autonomic nervous system. Peristalsis is far more intense when the parasympathetic nerves are stimulated, and less intense when the sympathetic nerves are stimulated. Strong stimulation of the sympathetic nervous system can totally block movement of food through the gastrointestinal tract.

Swallowing (deglutination) is a special type of propulsive movement. When food is pushed into the back of the mouth by the tongue, nerve receptors in the pharynx elicit an automatic swallowing process that occurs entirely in less than 2 seconds. Signals are transmitted from these receptors to the brain stem, integrated there, and a sequence of swallowing signals is then transmitted back through the pharyngeal and vagus nerves to the pharynx and upper esophagus. A peristaltic wave begins in the pharyngeal constrictors, the glottis closes so that food cannot pass into the trachea, and the upper esophageal constrictor muscle at the opening of the esophagus relaxes so that the food will enter the esophagus. Then, after this swallowing process is over, the peristaltic wave proceeds downward along the upper esophagus. All of this is controlled directly by nerve impulses from the brain stem. On reaching the lower half of the esophagus, the natural peristaltic process of the myenteric plexus in the gastrointestinal tract takes over and propels the food the rest of the way to the stomach.

"Retropulsion" is important for mixing the stomach contents. Peristaltic waves of constriction pass over the **antrum** at the lower end of the stomach toward the pylorus, yet the opening of the pylorus is too small to allow the antral contents to be expelled into the duodenum. Therefore, most of the contents are squirted backward through the peristaltic ring toward the body of the stomach, providing an intense type of mixing called retropulsion. The mixture of food and secretions that results is called **chyme.**

Segmentation movements provide mixing in the large and small intestine. Intermittent constrictive rings occur several times a minute, dividing the small intestine into segments. Then the first constrictions relax, and others occur at other points. In this way, the chyme is chopped again and again into small portions. These movements are called **segmenting contractions.** In the large intestine, similar but much slower contractions called **haustrations** occur. These slowly roll the fecal matter over and over, allowing almost complete absorption of the water and electrolytes.

Intestinal motility is regulated by neural reflexes. The **intestino-intestinal reflex** inhibits contraction of the remaining bowel when a portion of the bowel is severely distended. The **gastroileal reflex** results in the movement of ileal contents into the large intestine following a meal.

SECRETION

Salivary Secretion

Saliva has a digestive and lubricating function. Saliva is secreted principally by the parotid, submaxillary, and sublingual glands. Saliva is composed of a mucous secretion containing **mucin** and a serous secretion containing the enzyme **ptyalin,** which is an α-**amylase.** The mucus functions to provide lubrication for swallowing and the ptyalin functions to begin digestion of starches and other carbohydrates in the food. Saliva is also important for oral hygiene by virtue of its washing and bactericidal actions.

Salivary secretions are controlled by parasympathetic nervous signals from the salivatory nuclei. Salivatory nuclei in the brain can be stimulated by taste and tactile stimuli from the tongue and mouth. Salivation can also be stimulated from higher centers in the central nervous system, for example, when favorite foods are smelled or eaten. Drugs such as atropine and scopolamine block the acetylcholine receptors and abolish salivary secretions, whereas drugs such as neostigmine that inhibit acetylcholinesterase increase salivation.

Gastric Secretion

Gastric secretion is controlled by nervous and hormonal stimuli. The three phases of gastric secretion are shown in Table 8-3. The **cephalic phase** accounts for 30% of the response to a meal and is initiated by the anticipation of eating and the smell and taste of food. It is mediated entirely by the vagus nerve. The **gastric phase** accounts for 60% of the acid response to a meal. It is initiated by distention of the stomach, which leads to nervous stimulation of gastric secretion. In addition, partial digestion products of proteins in the stomach cause **gastrin** to be released from the antral mucosa. The gastrin then passes by way of the bloodstream to the **gastric glands** located in the upper three-quarters of the stomach, called the **fundus** and **body,** to cause secretion of a highly acidic gastric juice. The **intestinal phase** (10% of the response) is initiated by nervous stimuli associated with distention of the small intestine. The presence of digestion products of proteins in the small intestine can also stimulate gastric secretion by a humoral mechanism.

Pancreatic Secretion

Pancreatic secretions neutralize the duodenum and provide enzymes for digestion. The chyme, on entering the small intestine, causes the release of two different hormones from the intestinal mucosa, **secretin** and **cholecystokinin,** and these are carried in the

---

TABLE 8-3.

Summary of Mechanisms Stimulating Gastric Secretion

Phase of Secretion	Percent of Response	Initiating Stimulus	Stimulus at Parietal Cell
Cephalic	30%	Chewing	Acetylcholine released from vagus nerve*
		Swallowing	
Gastric	60%	Distention	Vagovagal and local reflexes lead to acetylcholine release
		Digested protein	Gastrin release from G cells
Intestinal	10%	Distention	Entero-oxyntin
		Digested protein	Circulating amino acids

Vagal stimulation also causes gastrin release.

blood to the pancreatic secretory cells. Secretin causes the pancreatic ducts to secrete large amounts of a highly alkaline, watery solution, and cholecystokinin causes the pancreatic acinar cells to release large quantities of digestive enzymes into this solution. Secretin is released from the intestinal mucosa mainly in response to acid emptied from the stomach into the duodenum; the alkaline pancreatic secretion in turn neutralizes the acid. Cholecystokinin is released mainly in response to the presence of fats and proteins in the chyme, and the secreted enzymes in turn help to digest the proteins and fats.

Intestinal Secretion

Intestinal fluid is secreted by epithelial cells in the crypts of Lieberkühn. Intestinal secretion normally results from either mechanical stimulation of the mucosa by the chyme or distention of the gut, and is mediated by intramural nervous reflexes in the submucosal and myenteric plexuses. Intestinal fluid is almost pure extracellular fluid, providing a watery vehicle for absorption of substances from the chyme.

Bile Secretion and Gallbladder

Bile is necessary for the digestion and absorption of lipids. The bile salts and other organic components of bile emulsify fat, rendering it soluble in water so that it can be digested by pancreatic lipase. The bile acids help to solubilize the digestion products into micelles that in turn aid in the transport and absorption of the fat digestion products. Bile also serves as a means for the excretion of waste products such as bile pigments (i.e., bilirubin), cholesterol, heavy metals, and various drugs.

Bile is secreted continuously by hepatocytes and flows toward the duodenum. The primary bile acids are synthesized by hepatocytes and secreted into bile canaliculi. The bile flows along a series of ducts where a second secretion is added consisting of a watery mixture of sodium and bicarbonate derived from epithelial cells lining the ducts. When the bile finally reaches the common bile duct, it can flow into the gallbladder during the interdigestive periods when the sphincter of Oddi is contracted and the gallbladder is distensible, or it can move directly into the duodenum following a meal when the gallbladder is contracted and the sphincter of Oddi is relaxed.

CCK causes the gallbladder to contract and the sphincter of Oddi to relax. CCK is released from the small intestine in response to amino acids and fatty acids. By relaxing the sphincter of Oddi (located at the junction of the common bile duct and duodenum) and contracting the gallbladder, CCK causes bile to be excreted into the duodenum so that fats can be emulsified and absorbed.

Mucus

Mucus protects the mucosa of the gastrointestinal tract. Mucus is secreted in all parts of the gastrointestinal tract, from the salivary glands all the way to the mucosal glands of the large intestine. Mucus is an excellent lubricant and is also very resistant to chemical destruction either by the gastric and intestinal juices or by different types of foods.

DIGESTION AND ABSORPTION

Hydrolysis is the basic chemical process of digestion. During hydrolysis the food molecule splits into two smaller molecules, and at the same time a hydrogen atom from a water molecule combines with one of the food products at the point of splitting while the remaining hydroxyl radical from the water combines with the other food product. The carbohydrates are hydrolyzed into monosaccharides, the fats into glycerol and fatty acids, and the proteins into amino acids.

Carbohydrate Digestion

Digestion of carbohydrates is begun by **ptyalin** secreted in the saliva and is carried still further by **pancreatic amylase** secreted in the pancreatic juice. After the action of these two enzymes, the carbohydrates will have been split principally into disaccharides. Then four enzymes in the epithelial cells of the small intestinal mucosa—**maltase, α-dextrinase, lactase,** and **sucrase**—split the disaccharides into monosaccharides, principally **glucose, galactose,** and **fructose.** Digestion of carbohydrates is summarized in Figure 8-3.

Protein Digestion

Protein digestion begins in the stomach under the influence of **pepsin** and **hydrochloric acid.** Then it continues in the upper small intestine under the influence of **trypsin, chymotrypsin, carboxypolypeptidase,** and **proelastase** secreted by the pancreas. At this point, the proteins will have been digested into large polypeptides. Several different **peptidases** secreted in the pancreatic juice or located in or on the surfaces of the intestinal epithelial cells split the polypeptides into amino acids. Digestion of proteins is summarized in Figure 8-4.

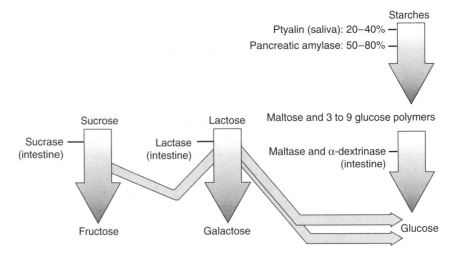

Figure 8-3.
Digestion of carbohydrates.

Fat Digestion

The most abundant fats in the diet are known as triglycerides. Each **triglyceride** molecule is composed of a glycerol nucleus and three fatty acids. Essentially all of the digestion of fats occurs in the small intestine. The fats are first broken into small globules by the emulsifying action of bile so that the water-soluble enzymes can act on the surfaces of the globules. The most important enzyme for fat digestion is **pancreatic lipase,** which splits the triglycerides into free fatty acids and 2-monoglycerides, the end-products of fat digestion, as shown in Figure 8-5.

Absorption by Intestinal Villi

Digestive products are absorbed by intestinal villi. Located on the mucosal surface of the small intestine are millions of small intestinal villi, each of which projects about 1 mm into the intestinal lumen, as shown to the left in Figure 8-6. These villi increase the surface area

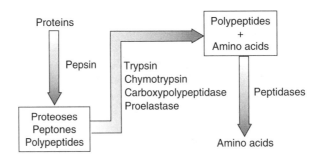

Figure 8-4.
Digestion of proteins.

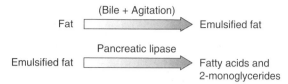

Figure 8-5.
Digestion of fats.

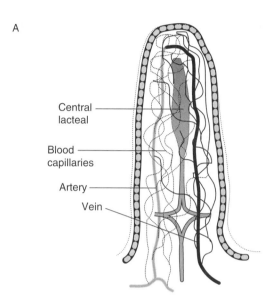

A

Central
lacteal

Blood
capillaries

Artery

Vein

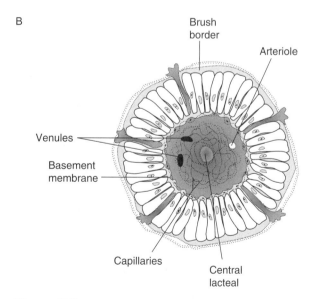

B

Brush
border

Arteriole

Venules

Basement
membrane

Capillaries

Central
lacteal

Figure 8-6.
Structure of the intestinal villus. *A:* Longitudinal
section; *B:* cross-section. (Adapted from Guyton
AC and Hall JE: *Textbook of Medical Physiology,*
9th ed. Philadelphia: WB Saunders, 1995, p 838.)

of the small intestine about 10-fold. The end-products of digestion are absorbed through the epithelial cells lining these villi into sublying villar blood vessels and lymphatics.

Microvilli increase the intestinal surface area by 20-fold. In addition to the increased surface area caused by the presence of the villi, each epithelial cell has approximately 600 microvilli, each 1 micron long. In all, the total area for absorption is about 250 m^2 for the entire small intestine.

Most digestive products enter the portal circulation. Almost all of the water-soluble substances are absorbed directly into the blood capillaries of the villi, and then pass with the portal blood through the liver sinusoids and into the general circulatory system. This is called the portal route of absorption, and the substances absorbed in this way are most of the **water and electrolytes,** the **carbohydrates,** and the **amino acids.**

Absorption of Water

Water diffuses freely through the intestinal membrane following osmotic gradients created by small molecules in the intestinal chyme and blood of the villus. Water is absorbed into the blood when the chyme is dilute, and water enters the chyme when the osmotic pressure of the chyme exceeds that of the blood. As small dissolved sub-

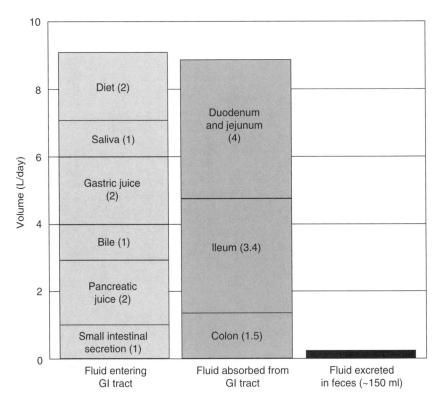

Figure 8-7.
Fluid balance in the gastrointestinal tract.

stances are absorbed by the blood, the osmotic pressure of the chyme tends to be reduced, but the water diffuses so rapidly that osmotic equilibrium occurs almost instantaneously.

Figure 8-7 shows the amounts of fluid entering the small intestine daily as well as the amounts of fluid absorbed by the small intestine and colon. Note that of the 9 L of fluid that enter the small intestine each day only about 150 ml are excreted in the feces.

Absorption of Ions

Approximately 25 to 35 grams of sodium are absorbed by the small intestine daily. This amounts to about one-seventh of the total amount of sodium in the entire body. The basic mechanism for sodium absorption is shown in Figure 8-8. Sodium is actively transported across the basolateral membrane into the paracellular spaces of the epithelial cells, causing the sodium concentration within the cells to decrease to about 50 mEq/L. Because the sodium concentration in the chyme is about 142 mEq/L (equal to that of plasma), a large concentration gradient causes sodium to move passively into the epithelial cells across the apical membrane (i.e., the brush border), which is the membrane surface facing the gut lumen. **Sodium moves into the epithelial cells by the following four passive processes: diffusion** through water-filled sodium channels, cotransport with sugars or amino acids, **cotransport** with chloride, and **countertransport** with hydrogen ions. Cotransport with sugars or amino acids, cotransport with chloride, and countertransport with hydrogen ions predominate in the small intestine, whereas diffusion through water-filled sodium channels predominates in the colon.

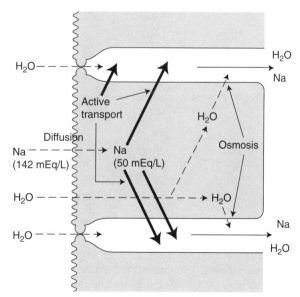

Figure 8-8.
Absorption of sodium through the intestinal epithelium.

Aldosterone greatly enhances sodium absorption in the colon in a manner similar to that which occurs in the renal distal tubules.

Chloride absorption accompanies sodium absorption to maintain electrical neutrality. Chloride absorption occurs by way of the following three mechanisms: passive diffusion by a paracellular pathway, cotransport with sodium, and countertransport with bicarbonate.

Carbohydrate Absorption

Most carbohydrates are absorbed by a sodium-dependent cotransport mechanism powered by the sodium-potassium pump. The primary digestion products of carbohydrates are the monosaccharides, glucose, galactose, and fructose. However, glucose accounts for more than 80% of the carbohydrate calories absorbed by the small intestine. **Glucose and galactose** are transported uphill across the brush border (apical membrane) by a sodium-dependent cotransport mechanism (secondary active transport) powered by the sodium-potassium pump on the basolateral membrane of the cell, as shown on Figure 8-9. The low sodium concentration inside the cell causes sodium in the intestinal lumen to diffuse through the brush border; however, the sodium must first combine with a protein carrier. The protein carrier will not transport the sodium until it also combines with an appropriate substance such as glucose (or galactose). Once glucose (or galactose) and sodium have joined with the protein carrier, they are both transported to the interior of the cell. The glucose (or galactose) is then trans-

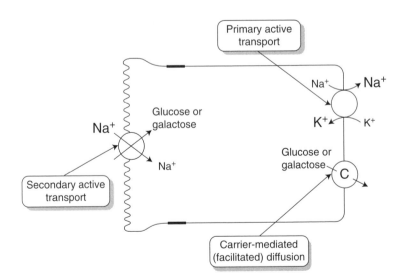

Figure 8-9.
Absorption of glucose and galactose through the intestinal epithelium. These substances are transported across the brush border (apical membrane) by a sodium-dependent cotransport mechanism (secondary active transport) and then across the basolateral membrane by facilitated diffusion.

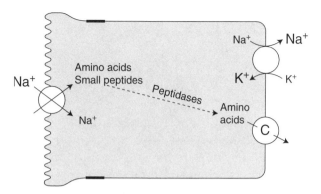

Figure 8-10.
Absorption of amino acids and small peptides through the intestinal epithelium. These substances are transported across the brush border (apical membrane) by a sodium-dependent cotransport mechanism (secondary active transport); the small peptides are digested into amino acids within the cell; and the amino acids then are transported across the basolateral membrane by facilitated diffusion.

ported across the basolateral membrane by facilitated diffusion. Fructose is not cotransported with sodium; instead, it is absorbed by facilitated diffusion and therefore cannot be absorbed against a concentration gradient.

Protein Absorption

Most proteins are absorbed by a sodium-dependent cotransport mechanism powered by the sodium-potassium pump. Free amino acids, as well as dipeptides and tripeptides, are absorbed along the brush border by a sodium-dependent cotransport mechanism, as shown in Figure 8-10. This mechanism is analogous to that described above for glucose and galactose. The dipeptides and tripeptides are hydrolyzed to amino acids within the cell by cytoplasmic peptidases. The free amino acids are then transported across the basolateral membrane by facilitated diffusion or by simple diffusion.

Fat Absorption

Chylomicrons are transported in the lymph. Fatty acids are not absorbed into the blood of the villi but, instead, while passing through the epithelium of the villus, recombine with glycerol to form triglycerides (neutral fat). This then passes into the lymphatics and is transmitted upward along the thoracic duct in the form of minute fatty globules called chylomicrons to be emptied into the veins of the neck.

GASTROINTESTINAL DISORDERS

Peptic ulcers are caused by gastric acid and pepsin. Peptic ulcers form when the damaging effects of acid and pepsin overcome the protective actions of bicarbonate secretion, mucus, and normal renewal of the cells. Because gastric and duodenal ulcers result from mucosal digestion by acid and pepsin, they are both categorized as peptic ulcers. However, their etiologies are quite different. **Duodenal ulcers** result from excessive secretion of acid and pepsin. Duodenal ulcer patients have higher than normal serum gastrin levels in response to a meal. Also, patients with **gastrinoma,** a gastrin-secreting tumor, always develop duodenal ulcers, never gastric ulcers. **Gastric ulcers** appear to result from a defect in the mucosa itself. Factors that can damage the so-called **gastric mucosal barrier** and thus promote gastric ulcer formation include abnormalities in mucosal blood flow, decreased mucus secretion, bacterial infection, and irritants such as alcohol and aspirin.

Irritation is the most common cause of diarrhea. Almost any irritation of the gastrointestinal mucosa greatly increases the local rate of secretion of intestinal juices and also the intensity of peristalsis. This causes a **"washout"** phenomenon, in which the material in the irritated area flows rapidly toward the anus. In addition to irritation and bacterial toxins, diarrhea can also result from excessive activity of the parasympathetic nervous system. Parasympathetic stimulation increases the secretory and propulsive activities of the entire gastrointestinal tract, especially the latter half of the large intestine.

Cholera toxin causes severe diarrhea. The rate of fluid secretion from intestinal epithelial cells can increase severalfold, causing as much as 5 to 10 L of diarrhea fluid to be lost during the first day of infection. Circulatory shock by dehydration will occur within a few hours if the body fluids are not replaced.

Gastric atrophy can cause pernicious anemia. Normal gastric secretions contain a glycoprotein called **intrinsic factor** that is required for absorption of vitamin B_{12} by the ileum. In the absence of intrinsic factor, **maturation failure** occurs in the production of red blood cells resulting in pernicious anemia.

Achalasia is characterized by failure of the lower esophageal sphincter to relax. This results in accumulation of food in the esophagus, sometimes causing an extreme enlargement of the esophagus called **megaesophagus.** It is caused by dysfunction of the myenteric plexus in the lower half of the esophagus. The musculature at the lower end of the esophagus remains tonically contracted and the myenteric plexus has lost its ability to transmit signals necessary for relaxation of the gastroesophageal sphincter.

Chapter 9

Metabolism and Energy

CARBOHYDRATE METABOLISM AND ATP SYNTHESIS

Carbohydrates are made available for metabolism in the form of glucose. The digestive products of essentially all carbohydrates are glucose, galactose, and fructose. Much of the fructose is converted to glucose as it is absorbed by the intestinal epithelium, and the remaining fructose and the galactose, after being absorbed, are transmitted by the portal blood mainly to the liver, where they, too, are almost completely converted into glucose. This glucose eventually passes back out of the liver cells into the blood to join the glucose that is absorbed directly from the gastrointestinal tract.

Insulin stimulates carrier-mediated (facilitated) diffusion of glucose into cells. Glucose is transported through the cell membrane by a facilitated diffusion mechanism. The activity of this mechanism for glucose transport is controlled by the amount of insulin secreted by the pancreas. Large amounts of insulin increase the rate of glucose transport to ten or more times the rate when no insulin is available. When the pancreas fails to secrete insulin, very little glucose can enter most cells, but when excess insulin is secreted, glucose enters the cells so rapidly that the blood glucose falls to a very low level.

Liver glycogen provides a blood glucosebuffering function. Large quantities of glucose, either that converted from fructose and galactose or that directly absorbed from the gastrointestinal tract, can be stored in the liver cells in the form of glycogen granules, glycogen being a polymer of glucose. When glucose enters the cells, it immediately becomes phosphorylated to glucose-6-phosphate, as shown in Figure 9-1. This phosphorylation of glucose serves to capture the glucose in cells that lack the enzyme glucose phosphatase, which includes most cells of the body with the exception of the liver and a few other types of cells. The glucose-6-phosphate can be used immediately for energy or can be polymerized into glycogen (glycogenesis). When the glucose level in the blood falls too low, the glycogen is split by still other enzymes back into glucose (glycogenolysis), which then passes into the blood to be utilized elsewhere in the body. In this way, extra glucose is removed from the blood when the

PHYSIOLOGY

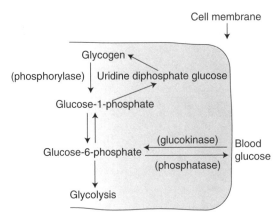

Cell membrane

Figure 9-1.
Chemical reactions of glycogenesis and glycogenolysis.

blood glucose concentration is too high, and then is returned when the blood glucose falls too low.

Glycolysis (splitting glucose to form pyruvic acid) is the initial step in the catabolism of glucose. After entry into the cells, glucose can be polymerized into glycogen and stored temporarily as glycogen granules, or it can be used immediately to provide energy for cellular functions. The initial process for providing energy is principally **glycolysis.** Each molecule of glucose is split to form pyruvic acid by a series of chemical reactions involving several stages of phosphorylation and several transformations, all catalyzed by protein enzymes in the cells.

Glycolysis provides a net yield of two molecules of adenosine triphosphate (ATP) for every molecule of glucose converted into pyruvic acid. In this way, a small portion (about 3%) of the energy stored in the glucose molecule is transferred to ATP molecules. The ATP in turn is very highly reactive and can provide immediate energy to other functional systems of the cells as needed.

Citric acid cycle: catabolism of pyruvic acid yields carbon dioxide and hydrogen. The two molecules of pyruvic acid formed from glucose in the glycolysis process still contain about nine-tenths of the energy originally in the glucose molecule. To make this available to the cell, each molecule of pyruvic acid is first split into carbon dioxide and hydrogen. This occurs principally by means of a series of chemical reactions called by various names: the **Krebs cycle,** the **tricarboxylic acid cycle,** or the **citric acid cycle.**

Chemical reactions in the citric acid cycle are catalyzed by decarboxylases (which remove carbon dioxide) and dehydrogenases (which remove hydrogen). The pyruvic acid is first decarboxylated to form **acetylcoenzyme A,** and this immediately combines with oxaloacetic acid to form citric acid. The citric acid is then progressively decomposed, liberating carbon dioxide and hydrogen atoms. The carbon dioxide diffuses out of the cells and is blown off by the lungs into the expired air. However, the hydrogen atoms are made available to react with oxygen, a process that provides tremendous amounts of energy to the cell, as will be discussed below. After the carbon dioxide and hydrogen atoms have been removed, the residue of the citric acid molecule is a new molecule of **oxaloacetic acid** that can be used over and over again in the citric acid cycle.

Oxidative phosphorylation is the overall process by which hydrogen ions are oxidized and the released energy is used to generate ATP. Most of the hydrogen that is released during the breakdown of pyruvic acid combines with nicotinamide adenine dinucleotide (NAD) and is then rapidly passed to another substance, a flavoprotein. The hydrogen then leaves the flavoprotein to become hydrogen ions, each hydrogen losing one electron in the process. The electrons removed from the hydrogen atoms are passed through a series of electron carriers, including cytochrome *b,* cytochrome *c,* cytochrome *a,* and cytochrome oxidase, and finally are combined with water and the dissolved oxygen in the fluids of the cell to convert these into hydroxyl ions. The presence of both hydrogen and hydroxyl ions in the same fluid allows immediate combination of the two to form water. The important feature about this oxidation of hydrogen is not the formation of water but, instead, the use of energy from the hydrogen and oxygen atoms for **synthesis of ATP.**

Hydrogen atoms are utilized to cause the formation of ATP. The first step is ionization of the hydrogen atoms and passage of the electrons removed during the ionization process through the electron carrier system. The electron carriers are large protein molecules that are integral parts of the inner wall of the mitochondrion. As the electrons pass from one carrier to the next, they give up energy, and the energy is used to pump the hydrogen ions from the central cavity of the mitochondrion into the outer chamber between the two mitochondrial walls. This creates a high concentration of hydrogen ions in this outer chamber.

The mitochondrial mechanism for synthesis of ATP is called the chemiosmotic mechanism. The entire process of oxidative phosphorylation occurs inside the mitochondria. The large hydrogen ion gradient across the inner wall of the mitochondrion causes the hydrogen ions to leak back through the inner wall toward the central cavity. This flow of hydrogen ions through the inner wall of the mitochondrion from the outer mitochondrial chamber to the central cavity occurs through very large protein molecules called ATP synthetase. Each of these molecules is an ATPase enzyme capable of using the energy derived from the flow of hydrogen ions through its molecular matrix to convert adenosine diphosphate (ADP) plus a phosphate radical into ATP—that is, to cause synthesis of ATP. Thus, in this roundabout way, the tremendous energy that was present in the glucose molecule is finally used to form ATP, which itself stores a large portion of this energy and uses it later to energize almost all intracellular reactions.

Oxidative phosphorylation accounts for more than 90% of ATP production. Thirty-six molecules of ATP are produced for each mol-

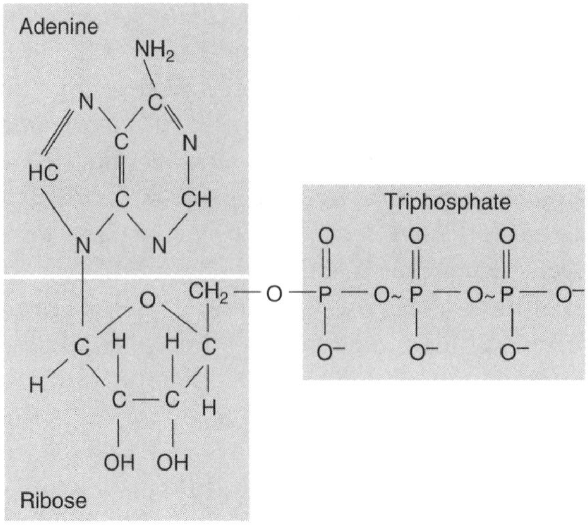

Figure 9-2.
Chemical structure of adenosine triphosphate.

ecule of glucose oxidized, 18 times as many molecules as are formed by the process of glycolysis alone.

ATP has high-energy phosphate bonds that can transfer energy to other chemical processes. ATP is a highly labile compound, the formula for which is shown in Figure 9-2. The points where the last two phosphate radicals attach (indicated by the curving bonds) are called "high-energy phosphate bonds." These bonds provide from 7,000 to 12,000 calories of energy for each mole of ATP, the exact amount depending on concentrations and other conditions of the reactants. The two phosphate radicals on the end of the molecule can split away with great ease and can transfer this energy to other chemical processes in the cells. Thus, ATP is almost an explosive compound that is ready to act immediately. After it begins to be consumed, more ATP is formed by the processes described above.

ATP is needed for almost all chemical reactions. Some of the specific functions of ATP are (1) to provide the immediate energy needed for contraction of muscle cells, (2) to provide the energy needed to pump sodium out of nerve and muscle cells so that action potentials can be transmitted along the membranes, (3) to provide the energy needed to synthesize new proteins by the ribosomes in the cytoplasm, and (4) to provide the energy needed for synthesis and secretion of almost all substances formed by the glands. These are only a few of the functions of ATP.

FAT METABOLISM

Chylomicrons transport fat from the gastrointestinal tract. Fatty acids absorbed from the small intestine immediately recombine in the intestinal epithelial cells with glycerol to form minute particles of **triglycerides** (neutral fat). The triglycerides then aggregate with small amounts of protein to form fat globules called chylomicrons that have an average size of 0.4 μm. These pass into the intestinal lymphatics and along the thoracic duct, eventually to empty into the blood. Within an hour or so after a meal has been completely absorbed, most of the chylomicrons will have been removed from the circulating blood, some of them being absorbed by the liver cells and a very large portion being split by an enzyme, **lipoprotein lipase,** into fatty acids that are either metabolized for energy or are resynthesized into fat by the fat cells and stored.

Ninety-five percent of the plasma lipids are in the form of lipoproteins. Most fatty substances are not soluble in the body fluids. Therefore, lipids are transported in the blood in the form of minute suspended particles called **lipoproteins.** Lipoproteins are composed of triglycerides, phospholipids, cholesterol, and protein. The fatty substances in the lipoproteins are loosely bound with varying amounts of protein. The proteins in turn, being miscible with water, increase the suspension stability of the lipoproteins in the plasma and prevent excessive adherence of them to each other or to the endothelial cells of the blood vessels.

Almost all fat is transported from one part of the body to another as free fatty acid. A very small amount of fatty acids, about 15 mg/100 ml, is present in the plasma in combination with the albumin of the plasma proteins. These are called free fatty acids. Despite this very small concentration, this free fatty acid is transferred extremely rapidly back and forth with the tissue fats, as much as one-half of it transferring every 2 to 3 minutes. Such rapid mobility makes this the principal means of transport of most of the fat from one area of the body to another.

Triglycerides are stored in fat depots. All of the fat in the fat cells of the body is known collectively as fat depots. When the amount of circulating fatty acid falls very low in the blood during the interdigestive period between meals, the stored triglycerides are split by tissue lipoprotein lipase to release free fatty acids, which are then transported in the blood to other cells of the body where they may be needed.

Fats provide energy in nearly the same manner as carbohydrates. The chemical processes for deriving energy from fats include the following stages:

1. The **triglycerides** are split inside the fat cells into glycerol and fatty acids. The glycerol, having a chemical composition very similar to that of certain glucose breakdown products, can easily be oxidized to energize the formation of ATP molecules.
2. The **fatty acid molecules** are partially oxidized, mainly in the liver cells, by a process called **beta carbon oxidation,** which causes the fatty acid to split and form many **acetylcoenzyme A** molecules. These in turn enter the mitochondria where they are decomposed by the citric acid cycle and then completely oxidized by the oxidative phosphorylation process to synthesize

large numbers of ATP molecules in exactly the same way that the acetylcoenzyme A derived from pyruvic acid is used for the same purpose.

Thus, fats provide energy in nearly the same manner as carbohydrates except that the initial stage for splitting carbohydrates is mainly the glycolysis mechanism while the initial stage for splitting fats is principally the beta carbon oxidation mechanism.

Acetylcoenzyme A formed in liver is transported to other tissues as acetoacetic acid. A large share of all fatty acid degradation begins in the liver, but only a small portion of the acetylcoenzyme A formed in the liver cells is used for energy in the liver itself. Instead, it is transported away from the liver by the blood in the form of acetoacetic acid, a condensation product of two molecules of acetylcoenzyme A. This in turn is absorbed by the other cells of the body, where it is split again into two molecules of acetylcoenzyme A, then it enters the citric acid cycle and is used for energy.

Phospholipids are a major constituent of cell membranes. Phospholipids are chemical substances derived mainly from fat and have certain of the physical and chemical characteristics of fats. All cells of the body synthesize phospholipids, but most of these are synthesized in the liver or intestinal epithelial cells. The phospholipids are used by the cells to form the most significant part of the cellular membrane, different intracellular membranes, and other intracellular structures.

Cholesterol is synthesized by all cells of the body. Cholesterol absorbed from the intestinal tract is called **exogenous cholesterol,** but an even greater quantity of cholesterol, called **endogenous cholesterol,** is formed in the cells of the body. Cholesterol is used along with phospholipids and small amounts of triglycerides in composing the different membranous structures of cells. An especially large amount of cholesterol is formed by the liver, and about 80% of this is then used by the liver to synthesize bile acids that are secreted in the bile into the small intestine. These in turn promote emulsification of fats in the gastrointestinal tract so that they can be digested, and they also promote fat absorption. Cholesterol is also a precursor for a number of steroid hormones.

Atherosclerosis is a major killer in the United States and Europe. Unfortunately, large quantities of **cholesterol** are occasionally either synthesized by the walls of the arteries or are deposited there from the blood. As a result, large plaques of cholesterol frequently develop in the arterial walls and protrude through the intima into the flowing arterial blood. This disease is called **arteriosclerosis.** Blood clots often develop on the protruding cholesterol plaques, at times becoming large enough to occlude completely the lumen of the vessel. This is the cause of most acute coronary occlusions that bring about heart attacks. Thrombosis or hemorrhage of arteries in the brain can lead to stroke and cause other problems in other organs such as the kidney, liver, gastrointestinal tract, limbs, and so forth.

Ingestion of saturated fats increases the severity of arteriosclerosis. The rate at which cholesterol is deposited in the walls of the arteries is directly proportional to the intake of calories in the diet, especially when this is above the daily requirements for energy and also when the diet contains a high percentage of cholesterol and fat. Atherosclerosis also seems to be more severe when most of the intake of calories is in the form of highly saturated fats rather than in the form of unsaturated fats, proteins, and carbohydrates. Certain persons with a hereditary disease called **familial hypercholesterolemia** are inclined to very severe arteriosclerosis.

PROTEIN METABOLISM

Amino acids absorbed into portal blood are transported to all cells of the body. The amino acids, which are the end-products of protein digestion in the gastrointestinal tract, are absorbed into the portal blood and then are disseminated into all parts of the body. Some of the amino acids are transported into the liver cells and are temporarily stored there, although this is of minor importance in comparison with the storage of glucose and fat by the liver. Most of the amino acids are rapidly absorbed directly into all the cells of the body by active transport or facilitated diffusion mechanisms.

Most amino acids are synthesized into proteins. This process is controlled by messenger ribonucleic acid (mRNA) molecules that act as templates for the formation of the protein molecules in the ribosomes. Some of these proteins include fibrous proteins such as collagen, elastins, keratins, actin, and myosin; plasma proteins such as albumin, globulin, and fibrinogen; and various enzymes, hormones, and growth factors. Amino acids can also be used to produce energy, as discussed below.

Essential amino acids cannot be synthesized in the cells. The usual proteins of the body contain 20 different amino acids, and these are available in almost all protein foods that are eaten. However, an occasional dietary protein has a deficiency of one or more of the usual amino acids. Often one of the available amino acids can be converted into the missing one, but 10 of the 20 amino acids cannot be formed this way and must be present in the diet in order for the cells to synthesize their normal complements of proteins. These 10 amino acids are called **essential amino acids.** The other 10 **nonessential amino acids** are necessary for the formation of proteins but it is not essential that they be present in the diet.

Catabolism of proteins means splitting of the proteins back into amino acids. A small concentration of amino acids, about 30 mg/100 ml, is always maintained in the plasma and interstitial fluid. When this amount decreases, amino acids are transported out of the cells, and this causes the proteins of the cells to begin to be catabolized into amino acids. This catabolism is catalyzed by intracellular enzymes called **cathepsins,** which are digestive enzymes released from lysosomes in the cells. In this way, a small concentration of amino acids is always maintained in the body fluids. Then, if a particular cell becomes damaged or for some other reason needs an immediate source of amino acids to repair its structural and enzyme systems, the amino acids are available. Amino acid mobilization is greatly accelerated by glucocorticoid hormones.

Excess amino acids are deaminated in the liver. If a person eats more protein than the amount needed to maintain adequate protein stores in the cells, all the excess amino acids are degraded and then used for energy as carbohydrates and fats are used. The first stage in this process is **deamination,** which means removal of the amino radical from the amino acid. This is accomplished by a specific enzyme system in the **liver cells.** The amino radical is then synthesized into urea, which is excreted through the kidneys into the urine.

Deaminated amino acids are oxidized to release energy. Many of the deaminated amino acids are similar to pyruvic acid and can enter the citric acid cycle either directly or after a few stages of minor alterations. In this way, the amino acids become **oxidized** in very much the same way as carbohydrates and fats, and large quantities of ATP are formed to be used as an energy source everywhere in the cells.

Proteins and amino acids in the blood and tissues are in reversible equilibrium. Figure 9-3 shows the reversible equilibrium

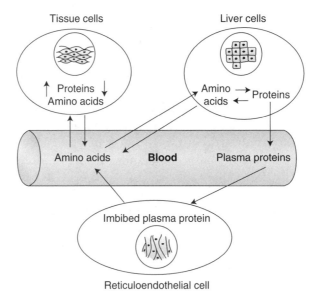

Figure 9-3.
Reversible equilibrium among the tissue proteins and amino acids and the plasma proteins and the amino acids. (Adapted from Guyton AC and Hall JE: *Textbook of Medical Physiology,* 9th ed. Philadelphia: WB Saunders, 1995, p 880.)

among the tissue proteins and amino acids and the plasma proteins and amino acids. Approximately 400 grams of protein are synthesized and degraded daily as part of the continual flux of amino acids between the different protein stores in the body. Because the amino acids can rapidly move between protein molecules in the plasma and various tissues of the body, one of the best therapies for severe protein deficiency is intravenous infusion of plasma protein.

NUTRITION

Metabolic Rate

The Calorie expresses the energy equivalent of foods. Energy is generally expressed in terms of Calories (spelled with a capital C), which are equivalent to kilocalories. A calorie (spelled with a lowercase c) is the amount of heat required to heat one gram of water 1° C. The different types of food are not equal in their capabilities for supplying energy. One gram of carbohydrate or 1 gram of protein supplies 4.1 Calories of energy to the body, while 1 gram of fat supplies 9.3 Calories of energy. Therefore, it is evident that over twice as much carbohydrate or protein must be eaten to provide the same amount of energy as a specified quantity of fat.

A normal adult burns about 1600 Calories daily simply to exist. To provide the energy needed for sitting, another 200 to 400 Calories are required; and for walking and working moderate amounts, still another 500 Calories. The total daily energy requirement for the average person adds up to about 2500 Calories. This is called the metabolic rate of the body. In a person performing very heavy physical labor, this requirement can occasionally be as high as 6000 to 7000 Calories daily.

The basal metabolic rate is the metabolic rate of an awake person whose body is as inactive as possible. To attain the basal state, a person must have had essentially no exercise for the previous 8 to 10 hours, no food for the previous 12 hours, and have quiet conditions and normal room temperature. The normal basal metabolic rate of the young male adult is about 40 Calories per meter squared of body surface area per hour or a total of about 70 Calories per hour for the whole body.

Regulation of Food Intake

Short-term food intake is regulated by signals from the gastrointestinal system. Whenever the gastrointestinal system becomes overly filled, nervous signals passing from the intestinal tract to the brain diminish one's desire for food and therefore diminish the intake of food. The degree of distention of the stomach and upper portions of the small intestine is particularly important. Because about 20 minutes or more are required for the nervous signals from a "full" stomach to make a person begin to lose his or her appetite for food, eating a meal very slowly can be an effective means of reducing caloric intake.

Long-term food intake is regulated by the hypothalamus. Located in the lateral nucleus of the hypothalamus is a feeding center, which, when stimulated, increases the food intake. Located in the ventromedial nucleus of the hypothalamus is a center that makes a person feel satisfied and therefore inhibits food intake; this center is called the satiety center. The degree of activity of the feeding center of the hypothalamus is determined by the metabolic status of the body. When a person becomes overweight, the activity of the feeding center diminishes. On the other hand, in starvation, the feeding center becomes greatly activated. The exact feedback signals to the feeding center that activate or inactivate it are not well understood, but it is known that increased levels of glucose and amino acids in the blood will inhibit feeding, while diminished quantities of these substances excite feeding. It is probable that the level of fatty acids in the circulating blood is also an important factor in controlling the activity of the feeding center.

Obesity can result from psychogenic factors, genetic factors, and childhood overnutrition. When more foods are ingested than are utilized each day, the extra amount is stored in the form of fat. Even excess carbohydrates and proteins are converted into fat and then stored. Thus, obesity actually develops from an excess intake of energy each day over utilization of energy. A common **psychological factor** that contributes to obesity is the notion that three meals are required every day and that each meal should result in complete satiation. This false notion often develops during childhood from overzealous parents. **Genetic factors** contributing to obesity can involve abnormalities of the feeding centers or psychic abnormalities that cause a person to eat as a "release" mechanism. **Childhood overnutrition** can increase the proliferation of fat cells in the first few years of life and thereby promote the development of obesity throughout the life of the individual.

Starvation

During starvation, carbohydrate stores are consumed before fat and protein stores. Figure 9-4 shows the effect of starvation on the food stores of the body. Sufficient glycogen is stored in the liver and muscle cells to provide significant amounts of energy for about one-half to one day. Beyond that, almost all the energy made available to the body at first comes from the stored fat, this sometimes lasting for as long as 3 to 8 weeks. However, during the entire process of starvation, a small amount of body protein is continually degraded and used for energy as well, and when the fat stores begin to run out,

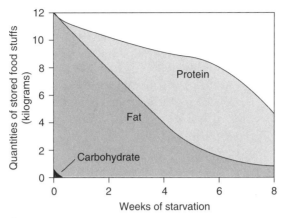

Figure 9-4.
The effect of starvation on the food stores of the body. (Adapted from Guyton AC and Hall JE: *Textbook of Medical Physiology,* 9th ed. Philadelphia: WB Saunders, 1995, p 894.)

tremendous amounts of protein then begin to be used for energy. When this happens, the functional state of the cells quickly deteriorates, and death soon follows. However, it is fortunate that most of the proteins are spared until the last.

Proteins must be ingested each day to prevent protein loss from the body. The body continues to degrade proteins into amino acids even when proteins are absent from the diet. This obligatory loss of proteins amounts to about 20 to 30 grams of protein each day. Because not all proteins contain amino acids in the same proportions as the bodily proteins and because not all proteins are used completely by the body, it is usually necessary to ingest 60 to 75 grams of protein each day to avoid significant protein loss.

Vitamins

Vitamins play key roles in the metabolic systems of the body. The functions of some of the important vitamins are the following:

- **Vitamin A** is used to synthesize rhodopsin, which is a chemical necessary for vision. It is also important for the normal growth of most cells, especially epithelial cells.
- **Thiamine (vitamin B_1)** functions as a cocarboxylase, mainly for decarboxylation of pyruvic acid. Thus, thiamine is necessary for normal metabolism of carbohydrates and many amino acids.
- **Niacin (nicotinic acid)** functions in the body as coenzymes in the forms of nicotinamide adenine dinucleotide and nicotinamide adenine dinucleotide phosphate (NADP). These act as hydrogen acceptors when hydrogen is removed from the foods; therefore, they are important in the oxidation of food.
- **Riboflavin (vitamin B_2)** is used to form flavoprotein, which is also a hydrogen carrier in the oxidation of foodstuffs.

- **Vitamin B$_{12}$ and folic acid** were discussed earlier in the chapter as substances needed for the maturation of red blood cells in the bone marrow.
- **Pantothenic acid** is a precursor of coenzyme A, which is needed for acetylation of many substances in the body. It is especially needed for the formation of acetylcoenzyme A prior to oxidation of both glucose and fatty acids.
- **Pyridoxine** is used as a coenzyme in many reactions involving amino acid metabolism, one of which acts to transfer amino radicals from one substance to another, and another of which causes deamination.
- **Ascorbic acid (vitamin C)** is a strong reducing compound that acts in several metabolic processes in which electrons are exchanged with oxidative chemicals. Although the precise nature of all these specific chemical reactions is not known, the major physiologic function of ascorbic acid is to maintain normal intercellular substances, including normal collagen fibers and normal intercellular cement substance between the cells.
- **Vitamin D** increases absorption of calcium from the gastrointestinal tract and also helps to control calcium deposition in the bone.
- **Vitamin K** is needed in the synthetic processes of the liver for formation of prothrombin, factor VII, and several other factors that are utilized in blood coagulation. Therefore, vitamin K deficiency retards blood clotting.

REGULATION OF BODY TEMPERATURE

The **temperature** of the body is determined by the balance between the **rate of heat production** and the **rate of heat loss.** If the rate of heat production exceeds the rate of heat loss, the body temperature will rise. If the rate of loss is greater, then the body temperature will fall.

The rate of heat production is controlled by the metabolic rate of the body. All of the **metabolic processes** of the body produce heat as a by-product, for almost all the energy in the food eventually becomes heat after it performs its other functions. Thus, in the average person about 2500 Calories of heat are formed daily.

Heat production increases greatly during exercise. Under basal conditions it is mainly the internal organs—the brain, the heart, the kidneys, the gastrointestinal tract, and especially the liver—that produce most of the heat. However, when a person uses muscles for various activities, these produce tremendous amounts of heat. During extreme muscular activity, the total heat production of the body can increase temporarily to as much as 15 to 20 times normal, over 90% of the heat then coming from the muscles.

Heat is lost from the body by convection, radiation, and evaporation. Figure 9-5 shows the basic mechanisms of heat loss from the body. Normally, about 60% of the heat loss from a nude person sitting in a room at 70°F (21°C) is by **radiation.** Heat loss in this manner results from heat waves—a type of electromagnetic radiation—that are transmitted from the surface of the body to surrounding objects. Approximately 18% of the heat that is lost by the nude person is by **conduction** to objects touching the body or to the surrounding air and then by convection of the heated air away from the body. Finally, a small amount of water continually diffuses through the skin, and evaporation of each gram of water removes about one-half calorie of heat from the body. **Evaporation** of water accounts for about 22% of the heat loss from the nude person.

Heat loss is controlled by blood flow to the skin. The skin is supplied with an abundant vasculature, but the amount of blood that flows through the skin vessels is controlled very exactly by the sympathetic nervous system so that when the body becomes overly heated skin blood flow will be tremendous, and when the body is underheated skin blood flow will be negligible. Rapid flow of blood heats the skin, allowing large amounts of heat to be lost to the surroundings. Also, sympathetic stimuli to the sweat glands increase sweat production and consequently greatly increase the evaporative loss of heat. Slow blood flow prevents heat loss because the internal heat of the body cannot be carried to the skin, and the skin temperature falls rapidly to approach that of the surroundings.

Body temperature is controlled by the "hypothalamic thermostat." The preoptic area of the anterior hypothalamus contains heat-sensitive neurons as well as cold-sensitive neurons. These neurons are thought to function as temperature sensors for controlling body temperature. The **heat-sensitive neurons** increase their rate of firing when the body becomes too hot, and the **cold-sensitive neurons** increase their firing rate when the body becomes too cold. The temperature sensory signals from the anterior hypothalamus are transmitted to the posterior hypothalamus where they are integrated with signals from the periphery to make appropriate adjustments in heat production and heat loss.

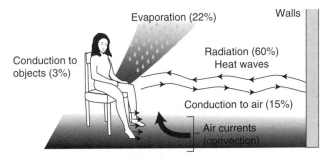

Figure 9-5.
Mechanism of heat loss from the body. (Adapted from Guyton AC and Hall JE: *Textbook of Medical Physiology,* 9th ed. Philadelphia: WB Saunders, 1995, p 913.)

Temperature-increasing mechanisms include skin vasoconstriction and increased heat production. Constriction of the skin blood vessels reduces the rate of heat loss because the skin temperature approaches the temperature of the surrounding air. **Heat production** is increased by **shivering,** which increases the muscle metabolic rate, and **sympathetic release of epinephrine,** which increases the metabolic rate of all the cells in the body. The result of these effects is an increase in the body temperature back toward normal.

Temperature-decreasing mechanisms include skin vasodilation, sweating, and decreased heat production. When the preoptic area becomes too hot, the temperature control system decreases heat production by reducing the tone of the skeletal muscles throughout the body and reducing the sympathetic release of epinephrine. More important, however, it increases the rate of heat loss in two ways: (1) **lack of sympathetic vasoconstriction** allows the blood vessels of the skin to become greatly dilated; the skin temperature rises, and increased amounts of heat are lost to the surroundings; and (2) **stimulation of a sweat center,** also located in the heat center of the hypothalamus, causes sweating over the entire body, with resultant evaporative loss of heat.

Summary of hypothalamic control of body temperature. If the preoptic area of the anterior hypothalamus becomes too hot, heat production is decreased while heat loss is increased, and the body temperature falls back toward normal. Conversely, if the preoptic area becomes too cold, heat production is increased, while heat loss is decreased so that the body temperature now rises toward normal.

Temperature signals from the skin also play a role in body temperature regulation. When the skin becomes too warm, nerve impulses are transmitted from warm receptors in the skin all the way to the preoptic area of the hypothalamus to increase sweating, and this obviously increases body heat loss. Conversely, when the skin becomes too cold, signals from cold receptors in the skin are transmitted to the posterior hypothalamus where they activate shivering, which in turn helps to increase the body temperature back toward normal. Therefore, body temperature control is vested in an integrated mechanism in which the hypothalamus is the central integrator, but reacts to signals both from the preoptic area and from the skin.

Pyrogens increase the hypothalamic set-point temperature to cause fever. Pyrogens consist of breakdown products of proteins, lipopolysaccharides released from bacteria cell membranes, or other substances released from breakdown of the body tissues. Pyrogens can increase the hypothalamic set-point temperature by an immediate direct action on the hypothalamus, or, more commonly, by an indirect mechanism requiring several hours to effect a change. This latter mechanism, which applies to **endotoxin** derived from gram-negative bacteria, involves increased production of interleukin-1 by leukocytes, tissue macrophages, and other cells that phagocytize the bacteria or bacterial products. The **interleukin-1** causes the anterior hypothalamus to increase its production of

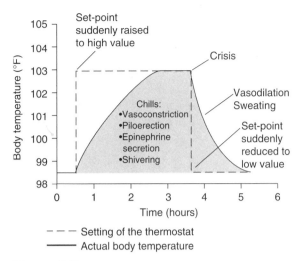

Figure 9-6.
Effect of changing the hypothalamic set-point temperature on body temperature. (Adapted from Guyton AC and Hall JE: *Textbook of Medical Physiology,* 9th ed. Philadelphia: WB Saunders, 1995, p 920.)

prostaglandins (mainly prostaglandin E_2), and the prostaglandins, in turn, increase the hypothalamic set-point temperature of the hypothalamus.

During the course of febrile illness, an individual experiences chills while the body temperature is rising and sweating while the body temperature is falling. Figure 9-6 shows the effect of changing the hypothalamic set-point temperature on body temperature. When the hypothalamic set-point temperature is suddenly raised to a high value, the person experiences chills and feels cold although the body temperature may already be above normal. The various temperature-increasing mechanisms discussed above continue to raise the body temperature until it reaches the set-point value of 103°F, as shown in Figure 9-6. The person feels neither hot nor cold, and the body temperature is regulated at this elevated temperature. If the factor causing the elevation in temperature is suddenly removed, the hypothalamic set-point temperature is suddenly decreased, and mechanisms are activated that cause heat loss (sweating and vasodilation). The person feels hot as the body temperature is falling back to normal.

Chapter 10

Endocrine Control Systems

The endocrine glands secrete **hormones** that control many of the body's functions. Hormones are chemical messengers that are carried by the blood from endocrine glands to the cells where they act—often referred to as **target cells** for that hormone.

In general, the hormone systems function to control transport of substances through cell membranes; to control the activity of specific cellular genes, which in turn determine the formation of specific enzymes and other cell factors; and to control directly some metabolic systems of the cells.

HORMONE STRUCTURE, SYNTHESIS, AND SECRETION

There are three general classes of hormones: steroids, such as cortisol and estrogen; proteins and polypeptides, such as adrenocorticotrophic hormone (ACTH) and growth hormone; and amino acid derivatives, such as epinephrine and thyroxine. Most hormones are water soluble. There are no hormones known that are polysaccharides or nucleic acids.

Steroid hormones are synthesized from cholesterol. They are lipid-soluble substances consisting of three cyclohexyl rings and one cyclopentyl ring combined into a single structure. Steroids are synthesized and secreted by the adrenal cortex and the gonads (including the testes and ovaries) as well as the placenta during pregnancy. Because the steroids are highly lipid soluble, once they are synthesized they simply diffuse across the cell membrane of the steroid-producing cell and enter the interstitial fluid and then the blood. The circulatory half-life of most steroids ranges between 60 and 100 minutes.

Protein and polypeptide hormones comprise the majority of hormones in the body. They range in size from small peptides of a few amino acids to small proteins. In many cases, they are synthesized first as larger proteins that are cleaved to form **prohormones,** which in turn must be modified by the secretory cells to produce biologically active hormones. Protein and polypeptide hormones are usually stored in subcellular membrane-bound secretory granules within the cytoplasm of the different endocrine cells. The hormones are released into the blood by **exocytosis,** which occurs when the secretory granule is fused with the cell membrane and the granular contents are then extruded into the interstitial fluid or directly into the bloodstream. In many cases, the stimulus for exocytosis is an increase in cytosolic calcium concentration caused by depolarization of the plasma membrane. The circulatory half-life of protein and peptide hormones is relatively short, lasting for as little as 5 to 6 minutes to as long as 60 minutes for some hormones.

Amine hormones are derived from the amino acid tyrosine and are sometimes called phenolic derivatives. These hormones include **epinephrine, norepinephrine, triiodothyronine** (T_3), and **thyroxine** (T_4). **The thyroid hormones** T_4 and T_3 are synthesized and stored in the thyroid gland and incorporated into molecules of the protein **thyroglobulin.** Hormone secretion occurs when T_4 and T_3 are split from the thyroglobulin, and the free hormones are then released into the bloodstream. After entering the blood, most of the T_4 and T_3 combine with plasma proteins, especially thyroxin-binding globulin, which slowly releases the hormones to the tissue cells.

The hormones epinephrine and norepinephrine are formed in the adrenal medulla, which normally secretes about four times more epinephrine than norepinephrine. Both of these catecholamines exist in the plasma in free form or in conjugation with other substances. Most of the circulating epinephrine is bound to plasma proteins, especially albumin, whereas norepinephrine does not significantly bind to plasma proteins. The circulatory half-life of norepinephrine and epinephrine is about 1 to 3 minutes.

Feedback Control of Hormone Secretion

Although many hormones fluctuate in response to various stimuli that occur throughout the day, all hormones that have been studied thus far appear to be closely controlled. In many instances, this control is exerted through **negative feedback mechanisms** that ensure a proper degree of activity of the hormone on the target tissue. After a physiologic condition stimulates the release of a hormone, conditions or products resulting from the action of the hormone tend to suppress its further release. In other words, the hormone (or one of its products) has a negative feedback effect to prevent oversecretion of the hormone or overactivity of the hormone on the target tissue.

The controlled variable in many cases is not the secretory rate of the hormone itself, but the degree of activity of the target tissue. Therefore, only when the target organ activity rises to an appropriate level will feedback to the endocrine gland become powerful enough to slow further secretion of the hormone. Feedback regulation of hormones can occur at all levels, including transcription and translational steps involved in synthesis, steps involved in processing the hormone, or release of the stored hormones.

In a few instances, **positive feedback** occurs when the biological action of the hormone either directly or indirectly causes additional secretion of the hormone. One example of this is the surge of **luteinizing hormone** (LH) that occurs prior to ovulation as a result of the stimulatory effect of estrogen on the anterior pituitary. The secreted LH then acts on the ovaries to cause more secretion of estrogen, which in turn causes additional secretion of LH. Eventually, the LH reaches an appropriate concentration, and typical negative feedback control of hormone secretion is then exerted.

Superimposed on negative and positive feedback control of hormone secretion are **cyclic variations of hormone release** that are influenced by seasonal variations, various stages of development and aging, diurnal (daily) cycle or by sleep. For example, the secretion of growth hormone is markedly increased during the early period of sleep but is markedly reduced during the later stages of sleep. In many cases, these cyclic variations in hormone secretion are due to variations in activity of neuropathways involved in controlling hormone release.

Mechanisms of Action of Hormones

The first step in a hormone's action is to bind to a specific **receptor** in the target cell. Cells that lack specific receptors for the hormone do not respond. For some hormones (e.g., peptide hormones), receptors are located on the target cell membrane. In other instances, the cell receptors are located in the cytoplasm (e.g., steroids) or at the nucleus (e.g., thyroid hormones).

Lipid-soluble hormones can cross the cell membrane. This enables them to interact with specific intracellular receptors, and increase or decrease protein synthesis by stimulating or inhibiting the production of messenger RNA (mRNA). These either bind to specific receptors in the cytoplasm or diffuse through the nuclear membrane and bind to specific receptors in the nucleus. Activation of these intracellular receptors then elicits a sequence of events that produces the hormone effects by regulating expression of specific genes in the cell nucleus.

Peptide hormones and catecholamines cannot penetrate cell membranes. They must interact with receptors on the exterior surface of the cell. Coupling of the receptor to the intracellular enzymes that control cell function results from generation of intracellular **signal transduction mechanisms** and **second messengers.** In some cases, the hormone receptor itself contains an ion channel, and activation of the receptor causes the channel to open, resulting in increased diffusion across the plasma membrane of ions specific for the channel. The altered ion concentration of the cell then elicits the cell responses to the hormone.

The adenylate cyclase/cAMP intracellular signaling mechanism. Figure 10-1 shows the adenylate cyclase/cAMP second messenger system for many hormones, including most polypeptides and catecholamines. Binding of the hormone to its receptor allows coupling of the receptor to a **G protein.** If the G protein stimulates adenylate cyclase/cAMP, it is called a **G_s protein,** denoting a stimulatory G protein. This stimulates adenylate cyclase, a membrane-bound enzyme, which then catalyzes the formation of cAMP inside the cell. This then activates cAMP-dependent protein **kinase,** which phosphorylates specific proteins in the cell, triggering the various biochemical reactions that ultimately lead to the cell's response to the hormone.

If binding of the hormone to its receptors is coupled to an **inhibitory G protein** (denoted G_I protein), adenylate cyclase will be inhibited, reducing the formation of cAMP and ultimately leading to an inhibitory action in the cell. Thus, depending on the coupling of the hormone receptor to inhibitory or stimulatory G proteins, a hormone can either increase or decrease the concentration of

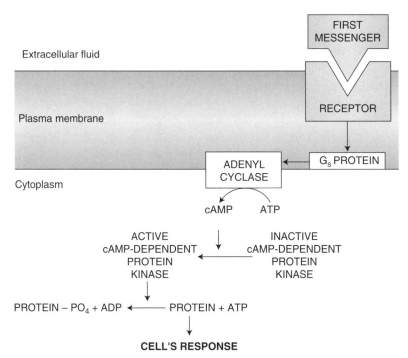

Figure 10-1.

cAMP second messenger system. A combination of the first messenger (e.g., hormone) with its specific receptor allows the receptor to bind to a membrane-stimulating G protein (G_s). This protein binds to adenyl cyclase, activating it and causing it to catalyze the formation of cAMP inside the cell. This cAMP activates cAMP-dependent protein kinase, which then phosphorylates specific proteins in the cell, ultimately leading to the cell's response.

cAMP in the cell and thereby reduce the phosphorylation of key proteins inside the cell.

Membrane phospholipid breakdown products serve as second messengers. Some hormones activate transmembrane receptors that then activate the enzyme **phospholipase C**. This enzyme catalyzes the breakdown of some phospholipids in the cell membrane, especially **phosphatidylinositol biphosphate**, into two important second messenger products, **inositol triphosphate** (IP_3) and **diacylglycerol** (DAG). The IP_3 mobilizes calcium ions from mitochondria and the endoplasmic reticulum, and the calcium ions then have their own second messenger effects, such as smooth muscle contraction and changes in cell secretion.

Diacylglycerol, the other lipid second messenger, activates the enzyme **protein kinase C** (PKC), which then phosphorylates a large number of proteins leading to the cell's response (Fig. 10-2).

Calcium ions and calmodulin act as a second messenger system. Another second messenger system operates in response to entry of calcium into the cells. The calcium entry may be initiated by changes in membrane potential, which open calcium channels, or by hormones interacting with membrane receptors that open calcium channels. On entering the cell, calcium ions bind with a protein called **calmodulin.** After binding with calcium, the calmodulin changes its shape and initiates multiple effects inside the cell including activation or inhibition of protein kinases. Activation or inhibi-

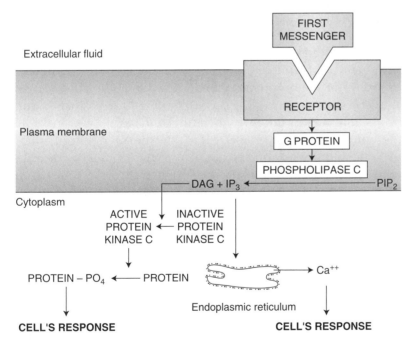

Figure 10-2.
Mechanism by which activation of a cell receptor stimulates the breakdown of phosphatidylinositol biphosphate into inositol triphosphate (IP_3) and diacylglycerol (DAG). IP_3 then causes the release of calcium ions from the endoplasmic reticulum, and DAG activates protein kinase, which then phosphorylates large numbers of proteins to the cell's response.

tion of **calmodulin-dependent protein kinases** causes, via phosphorylation, activation or inhibition of proteins involved in the cell's responses to the hormone. For example, one specific function of calmodulin is to activate **myosin kinase,** which acts directly on the myosin of smooth muscle to cause smooth muscle contraction.

THE HYPOTHALAMUS–PITUITARY HORMONE SYSTEM

The **pituitary gland** (also called the **hypophysis**), located at the base of the brain just below an area called the **hypothalamus,** secretes a large number of peptide hormones that regulate many body functions. The pituitary is connected to the hypothalamus by a stalk that contains nerve fibers and blood vessels and is divided into two major parts: the **anterior pituitary** and the **posterior pituitary** gland.

Six well-known hormones are secreted by the anterior pituitary gland, including growth hormone, thyroid-stimulating hormone (TSH), ACTH, prolactin, luteinizing hormone (LH), and follicle-stimulating hormone (FSH). The posterior pituitary gland secretes two primary hormones: antidiuretic hormone (ADH) and oxytocin.

Anterior pituitary hormone secretion is controlled by hypothalamic-releasing hormones. If the anterior pituitary gland is separated from the hypothalamus, secretion of most of its hormones, with the exception of prolactin, decreases to very small amounts. This occurs because secretion of the anterior pituitary gland is controlled mainly by additional hormones formed in the hypothalamus, which are then conducted to the pituitary through the **hypothalamic-hypophyseal venous portal system,** shown in Figure 10-3. The system consists of small veins that carry blood (and therefore hormones) from capillaries in the lower hypothalamus to the venous sinuses of the anterior pituitary gland. Here, the hormones from the hypothalamus act directly on the anterior pituitary cells to control secretion of the pituitary hormones. Most of the hormones from the hypothalamus stimulate hormone secretion from anterior pituitary cells, but some of them inhibit release of hormones.

Some of the most important hormones that control the function of the anterior pituitary gland are the following:

- **Growth hormone-releasing hormone** (GHRH), which causes release of growth hormone
- **Corticotropin-releasing hormone** (CRH), which causes release of ACTH
- **Thyrotropin-releasing hormone** (TRH), which causes release of TSH
- **Gonadotropin-releasing hormone** (GnRH), which causes release of two gonadotropins, LH and FSH
- **Prolactin inhibitory hormone,** believed to be **dopamine,** which

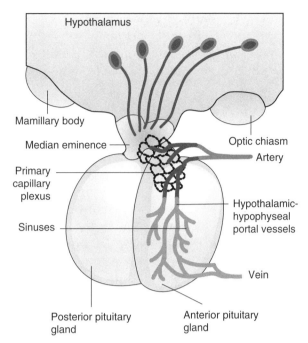

Figure 10-3.
Hypothalamic-hypophyseal portal system.

inhibits the release of prolactin. In the absence of this hormone, the rate of prolactin secretion increases to about three times normal

- **Somatostatin or somatropin-releasing inhibiting factor** (SRIF), which inhibits the secretion of growth hormone, TSH, and possibly other hormones

Growth hormone stimulates growth during childhood and adolescence. The anterior pituitary gland secretes growth hormone throughout life, not only while a person is growing. However, growth hormone does not markedly influence fetal growth and is not a major growth factor during the first few months after birth. On the other hand, growth hormone is essential for normal body growth during childhood and adolescence. It causes enlargement and proliferation of cells in all parts of the body, resulting in progressive growth of the body stature until adolescence. At this time, the epiphyses of the long bones unite with the shafts of the bones so that further increase in height of the body cannot occur. However, certain of the "membranous" bones such as the bones of the nose and certain bones of the skull, as well as the solid tissues of some of the internal organs, can continue to grow under the influence of excess growth hormone.

In addition to its growth-promoting action, growth hormone also has effects on lipid, carbohydrate, and protein metabolism. For example, growth hormone is believed to modulate some of the actions of insulin on the liver and peripheral tissues, inhibiting glucose used by muscle and fat and increasing glucose production by the liver. In addition, growth hormone also stimulates mobilization of triglycerides from fat depots in the body. For this reason, excess secretion of growth hormone can produce **metabolic disturbances** much like those in individuals with non–insulin-dependent diabetes mellitus.

The precise mechanism by which growth hormone exerts its effects on cells is not clear. However, it is known to act on the liver to cause formation of several small protein substances called **somatomedins,** one of which is called **insulin-like growth factor I** (IGF-1). IGF-1 is produced by many cells of the body and released in the bloodstream, but the liver is the primary source of IGF-1 in the blood. IGF-1, in turn, has a negative feedback effect to decrease the secretion of growth hormone.

Growth hormone specifically promotes transport of some amino acids through cell membranes, thereby making more of these available to the cells. In addition, (1) it **activates the RNA translation process** to stimulate formation of proteins by the ribosomes; (2) it **increases the rate of DNA transcription** to increase the amount of mRNA; (3) it **increases the replication of DNA,** which causes increased production of the cells themselves; and (4) it **decreases the rate of breakdown of proteins in the cells.** Thus, growth hormone has the potent effect of enhancing all aspects of protein synthesis and storage in the cells of the body.

Growth hormone secretion is controlled by **growth hormone-releasing hormone** (GH-RH), which is secreted in the hypothalamus and transported to the anterior pituitary through the hypothalamic-hypophyseal portal system. When nutritional debility causes protracted hypoglycemia or reduces the body's protein storage, the secretion of growth hormone-releasing hormone and, therefore, the rate of growth hormone secretion are increased. Also, other types of physical or mental stress often greatly increase the rate of growth hormone secretion.

ACTH regulates adrenocortical function. ACTH strongly stimulates **cortisol** production of the adrenal cortex, and to a lesser extent the production of other adrenocortical hormones. Cortisol, in turn, has many different metabolic effects on the body, including degradation of proteins in the tissues, release of amino acids into the circulating blood, conversion of many of these amino acids into glucose (the process of **gluconeogenesis**), and decreased utilization of glucose by the tissues. These effects will be discussed later in relation to the adrenal hormones.

Figure 10-4 shows that the secretion of ACTH by the anterior pituitary is controlled mainly by **corticotropin-releasing hormone** (CRH) released by the hypothalamus. The rate of secretion of CRH, in turn, is strongly stimulated by stressful states such as disease, trauma to the body, and even emotional excitement. When excess quantities of cortisol are present in the blood, these feed back on the hypothalamus and anterior pituitary cells to decrease the rate of secretion of ACTH, thus providing a negative feedback mechanism for controlling cortisol concentration in the blood.

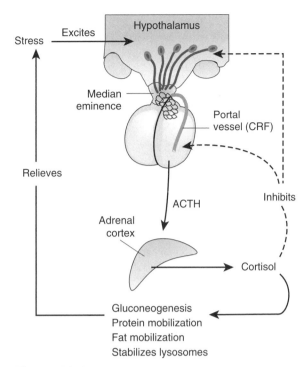

Figure 10-4.
Mechanism for regulation of glucocorticoid secretion.

TSH regulates the function of the thyroid gland. TSH, released by the anterior pituitary, stimulates the thyroid gland follicles in several ways: (1) It increases the rate of synthesis of thyroglobulin; (2) it increases the uptake of iodide ions from the blood by the glandular cells; and (3) it activates all of the chemical processes that cause T_4 production and release by the thyroid gland. Therefore, TSH indirectly **increases the overall rate of metabolism of the body** through increased formation and release of T_4.

The rate of TSH secretion by the anterior pituitary is controlled mainly by the negative feedback effect of T_4. When T_4 concentrations are high, this decreases the rate of secretion of TSH by the anterior pituitary and therefore reduces production of T_4.

Prolactin stimulates milk synthesis. Prolactin is important in the development of the breast during pregnancy and in promoting milk secretion by the breast after the birth of the infant. These functions will also be discussed in more detail in relation to milk production.

The secretion of prolactin is controlled mainly by **prolactin inhibitory hormone** in the hypothalamus, which reduces the secretion of prolactin. There are also **prolactin-releasing factor(s)** released by the hypothalamus, but these have not been fully characterized. Prolactin secretion increases about 10-fold during pregnancy. This rate of secretion is also increased by suckling of the nipples by the infant, which in turn causes production of more milk.

The gonadotropic hormones: FSH and LH. The anterior pituitary secretes two hormones that regulate many male and female sexual functions: FSH and LH influence virtually every aspect of the reproductive process. The rates of secretion of these hormones are controlled mainly by gonadotropin-releasing hormone from the hypothalamus, but feedback from the sex hormones also helps to control their rates of secretion. The gonadotropic hormones will be discussed in more detail later in relation to reproduction.

ADH and oxytocin are secreted by the posterior pituitary. The posterior pituitary is, to a large extent, a neural extension of the hypothalamus (Fig. 10-5). The hormones of the posterior pituitary

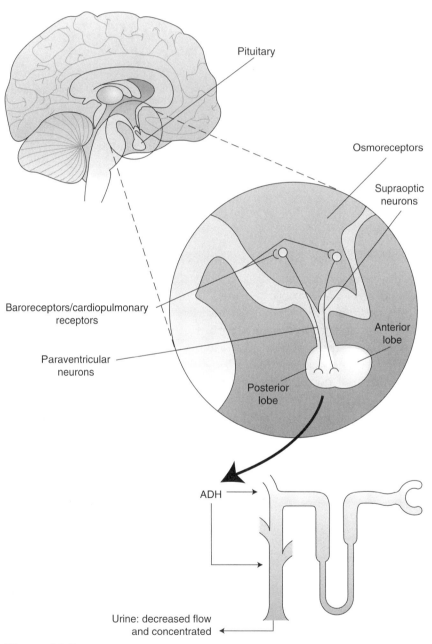

Figure 10-5.
Neuroanatomy of the hypothalamus, where ADH is synthesized, and the posterior pituitary gland, where ADH is released, and action of ADH on the distal and collecting tubules of the kidney.

are antidiuretic hormone (ADH) and oxytocin, which are synthesized in the cell bodies of the hypothalamus whose axons pass through the **median eminence** and enter the posterior pituitary. These hormones move down the neural axons to accumulate at the axon terminals in the posterior pituitary. When the hypothalamus neurons receive input signals for ADH or oxytocin secretion, action potentials are generated in these cells and travel down the axons to trigger the release of ADH or oxytocin from the axon terminal. These substances then diffuse into the capillaries and enter the systemic circulation.

Secretion of **ADH,** discussed previously in relation to water reabsorption by the renal tubules, is stimulated when the body fluids become excessively concentrated. Neuronal cells in or near the **supraoptic nucleus** of the hypothalamus, called **osmoreceptors,** are stimulated by increased osmolality or cell shrinkage and cause the release of ADH from the posterior pituitary gland (see Fig. 10-5). On reaching the kidney, ADH increases the rate of water reabsorption from the renal tubules and tends to correct the overconcentration of the body fluids.

ADH secretion is also stimulated by decreased blood pressure or decreased blood volume, through baroreceptor and cardiopulmonary reflexes, which send signals to the hypothalamus. ADH, in addition to causing water retention, also constricts the arterioles and causes the arterial pressure to rise—the primary reason it is also called **vasopressin.**

Oxytocin causes contraction especially of the uterus and to a lesser extent other smooth muscles of the body. Oxytocin is released by the posterior pituitary gland in increased amounts during parturition and may play a significant role in initiating the birth of the infant. Oxytocin also plays a role in lactation in the following ways: sucking on the breast initiates nerve impulses that pass all the way from the nipple to the hypothalamus, which then sends signals to the posterior pituitary gland to stimulate release of oxytocin. The oxytocin, in turn, stimulates **myoepithelial cells** in the breast that constrict the alveoli of the breast in a manner that makes the milk flow into the ducts. This is called **milk ejection** or **milk let-down.**

ADRENOCORTICAL HORMONES

The adrenal cortex secretes three different types of steroid hormones that are chemically similar but physiologically very different. These are (1) **mineralocorticoids,** represented principally by aldosterone; (2) **glucocorticoids,** represented mainly by cortisol; and (3) several **androgens,** which have masculine sexual effects.

The adrenal cortex of the adult human consists of three distinct layers (Fig. 10-6): (1) the **zona glomerulosa,** the outer zone, which

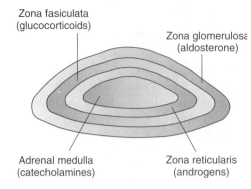

Figure 10-6.
Secretion of adrenocortical hormones by the different zones of the adrenal cortex.

produces aldosterone; (2) the **zona fasciculata,** the middle layer, which produces glucocorticoids and androgens; and (3) the **zona reticularis,** the inner layer of the cortex, which also produces glucocorticoids and androgens.

Mineralocorticoids control electrolyte balance. Although several mineralocorticoid hormones are secreted by the adrenal cortex, about 95% of the total mineralocorticoid activity is due to **aldosterone.** As discussed in Chapter 2, aldosterone enhances sodium reabsorption from the renal tubules into the blood and at the same time increases potassium secretion from the blood into the tubules. Therefore, aldosterone causes the body to conserve sodium while increasing the excretion of potassium in the urine.

Aldosterone also increases reabsorption of chloride ions and water from the tubules in parallel with increased sodium reabsorption, as explained previously. As a result, increased aldosterone secretion causes the kidneys to retain water and chloride and to increase extracellular fluid volume, while decreasing extracellular potassium concentration.

Although aldosterone mainly causes potassium to be secreted into the tubules in exchange for sodium reabsorption, it also causes tubular secretion of hydrogen ions in exchange for sodium ions in the collecting tubule. Thus, with excess aldosterone secretion, a moderate degree of alkalosis can develop as a result of excess hydrogen ion secretion.

The main factors that control aldosterone secretion are as follows:

- **Increased potassium ion concentration stimulates aldosterone secretion by the adrenal gland.** This provides an important negative feedback control system for extracellular-fluid potassium ion concentration because aldosterone then promotes excretion of excess potassium from the extracellular fluid into the urine.
- **Angiotensin II also stimulates secretion of aldosterone.** This effect provides important feedback control of extracellular fluid volume. When extracellular fluid volume or blood pressure decreases, this stimulates the kidney to release renin, which in turn stimulates angiotensin II formation and aldosterone secre-

tion. The increased aldosterone concentration causes the renal tubules to reabsorb more sodium, chloride, and water, thus returning extracellular fluid volume and blood pressure toward normal.

- **Decreased sodium ion concentration in the extracellular fluid also stimulates aldosterone secretion.**
- **ACTH transiently increases aldosterone secretion.** ACTH also plays a permissive role in allowing other stimuli, such as potassium and angiotensin II, to exert their long-term effects on aldosterone secretion.

A deficiency of aldosterone secretion, such as occurs in **Addison's disease,** is associated with a tendency toward decreased extracellular fluid volume and accumulation of potassium. Excess secretion of aldosterone, as occurs in **primary aldosteronism (Conn's syndrome),** causes sodium and water retention, mild expansion of extracellular fluid volume, and hypokalemia (decreased plasma potassium concentration).

Glucocorticoids affect metabolism and suppress inflammation. Several different glucocorticoids are secreted by the adrenal cortex, but most of the glucocorticoid activity is caused by **cortisol,** also called **hydrocortisone.** Some of the main actions of cortisol are the following:

- **Cortisol reduces glucose uptake into cells** and also reduces the rate of utilization of glucose by the cells.
- **Cortisol increases the rate of gluconeogenesis in liver cells,** which convert amino acids into glucose. Thus, several different effects of glucocortisol cause the blood concentration of glucose to increase.
- **Cortisol causes degradation of proteins and decreases protein synthesis in most tissues of the body.** This causes amino acids to be released from the cells and therefore increases their concentration in the blood. In the liver, however, cortisol has the opposite effect—it increases amino acid uptake. These amino acids are used to synthesize large quantities of plasma proteins, to provide metabolic energy for the liver, and to be converted into glucose by the process of gluconeogenesis as discussed previously.
- **Cortisol increases the use of fat for energy.** This results mainly from an effect of cortisol to activate **hormone-sensitive lipase** in fat cells, which causes splitting of the fat and release of amino acids into the circulating blood.
- **Cortisol can also suppress inflammation by stabilizing cellular lysosomes.** This prevents them from rupturing and releasing their digestive enzymes as well as histamine, bradykinin, and other factors that promote inflammation. Cortisol also reduces the permeability of capillary membranes, which minimizes leakage of protein and fluid into the tissues during inflammation. Because of the anti-inflammatory effect of cortisol, it is often administered clinically to reduce inflammation in certain aller-

gic reactions, to treat bursitis and arthritis, and in some types of infection.

ACTH is a primary stimulus for glucocorticoid secretion. Secretion of both cortisol and ACTH usually follow a circadian pattern, dependent on the sleep/wake cycle, with levels of both hormones increasing during waking hours. Other important stimuli for ACTH secretion, and therefore cortisol secretion, include emotional and physical stress, trauma, shock, infection, and hypoglycemia. There is also feedback inhibition of ACTH secretion by glucocorticoids acting at the pituitary as well as the hypothalamus to inhibit corticotropic-releasing hormone.

Deficiency of glucocorticoid production (as occurs in **Addison's disease**) reduces the ability to cope with stress, to form glucose during fasting, and to mobilize lipids for use by peripheral tissues. Conversely, excess glucocorticoid production, as occurs in **Cushing's syndrome,** is associated with increased production and decreased utilization of glucose and therefore a tendency toward hyperglycemia. In addition, prolonged exposure of the body to large amounts of glucocorticoids causes breakdown of peripheral tissue proteins, increased mobilization of lipid from fat stores, and a redistribution of fat on the abdomen, shoulders, and face. Although Cushing's syndrome can be caused by primary hypersecretion of cortisol by the adrenal gland, it usually occurs in response to hypersecretion of ACTH by the pituitary.

Although the most important androgen in the body is testosterone secreted by the testes, the adrenal cortex also secretes several **androgenic hormones.** These are normally of minor importance, but when an adrenal tumor develops or when adrenal glands become hyperplastic and secrete excess quantities of hormones, the amounts of androgens then secreted occasionally become great enough to cause even a child or an adult female to take on adult masculine characteristics.

THYROID HORMONES

The thyroid gland, located immediately below the larynx on either side and to the front of the trachea, secretes the two hormones T_4 and T_3 that have marked effects on the metabolic rate of the body. The thyroid gland also secretes **calcitonin,** a hormone that is important for calcium metabolism and will be considered later.

Synthesis and secretion of T_4 and T_3. The thyroid gland is composed of follicles lined with thyroid glandular cells, shown in Figure 10-7. These cells secrete a very large glycoprotein called **thyroglobulin** to the inside of the follicles. They also absorb iodide ions from the circulating blood and secrete these in an oxidized form into the

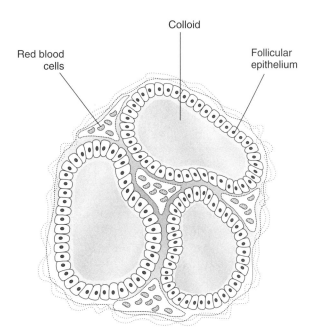

Colloid

Red blood cells

Follicular epithelium

Figure 10-7.
Histology of the thyroid gland. Thyroglobulin is secreted into the follicles.

follicles along with the thyroglobulin. This oxidized iodine combines with tyrosine amino acid molecules that are integral parts of the thyroglobulin molecule. In this manner, large quantities of T_4 and smaller amounts of T_3 are formed within the thyroglobulin molecule, and the thyroglobulin then remains stored in the thyroid follicles for an average of about 6 weeks.

As thyroid hormones are needed in the circulating blood, some of the thyroglobulin is reabsorbed back into the glandular cells by the process of **pinocytosis.** Then the thyroglobulin is digested by proteases formed by lysosomes in the cells, thus releasing T_4 and T_3 into the blood.

The rate of formation of the thyroid hormones, and especially their rate of release from thyroglobulin, is controlled to a large extent by **TSH** released from the anterior pituitary gland, as discussed earlier. The thyroid hormones, in turn, regulate their own secretion by exerting an inhibitory effect on TSH secretion. Consequently, when circulating concentrations of thyroid hormones are too high, the rate of TSH secretion is reduced and the secretion of thyroid hormones returns toward normal.

Once the thyroid hormones have been released into the blood, they combine with several different plasma proteins. Then during the following week, they are slowly released from the blood into the tissue cells.

T_4 and T_3 increase metabolic activity of the cells. The overall action of the thyroid hormones is to increase the metabolic activity in almost all cells of the body. They also increase the breakdown of all cell foodstuffs and the rate of release from the cells. T_4 is the major circulating form of thyroid hormone, although T_3 has the same effects as T_4 except that it acts more rapidly.

Thyroid hormones influence growth and differentiation. Fetal or neonatal thyroid deficiency leads to **cretinism,** a condition associated with abnormal structure of the face, reduced neuronal development, and delayed skeletal maturation. Unfortunately, the exact mechanisms by which the thyroid hormones perform their functions in the cells are still unknown. However, they increase the rate of synthesis of proteins in almost all cells, and especially the rate of synthesis of the different intracellular enzymes that are the basis for increased metabolic activities of the cells. They also increase the sizes and numbers of mitochondria in the cells and these, in turn, increase the rate of production of ATP, which also promotes enhanced cellular metabolism.

Hyperthyroidism produces metabolic and nervous disorders. One of the most common causes of excess thyroid hormone production in the human is **Graves' disease,** an autoimmune disease believed to be caused by antibodies formed against certain components of follicular cell membranes, resulting in enlargement of the thyroid gland and oversecretion of the thyroid hormones. The main effects caused by excess thyroid activity include:

- Greatly **increased rate of metabolism** throughout the body
- **Increased heart rate** and **cardiac output**
- **Increased gastrointestinal secretion and motility**
- **Increased activity of the nervous system,** sometimes causing a fine tremor of the muscles
- **Increased respiratory rate**
- Often, **abnormal glandular secretion** by the endocrine systems
- **Severe weight loss** in extreme cases. In general, hyperthyroid individuals are nervous and irritable and experience physical weakness and fatigue, gradual destruction of body tissue, and weight loss despite increased food intake.

Hypothyroidism decreases metabolic activity. A deficiency of thyroid hormone greatly inhibits activity of almost all functional systems, causing (1) a decrease in metabolic rate to as low as 40% below normal; (2) lethargy so that a person may sleep 14 to 16 hours a day and have difficulty cerebrating even when awake; and (3) collection of mucinous fluid in the tissue spaces between the cells, creating an edematous state called **myxedema.**

THE PANCREATIC HORMONES INSULIN AND GLUCAGON

The pancreas has thousands of small clusters of cells, called **islets of Langerhans,** one of which is illustrated in Figure 10-8. These islets are composed of four main cell types including:

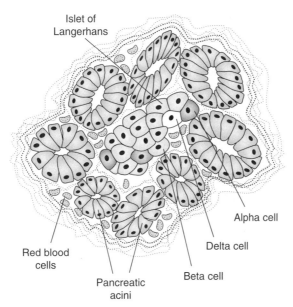

Islet of
Langerhans

Alpha cell

Delta cell

Red blood
cells

Beta cell

Pancreatic
acini

Figure 10-8.
Histology of islets of Langerhans in the pancreas.

- **Alpha cells, which secrete glucagon**
- **Beta cells, which secrete insulin**
- **Delta cells, which secrete somatostatin**
- **F cells, which secrete pancreatic polypeptide**

The precise roles of pancreatic somatostatin and pancreatic polypeptide have not been elucidated, but both **insulin** and **glucagon** are known to be important in the storage and use of the fuels and in regulating metabolism.

Insulin is a small protein with a molecular weight of about 6000 and is often referred to as the "**storage hormone**" because its secretion is greatly increased immediately after a meal and it causes cellular storage of all the different foodstuffs, including carbohydrate, fat, and protein. Glucagon, on the other hand, has effects that are in many ways opposite to those of insulin as will be discussed later.

Insulin regulates blood glucose concentration. Insulin is perhaps the most important hormone for maintaining normal blood glucose concentration. After a large meal, increased secretion of insulin prevents the glucose concentration from rising too high by causing storage of approximately 60% of the absorbed glucose in the liver and 15% in the skeletal muscle; the remaining glucose is used for energy. Then, between meals the secretion of insulin is markedly decreased and most of the glucose returns to the blood because of continual breakdown of glycogen in the liver when insulin is not present. This maintains a relatively constant blood glucose concentration in the fed and fasting states, which ensures a relatively constant rate of delivery of glucose to the brain. Because the brain cells utilize fats and proteins very poorly for energy, the precise regulation of blood glucose concentration helps to maintain a steady rate of neuronal activity.

Insulin promotes glucose storage in the liver. In the liver, glucose can rapidly diffuse across the cell membrane without the action of insulin; once glucose diffuses across the cell membrane, it is trapped in the hepatic cells by being converted first into **glucose-6-phosphate** and then into **glycogen.** Insulin promotes this effect by greatly increasing the activity of two liver enzymes, **glucokinase** and **glycogen synthetase.** Insulin also reduces hepatic glucose output by inhibiting **glycogenolysis** (breakdown of glycogen) rapidly by decreasing glycogen phosphorylase activity and gradually by decreasing glucose-6-phosphatase levels. In addition, insulin also inhibits **gluconeogenesis** by decreasing hepatic uptake of precursor amino acids, which are used to synthase glucose.

In skeletal muscle and in most other cells of the body besides the liver and the brain, insulin greatly stimulates glucose transport across the cell membrane. In the presence of large amounts of insulin, the cell membrane of resting skeletal muscle is about 15 times as permeable to glucose as it is when there is no insulin. Insulin also activates **glycogen synthetase** and **phosphofructokinase,** which cause glycogen synthesis and glucose utilization, respectively, in skeletal muscle. And finally, insulin also increases muscle blood flow due to its metabolic effects; in turn, increased blood flow causes more delivery of glucose to the skeletal muscle.

In adipose tissue, insulin stimulates the transport of glucose into the cells. This glucose is then used for esterification of fatty acids and permits their storage as triglycerides. In the brain, insulin has little or no effect on transport of glucose across the cell membrane. Instead, the rate of glucose diffusion to these cells is directly proportional to the blood glucose concentration.

Insulin reduces lipolysis and promotes storage of triglycerides in fat tissues. Insulin has both direct and indirect effects on fat metabolism. The direct effect is to reduce the rate of fatty acid release (lipolysis) from fat tissues in the body fluids. The mechanism of this is mainly intense depression of **hormone-sensitive lipase** by insulin, preventing the hydrolysis of triglycerides in the fat tissue and therefore preventing fatty acid release. When there is very little insulin, this enzyme becomes active and causes large quantities of fatty acids to be released into the tissue fluids. In this way, fatty acids become mobilized and are utilized for energy in place of glucose that cannot be effectively used in most tissues of the body without insulin.

The indirect effect of insulin on fat metabolism occurs secondarily to insulin-induced changes in carbohydrate metabolism. As discussed above, insulin has the same effect on fat cells that it has on muscle cells—to cause increased glucose transport into the cells. This causes some increase in formation of fatty acids in these cells and then storage of these in the form of triglycerides. However, the most important effect is that the glucose inside the fat cells is used to form the glycerol portion of the stored triglycerides.

In the liver, fatty acids are also synthesized from glucose under the influence of insulin. These fatty acids are then transported to the fat cells, where they combine with the glycerol to form still more triglycerides that are stored. In the absence of insulin, large amounts of fatty acids are released from the fat cells, and many of these are transported to the liver where they form excessive quantities of (1) stored fat in the liver; (2) cholesterol, phospholipids, and triglycerides that are released into the blood; and (3) aceto-acetic acid, which is also released into the blood. In prolonged periods of low insulin secretion (e.g., diabetes mellitus), the aceto-acetic acid can become so great that it causes severe acidosis.

Insulin is essential for normal growth and protein metabolism. Even growth hormone will not cause significant growth in an animal in the absence of insulin. Insulin probably has this effect because it increases the formation of protein in cells. It does this by promoting active transport of some of the amino acids through the cell membranes to the interior of the cell, by increasing the number of functional ribosomes for forming proteins, and by increasing the activity of the DNA-RNA system that controls protein formation. Thus, insulin is an anabolic hormone.

Insulin secretion is stimulated by increased glucose and increased amino acids in the blood. The rate of secretion of insulin by the pancreas is controlled mainly by the concentration of glucose in the circulating blood, but also by the concentration of amino acids. When glucose concentration increases, large quantities of insulin are secreted and the insulin, in turn, promotes the storage of glucose in the body cells, especially in the liver. When blood glucose concentration decreases, the rate of insulin secretion also decreases and glucose is then transported out of the liver back into the blood, helping to prevent excessive decreases in plasma glucose concentration.

Glucagon has effects opposite to those of insulin in the liver. In almost all respects, the actions of **glucagon** are exactly opposite those of insulin. Glucagon promotes mobilization rather than storage of fuels, especially glucose. The most important site of action of glucagon is in the liver where it has a hyperglycemic action resulting mainly from stimulation of **liver glycogenolysis.** Over a longer period of time, glucagon also causes marked increases in **liver gluconeogenesis,** which is the process by which amino acids are converted into glucose. Glucagon causes glucose release in the liver cells by the following mechanism. Glucagon activates adenylate cyclase in the liver cell membranes. This in turn causes formation of cAMP in the cells, which then activates phosphorylase, the enzyme that causes glycogen to split into glucose molecules.

When the blood concentration of glucose falls below normal, the pancreas secretes large amounts of glucagon. The blood glucose-raising effect of glucagon then helps to correct the hypoglycemia. Therefore, glucagon, like insulin, is also important for control of blood glucose concentration.

Diabetes Mellitus Is a Syndrome of Impaired Carbohydrate, Protein, and Fat Metabolism

There are two types of diabetes mellitus:

- **Insulin-dependent diabetes mellitus (IDDM, or type I)**, caused by insufficient secretion of insulin
- **Non–insulin-dependent diabetes mellitus (NIDDM, or type II)**, caused by resistance to the metabolic effects of insulin on the target tissue

In both of these types of diabetes mellitus, metabolism of all the basic foodstuffs is altered. The basic effect on glucose metabolism is to prevent the efficient uptake and utilization of glucose by most of the cells of the body except those of the brain. As a result, blood glucose concentration increases, cell utilization of glucose falls increasingly lower, and fat utilization increases.

IDDM is caused by impaired insulin production by the beta cells of the pancreas. Any disease or injury to the beta cells that impairs insulin production can lead to IDDM. Most often, beta cell injury is caused by an **autoimmune disorder** in which the beta cells are destroyed by the immune system, or in some cases by **viral infections** that destroy the beta cells. The usual onset is at about 12 years of age; for this reason IDDM is often called **juvenile diabetes mellitus.** There is considerable evidence that IDDM has a genetic basis, but environmental factors, especially viral infections, may also be involved.

The lack of insulin secretion results in four principal sequelae:

- **Blood glucose concentration rises to very high levels** because the efficiency of peripheral glucose utilization is reduced and glucose production is augmented. Increased plasma glucose then has multiple effects throughout the body, especially injury to the blood vessels and the kidneys. Diabetes is one of the leading causes of **end-stage kidney disease** and an important contributor to the mortality and morbidity associated with **cardiovascular diseases.** The high blood glucose also causes more glucose to filter into the renal tubules than can be reabsorbed. This creates an osmotic pressure in the tubules that prevents reabsorption of much of the tubular water, promoting very rapid diuresis and requiring the person to drink large amounts of water to maintain fluid balance. Thus, one of the classic symptoms of diabetes mellitus is **polyuria** (excessive urine excretion) and **increased thirst**.
- **Increased utilization of fats for energy** is a second effect of diabetes mellitus. This causes the liver to release acetoacetic acid into the plasma more rapidly than it can be taken up and oxidized by the tissue cells. As a result, the patient develops **severe acidosis** from the excessive acetoacetic acid, which, in association with dehydration due to the excessive urine formation, can cause **diabetic coma.** This leads rapidly to death unless the condition is treated immediately with large amounts of insulin.

- **Depletion of the body's proteins** is a third effect that occurs over a prolonged period of time. Also, failure to utilize glucose for energy leads to decreased storage of fat as well. Therefore, a person with severe untreated diabetes mellitus suffers rapid weight loss and often death within a few weeks without treatment.
- **Excess fat utilization in the liver** occurring over a long period of time causes large amounts of cholesterol in the circulating blood and increased deposition of cholesterol in the arterial walls. This leads to severe **arteriosclerosis** and other **vascular lesions.**

NIDDM is caused by resistance to the metabolic effects of insulin. NIDDM is by far more common than IDDM, accounting for about 80% to 90% of all diabetics. This syndrome is characterized by impaired ability of target tissues to respond to the metabolic effects of insulin, a condition referred to as **insulin resistance.**

There are multiple causes of insulin resistance, but the most common cause is **obesity.** In most cases, the onset of IDDM is gradual with relatively mild hyperglycemia occurring after ingestion of carbohydrates. However, higher rates of insulin secretion by the pancreas are required to maintain normal blood levels of glucose because of the insulin resistance. Because the average age of onset of NIDDM is 50 to 60 years, this syndrome has often been referred to as **adult-onset diabetes.**

In many instances, NIDDM can be effectively treated, at least in the early stages, with caloric restriction and weight reduction, and no exogenous insulin administration is required. However, in the later stages when the pancreatic beta cells can no longer secrete enough insulin to prevent hyperglycemia, insulin administration is required.

PARATHYROID HORMONE, CALCITONIN, VITAMIN D, PHOSPHATE METABOLISM, AND BONE

The control of calcium and phosphate metabolism, bone formation, and regulation of parathyroid hormone, calcitonin, and vitamin D are closely intertwined. Extracellular calcium ion concentration, for example, is determined by the interplay of potassium reabsorption, renal excretion, and bone uptake and release of calcium, each of which is regulated by hormones (Figure 10-9).

Almost all the calcium in the body (99%) is stored in the bone, with only about 1% in the body fluids. The bone, therefore, acts as a large reservoir for storing and releasing calcium in response to hormonal signals. The bones, however, do not have an inexhaustible supply of calcium. Therefore, over the long term, the intake of calcium must be in balance with calcium excretion by the gastroin-

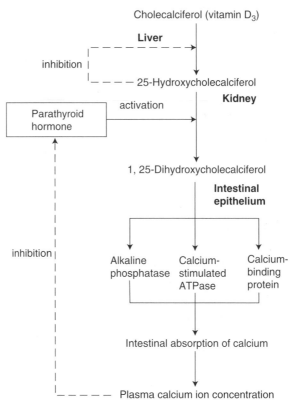

Figure 10-9.
Activation of vitamin D_3 to form 1,25-dihydroxycholecalciferol, and control of plasma calcium concentration by vitamin D.

testinal tract and the kidneys. The most important hormone regulators of calcium reabsorption at both of these sites are **parathyroid hormone (PTH)** and **vitamin D.**

Vitamin D increases calcium absorption from the intestinal tract. Although vitamin D has important effects on calcium absorption from the intestinal tract as well as storage and release of calcium from bones, vitamin D itself is not active in causing these effects until it is first converted through a succession of reactions in the liver and the kidney to the final active product, **1,25-dihydroxycholecalciferol:**

- **Vitamin D_3,** formed in the skin, is converted to 25-dihydroxycholecalciferol in the liver.
- **25-Hydroxycholecalciferol** is converted in the kidneys to 1,25-hydroxycholecalciferol, the active form of vitamin D. This conversion requires PTH, and in the absence of PTH almost none of the 1,25-dihydroxycholecalciferol is formed. Therefore, PTH has a potent effect in determining the functional effects of vitamin D, which in turn has several effects that increase gastrointestinal absorption of calcium.

Parathyroid hormone regulates plasma calcium concentration. PTH is secreted by the **parathyroid glands,** which are embedded in the surface of the thyroid gland, but are distinct from it. PTH secretion is controlled primarily by changes in plasma calcium concentration; decreased plasma calcium concentration stimulates PTH secretion whereas increased calcium concentration reduces PTH secretion (Fig. 10-10).

PTH, in turn, has several actions that tend to increase plasma calcium concentration:

- **PTH increases the activity of osteoclasts,** present in the bone cavities, which secrete hydrogen ions and enzymes that cause breakdown (resorption) of bone. This results in movement of calcium and phosphate from bone into the extracellular fluid.
- **PTH stimulates the activation of vitamin D,** which increases intestinal calcium reabsorption.
- **PTH increases renal tubular calcium reabsorption,** thereby reducing urinary excretion of calcium.
- **PTH reduces phosphate reabsorption by the kidney,** thereby increasing excretion of phosphate and lowering extracellular phosphate concentration. Although PTH releases phosphate from the bones, PTH does not increase the plasma phosphate concentration because of the increased loss of phosphate by the kidneys.

The regulation of calcium ion concentration by PTH is very important because all of the excitable tissues in the body, including nerves, skeletal muscles, the heart, and smooth muscles, depend on well-regulated calcium ion concentration for normal function. For instance, low calcium ion concentration (**hypocalcemia**) causes extreme increase in irritability of the peripheral nerves so that they begin to emit impulses spontaneously, causing a state of continual muscle contraction called **tetany.** Also, greatly decreased calcium ion concentration reduces the contractility of cardiac muscle. PTH secretion is the principal method by which normal calcium ion concentration is maintained in the body fluids, and the bones act as a large reservoir of calcium to be used for this purpose.

Calcitonin decreases calcium ion concentration. The hormone calcitonin, when injected into an animal, causes rapid deposition of

Figure 10-10.
Compensation for decreased plasma ionized calcium concentration by parathyroid hormone and vitamin D.

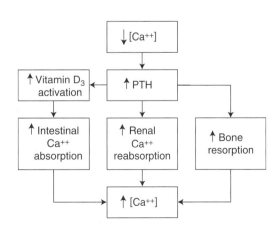

calcium into the bones and therefore rapid decrease in calcium ion concentration of the body fluids. This hormone is secreted by special cells called **parafollicular cells,** located between the follicles of the thyroid gland in the human. Its rate of secretion increases when the calcium ion concentration rises above normal. Therefore, it functions in exactly the opposite manner to PTH, returning calcium ion concentration back towards normal when this concentration rises too high. Further, it responds more rapidly than does PTH. However, the quantitative role of calcitonin in calcium ion regulation is far less than that of parathyroid hormone. Also, its effect usually does not continue for more than a few hours to a few days.

Formation of bone by osteoblasts. Bone is formed by osteoblasts, which line the outer surfaces of all bones and are also present inside most of the bone cavities. The osteoblasts secrete a very strong **protein bone matrix,** comprised mainly of collagen fibers, which gives the bone its toughness. This matrix has the special property of causing phosphate ions to combine with calcium ions, precipitating a complicated salt of calcium and phosphate called **hydroxyapatite** in the protein matrix to make it extremely hard. The strength of the collagen fibers gives the bone tremendous tensile strength, while the hardness of bone gives it compressive strength.

When calcium is not available in large quantities in the body fluids, bone is poorly formed. Calcium is not well absorbed from the gastrointestinal tract, and when it fails to be absorbed, phosphate is also poorly absorbed because the two substances form insoluble compounds in the gut. However, as explained previously, **vitamin D** greatly increases calcium absorption, which in turn allows increased phosphate absorption. Therefore, lack of vitamin D reduces the amount of available calcium and phosphate for bone formation and can result in poor mineralization of the bones so that they no longer resist compressive forces. This is the disease known as **rickets.**

Because vitamin D must be converted by the liver and kidney from its natural form into the substance 1,25-dihydroxycholecalciferol before it is active, severe liver disease, kidney disease, and lack of parathyroid hormone can all lead to diminished calcium absorption by the intestinal tract and weakened bones.

Resorption of bone by osteoclasts. Bone is continually being broken down (resorbed) by large numbers of **osteoclasts** present in the bone cavities. The osteoclasts are large multinucleated cells that secrete hydrogen ions (which dissolve the crystals) and hydrolytic enzymes that digest the collagen matrix. This resorption of bone has two major functions: (1) In all bones the protein matrix becomes aged and loses its toughness, thereby allowing the bones to become brittle. The osteoclastic resorptive process removes the aging bone, which is then continuously replaced through new bone formation by the osteoblasts. (2) Resorption of bone also provides a means by which calcium ions can rapidly be made available to the extracellular fluids.

REPRODUCTIVE FUNCTIONS OF THE MALE

Spermatogenesis—the formation of sperm by the testes. A basic reproductive function of the male is the formation of sperm by the testes. The seminiferous tubules of the testes contain a basal layer of **germinal epithelium**, the cells of which divide through several stages and gradually form the sperm (Fig. 10-11). During sperm formation, essentially all of the cytoplasm is lost from the cell and the cell membrane elongates in one direction to form a tail. Also, at one stage of division, the 23 chromosome pairs (46 chromosomes) split into two unpaired sets of 23 chromosomes, one of these sets going to one sperm and the other to the second sperm. One pair of chromosomes, called the **XY pair**, is known as the sex pair. After separation of this chromosome pair in the process of sperm formation, half of the sperm carry an X chromosome and the other half a Y chromosome. The X chromosome causes a female child to be formed while the Y chromosome causes a male child to be formed. Thus, the sex of the offspring is determined by the type of sperm that fertilizes the ovum. Furthermore, this division of chromosomes allows half of the genes of the father to be inherited by the child while the other half comes from the mother.

In addition to the germinal cells, the seminiferous tubules contain large **Sertoli cells** from which the developing sperm obtain nutrient substances during their development.

Transport of sperm, storage, and ejaculation. After sperm are formed in the seminiferous tubules, they pass into the **epididymis** where they remain for approximately 1 day while they mature within this new environment. From there, they pass into the **vas deferens** and the **ampulla,** where they are stored. During coitus, sexual stimulation transmits impulses to the spinal cord that cause reflex erection of the penis mediated by parasympathetic impulses. Then at the height of sexual stimulation, rhythmic peristalsis, which is mediated by the sympathetic nervous system, begins in the epididymis and spreads up the vas deferens, into the ampulla and seminal vesicles, and finally through the prostate gland. This expels the semen into the posterior urethra, a process called **emission.** Then, rhythmical contractions of the bulbocavernosus muscle cause rhythmic compression of the urethra, and about 3 ml of semen are expelled. This process is called **ejaculation.**

The **ejaculate** is composed of a mixture of (1) **sperm** from the testes; (2) a highly **mucid fluid** from the seminal vesicles; and (3) a highly **alkaline fluid** from the prostate gland. The alkalinity of the prostate fluid causes the sperm to become immediately motile by activating movement of the sperm tail, permitting the sperm to travel at a velocity of as much as 1 to 4 mm/min, and to pass through the uterus and the fallopian tubes to fertilize the ovum.

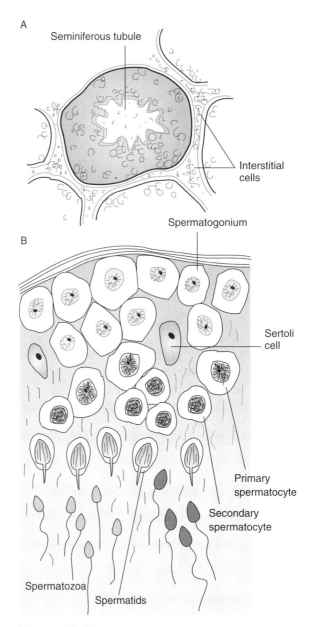

Figure 10-11.
A: Cross-section of a seminiferous tubule; *B:* stages
in the development of sperm from spermatogonia.

Failure of the male to expel more than 60 to 80 million sperm in each ejaculate usually results in sterility. This is probably caused by the lack of sufficient hyaluronidase and various proteolytic enzymes that are secreted by the sperm. These enzymes dissolve the mucus plug of the female cervix and possibly also help to break granulosa cells away from the surface of the ovum to allow the sperm to penetrate the ovum.

Testosterone is formed by the interstitial cells of the testis. Large amounts of testosterone are formed by the **interstitial cells of**

Leydig, which lie in the interstitial spaces between the seminiferous tubules of the testes. This testosterone has the local effect in the testes of promoting sperm production by the seminiferous tubules; without it spermatogenesis cannot occur. However, testosterone is also secreted into the blood and causes development of the sexual and secondary sexual characteristics of the male, including the following:

1. Formation of the penis when the fetus is developing
2. Descent of the testes into the scrotum in the adult male
3. Development of the enlarged musculature of the male
4. Increase in the thickness of the skin
5. Deepening of the voice resulting from growth of the larynx
6. Growth of the beard on the face and hair in many areas of the body
7. Baldness in those male individuals who are genetically predisposed to this condition

Testicular function is regulated by the anterior pituitary gland. At least two gonadotropic hormones, **FSH** and **LH,** are secreted by the pituitary gland to help control spermatogenesis and testosterone secretion by the testes. During childhood, almost no gonadotropic hormones are secreted by the anterior pituitary gland, but at puberty both FSH and LH begin to be secreted. The FSH promotes division of the germinal cells to initiate spermatogenesis, while LH stimulates the interstitial cells to produce testosterone. The testosterone, in turn, is required for proper development and maturation of the sperm. The testes continue to produce both sperm and testosterone from puberty until death, although beyond approximately age 40, the rates of production gradually decline.

GnRH controls gonadotropic hormone (LH and FSH) secretion. The rate of secretion of gonadotropic hormones by the anterior pituitary is controlled mainly by GnRH. This hormone causes both LH and FSH to be released from the anterior pituitary gland.

Testosterone secreted by the testes inhibits the formation of LH-releasing hormone (LHRH) by the hypothalamus, thus providing a negative feedback mechanism for control of testosterone secretion. In addition, a substance called **inhibin** is secreted by the Sertoli cells and suppresses FSH secretion by the anterior pituitary, thus providing a feedback mechanism for controlling sperm formation.

During childhood, the rate of secretion of LHRH is very low because the hypothalamus is extremely sensitive to the inhibitory effects of even small amounts of circulating testosterone. However, at approximately the age of 12 years, the hypothalamus loses most of this inhibitory sensitivity and large amounts of LHRH then begin to be secreted, and the sex life of the male begins. This initiates the period of **puberty.**

REPRODUCTIVE FUNCTIONS OF THE FEMALE

Ovarian cycle and oogenesis—the formation of ova. The newborn female has about two million **primordial ova** in her two ovaries. Many of these degenerate during childhood and only about 300,000 to 400,000 remain at puberty. At that time, under the stimulation of gonadotropic hormones from the anterior pituitary, the rhythmic monthly sexual cycle begins. At the beginning of each month, the cells surrounding a few of the ova, the **granulosal** and the **thecal cells,** begin to proliferate, and these secrete large quantities of **estrogen,** one of the female sex hormones. Fluid is secreted by the granulosa and thecal cells, forming cavities around these few ova called **follicles.** After approximately 14 days of growth, one of the growing follicles breaks open and expels its ovum into the abdominal cavity. Then, all of the other growing follicles begin to degenerate within a few hours, a process called **atresia.** Presumably, this results from an inhibitory hormone action on the ovaries after ovulation from the first follicle occurs. The result is normally the release of one single ovum each month at approximately the 14th day of the female sexual cycle.

Immediately after the ovum has been expelled, the granulosal and thecal cells undergo rapid fatty changes and considerable swelling, a process called **luteinization,** and they begin to secrete large amounts of progesterone in addition to estrogens. This modified mass of cells, now called a **corpus luteum,** persists for another 14 days, after which time it degenerates. Then a new set of follicles begins to develop and at the end of another 14 days another ovum is expelled into the abdominal cavity, with the cycle continuing on and on.

At about the same time the ovum is expelled from the follicle, the **nucleus of the ovum** divides two times in rapid succession. During one of these divisions, the pairs of chromosomes separate and half of them are expelled from the ovum, leaving **23 unpaired chromosomes** in the final mature ovum, which is then ready for fertilization.

Estrogens and progesterone cause proliferation of the uterine endometrium. The estrogens and progesterone secreted by the ovaries prepare the uterine endometrium for implantation of the fertilized ovum. The **estrogen** secreted during the first half of the monthly ovarian cycle causes very rapid proliferation of the endometrial stroma and glandular cells. Then during the second half of the monthly cycle, **progesterone** causes the stromal and glandular cells to begin secreting a serous fluid while the stromal cells store large quantities of protein and glycogen in preparation for supplying nutrition to the developing ovum.

Menstruation. When the corpus luteum degenerates at the end of the monthly cycle, very little estrogen or progesterone is secreted by the ovaries for the next few days. Lack of the normal stimulatory effect of these hormones causes the endometrial cells to lose their stimulus for increased activity. This results in rapid necrosis of the superficial two-thirds of the endometrium, with the dead tissue sloughing away and being expelled through the vagina along with about 40 ml of blood and at least as much additional serous exudate. This process, called menstruation, normally lasts about 4 days. By the end of **menstruation,** new follicles have begun to develop and the ovaries have begun to secrete estrogens once again. Under the influence of these estrogens, the endometrium begins a new cycle of development.

Effects of estrogens and progesterone on other tissues. Estrogens and progesterone, in addition to their effects on the endometrium, have other effects throughout the body, including the following:

1. Estrogens cause proliferation and enlargement of the smooth muscle cells in the uterus, **increasing the uterine size** after puberty to about double the childhood size.
2. Estrogens also cause **proliferation of the glandular cells of the breast and deposition of fat** in the breast tissue, the hips, and other points that give the **characteristics of the adult female.**
3. Estrogens and progesterone cause very **rapid growth of bones** immediately after puberty, but also promote early uniting of the epiphyses with the shafts of the long bones so that the final height of the female, despite her rapid growth immediately after puberty, is less than it otherwise would have been.
4. Estrogens and progesterone cause **enlargement of the external genitalia.**

Progesterone has very much the same effect on the breast that it has on the uterine endometrium, causing the glandular cells to increase in size and to develop secretory granules in their cells. In addition, progesterone causes accumulation of fluid and electrolytes in the breast tissue, making them swell during the latter half of each monthly sexual cycle.

Regulation of the female sexual cycle by the hypothalamus and anterior pituitary gland. Until the female is about 12 years of age, the **anterior pituitary** secretes no gonadotropic hormones, as is also true in the male. This is thought to be caused by a very high sensitivity of the **hypothalamus** to the inhibitory effect of even small amounts of estrogen and progesterone secreted by the ovaries. However, at puberty, the hypothalamus loses this inhibitory sensitivity and begins to secrete LHRH in the same manner that this occurs in the male. This hormone, in turn, stimulates the secretion of both LH and FSH in the monthly cycles by the anterior pituitary gland as shown in the lower curve of Figure 10-12. The FSH causes initial growth of the ovarian follicles during the first few days of the monthly cycle. Then this hormone, aided by LH as well, causes the thecal cells and possibly also the granulosal cells to secrete estrogen (the "estradiol" curve in the upper part of the figure) plus large quantities of fluid into the developing follicles.

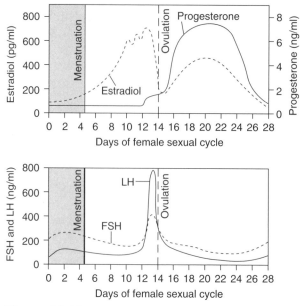

Figure 10-12.
Approximate plasma concentration of the gonadotropins and ovarian hormones during the normal female sexual cycle.

At about the **13th day** of the ovarian cycle, a very large amount of LH is secreted by the anterior pituitary, which is called the **LH surge.** The excess LH, in some way not completely understood, causes ovulation about 24 hours later. The LH also causes the granulosal and thecal cells to change into **lutein cells,** which in the aggregate become the corpus luteum. LH then stimulates the corpus luteum to produce large amounts of both progesterone and estrogen during the latter half of the female sexual cycle. Finally, when the corpus luteum degenerates at the end of the cycle, the resulting lack of progesterone and estrogen leads to menstruation as described above.

The exact mechanisms that control secretion of FSH, LH, estrogen, and progesterone during the female sexual cycle are not completely understood. However, it is known that estrogen and progesterone normally cause feedback inhibition of LHRH secretion by the hypothalamus. Therefore, during the latter part of the ovarian cycle, when large amounts of progesterone and estrogen are secreted by the corpus luteum, secretion of both FSH and LH by the anterior pituitary decreases markedly. This, in turn, leads to degeneration of the corpus luteum and greatly decreased production of progesterone and estrogen. Next, lacking the feedback inhibition of these two hormones, the hypothalamus and pituitary gland become active once again during the next few days, and the rates of secretion of FSH and LH rise again, thus beginning a new cycle.

Menopause. At 40 to 50 years of age, essentially all of the ova in the ovaries have been used up, a few expelled into the abdominal cavity by ovulation and large numbers degenerated in situ in the ovaries. Therefore, no follicles or any corpus luteum can develop in the ovaries to secrete either estrogens or progesterone. The anterior pituitary gland continues to secrete large amounts of gonadotropic hormones, but since no estrogen or progesterone can be secreted to inhibit the hypothalamus or the pituitary, no monthly sexual cycle occurs thereafter.

PHYSIOLOGY OF PREGNANCY

Fertilization of the ovum and implantation in the endometrium. After coitus, millions of motile sperm make their way up to the uterus and fallopian tubes. The sperm are capable of living in the genital tract of the female for as long as 72 hours but are very fertile for only about 24 hours. If during this time an ovum is expelled from the ovary, or if an ovum has been expelled up to 24 hours prior to coitus, then a sperm can cause fertilization.

In the process of fertilization, the head of the sperm combines with the nucleus of the ovum. Since each of these contains 23 unpaired chromosomes, the combination restores the normal cellular complement of 23 pairs of chromosomes. The fertilized ovum then contains 44 autosomal chromosomes and either 2 X chromosomes, causing a female child to develop, or an X and a Y chromosome, causing a male child to develop. After fertilization, the process of division begins, the first division occurring approximately 30 hours after fertilization. Subsequent divisions occur at a rate of about once every 18 to 24 hours.

Shortly before fertilization the ovum usually passes to one of the two fallopian tubes, the fimbriated ends of which lie in approximation to the ovaries. The cilia that line the fallopian tube beat toward the uterus and slowly move the dividing ovum downward along the tube to reach the uterus in about 3 days.

The dividing ovum develops an outer layer of **trophoblast cells.** These are capable of phagocytizing nutrient materials from the secretions of the fallopian tube and uterus, thus making nutrients available to the developing mass of cells. The trophoblast cells also secrete proteolytic enzymes that allow the developing cell mass to eat its way into the endometrium and thereby implant itself. Once implantation occurs, the trophoblastic and underlying cells proliferate rapidly.

Chorionic gonadotropin is secreted by the trophoblast cells. When the corpus luteum degenerates at the end of the normal monthly menstrual cycle, the endometrium of the uterus sloughs away and menstruation occurs. However, when the ovum becomes fertilized, it is important that the endometrium remain intact for the early developing fetus to implant and grow. Fortunately, the trophoblast cells secrete a hormone called **human chorionic gonadotropin (HCG),** which has the same effects on the corpus luteum as LH secreted by the pituitary gland. Therefore, this hormone keeps the corpus luteum from degenerating and keeps it secreting large quantities of estrogens and progesterone; as a result, menstruation does not occur. Instead, the endometrium actually grows thicker and the large endometrial cells, called **decidual cells,** are gradually phagocytized by the growing fetal tissues, providing the major portion of the nutrition for the fetus during the first 8 to 12 weeks of pregnancy.

After the first 2 to 4 months of pregnancy, the placenta begins to secrete large amounts of estrogen and progesterone (Fig. 10-13). From then on, the hormones from the corpus luteum are not needed, and it degenerates.

The placenta serves as an organ of exchange between the mother and the fetus. During the early weeks of pregnancy, the trophoblast cells and other fetal tissues gradually develop the placenta. This organ contains multiple large chambers filled with the mother's blood and into these chambers project millions of small villi containing blood capillaries from the fetus. Trophoblast cells cover the surfaces of the villi, and these actively reabsorb many nutrients from the mother's blood and transport them into the fetal blood during the early weeks of pregnancy. However, after 4 to 8 weeks most of the necessary nutrients are absorbed passively from the mother's blood into the fetal blood. This occurs because the concentrations of the nutrients are greater in the mother's blood than in the fetal blood, and as a result they can diffuse through the placental membrane into the fetal blood. Conversely, waste products of metabolism, such as urea, uric acid, and creatinine, accumulate in higher concentrations in the fetal blood and diffuse through the placental membrane into the mother's blood and then are excreted by the mother's kidneys.

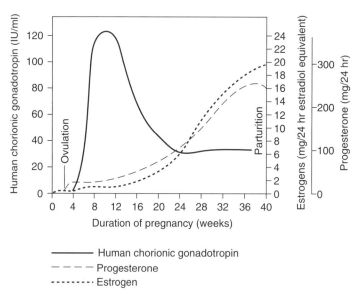

Figure 10-13.
Rates of secretion of estrogens, progesterone, and chorionic gonadotropin at different stages of pregnancy.

The placenta secretes HCG, estrogen, and progesterone. In addition to secreting HCG, the placenta also secretes other important hormones, especially estrogens and progesterone. After approximately the third month of pregnancy, the secretion of HCG becomes greatly reduced and the corpus luteum begins to degenerate. From that time onward, the estrogens and progesterone from the placenta are essential for the maintenance of pregnancy. Toward the end of pregnancy, the rate of secretion of active estrogens is about 30 times that during the normal ovarian cycle, and the rate of secretion of progesterones is about 10 times as great (see Fig. 10-13). The estrogens and progesterone are essential for growth and development of the fetus.

The progesterone secreted by the placenta is formed from cholesterol derived from the mother's blood. However, secretion of estrogens by the placenta requires a two-stage process. The first stage is the formation of large quantities of androgens by greatly enlarged adrenal cortices in the fetus. These androgens are then carried into the blood to the placenta where they are converted into several different types of estrogens, including **estradiol,** the most potent of all of the estrogens.

The placenta produces large amounts of human chorionic somatomammotropin. This hormone has several important effects: (1) It promotes growth of the fetus. (2) It causes increased use of fatty acids by the mother for energy and decreased use of glucose; this makes the excess glucose of the mother available for use by the fetus. (3) It aids in the growth and development of the breasts during pregnancy, thus preparing the breast for lactation following birth of the infant.

Growth of the fetus is slow during the first 4 weeks and rapid during the last 3 months of pregnancy. During the first few weeks of pregnancy, the fetus hardly grows, although the surrounding fetal membranes, especially the placenta, develop very rapidly. After 4 to 5 weeks, however, the length of the fetus increases approximately in proportion to the time of gestation, and the weight increases with the cube of the time. Thus, at 6 months, the length is approximately six-ninths the final length, but the weight is still only about one-fourth the final weight. Thus, the greatest growth in weight of the fetus occurs in the last 3 months, and during this time pregnancy makes many demands on the mother for nutritive substances needed by the infant, including especially proteins, vitamins, large amounts of calcium for the bones, and iron for the red blood cells.

PARTURITION

Parturition is initiated by uterine contractions. When the fetus is fully formed, approximately 9 months after fertilization, the uterus becomes more excitable than usual, labor begins, and the infant is expelled. This is called **parturition.** The precise factors that initiate parturition are uncertain, but a combination of several different factors progressively increases the excitability of the uterine musculature as follows:

- Near term, the placenta begins to secrete a progressively higher ratio of estrogens to progesterone. Since estrogens normally excite uterine activity while progesterone inhibits it, this change in ratio increases the excitability of the uterine musculature.
- The fetus itself increases in size, which stretches the uterus, thus increasing its excitability.
- The head of the fetus presses downward against the cervical opening of the uterus and begins to stretch the cervix; this also increases the excitability of the uterus.
- The posterior pituitary gland begins to secrete additional amounts of **oxytocin** and at the same time the sensitivity of the uterus to oxytocin increases greatly, both of which increase the excitability of the uterine musculature.

As a result of all of these factors, the rhythmic contractions of the uterus become stronger and stronger. Finally, they become strong enough to begin pushing the baby into the birth canal. This stretches the cervix very rapidly, causing still greater increase in the excitability of the uterus and making it contract even harder. Also, sensory signals from the cervix to the hypothalamus cause progressively increasing secretion of oxytocin, which excites the uterus even more. Thus, a positive feedback cycle is set up as follows: Strong uterine contraction stretches the cervix, which stimulates even stronger uterine contraction, causing still more stretch of the cervix, and so forth until the baby is expelled.

Changes occur in the infant immediately after birth. Prior to birth, the infant receives its nutrition and oxygen through the placenta. Normally, the first function performed by the newborn is rapid expansion of its lungs and onset of respiration to oxygenate its own blood. An infant can usually go as long as 4 to 6 minutes without breathing before neuronal cells in the brain are damaged.

In the fetus, blood bypasses the lungs by two routes (Fig. 10-14): (1) Some blood flows directly from the right atrium through the **foramen ovalae** directly into the left atrium. (2) Most of the remaining blood that does not take this route is pumped by the right ventricle into the pulmonary artery and then through the **ductus arteriosus** directly into the aorta rather than through the lungs.

Growth of the infant changes these directions of blood flow in the following ways: (1) **Loss of blood flow through the placenta** after birth greatly increases the total peripheral resistance in the infant's systemic circulation. (2) **Expansion of the lungs** expands the pulmonary blood vessels, and this greatly reduces the resistance to blood flow through the pulmonary circulation. As a result, the ratio of resistance in the systemic circulation to resistance in the pulmonary circulation increases severalfold, allowing much easier flow of blood through the lungs but considerably more difficult flow through the systemic circu-

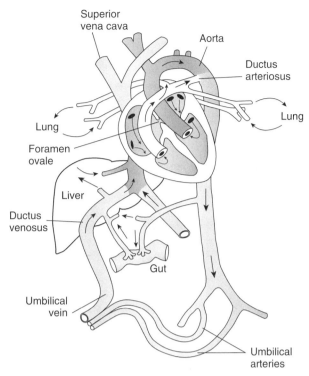

Figure 10-14.
The fetal circulation.

lation. Because of this, the pulmonary artery pressure falls while the systemic arterial pressure rises so that blood now begins to flow backward from the aorta through the ductus arteriosus rather than forward. This brings arterialized blood, containing a high oxygen concentration, into contact with the ductus, and the oxygen constricts the ductus, causing functional closure within a few hours. Then fibrous tissue grows into the ductus and causes **permanent closure in 1 to 2 months** in all infants, except one in several thousand.

Also, immediately after birth, the increased resistance in the systemic circulation raises the load on the left heart and therefore increases the left atrial pressure. At the same time, the decreased resistance in the lungs reduces the right atrial pressure. This higher pressure in the left atrium than the right atrium closes a valve-like structure over the foramina ovalae preventing backflow through this route. Thus, these two changes in the circulatory system provide normal blood flow through the lungs.

LACTATION

During pregnancy, large amounts of estrogens and progesterone are secreted either by the corpus luteum or the placenta. **Estrogens** cause proliferation of the glandular tissues of the breasts and **progesterone** causes development of the alveoli and storage of nutrient materials in the glandular cells.

Other hormones that help to promote breast development during pregnancy include **prolactin** and **growth hormone** from the mother's anterior pituitary gland, **insulin** from her pancreas, **glucocorticoids** from her adrenal glands, and **human chorionic somatomammotropin** from the placenta. However, progesterone and estrogens also inhibit milk production despite their effects on breast tissue proliferation. Therefore, before birth of the infant, the mother does not secrete milk. Loss of the placenta from the mother's body when the infant is born removes the source of progesterone and estrogen so that the breasts are no longer inhibited and within 24 to 48 hours milk begins to flow.

During pregnancy, the mother's anterior pituitary produces increased amounts of prolactin, increasing to about ten times the normal rate of secretion. This hormone is required to cause final development of the breasts and also to cause them to secrete milk. After the birth of the infant, continued suckling of the breast stimulates the anterior pituitary gland to continue secreting large quantities of prolactin and this, in turn, causes the breasts to continue producing milk. When milk is no longer needed by the child and is no longer removed from the breast, the anterior pituitary gland stops producing prolactin and milk production ceases within a few days.

Oxytocin secreted by the posterior pituitary gland during the suckling process is also important for lactation, causing milk ejection from the breast alveoli.

PHYSIOLOGY QUESTIONS

DIRECTIONS: Each of the numbered items or incomplete statements in this section is followed by answers or by completions of the statement. Select the ONE lettered answer or completion that is BEST in each case.

1. A normal individual on a diet high in K^+ exhibits increased K^+ excretion. The major cause of this increased renal excretion of K^+ is

   (A) decreased aldosterone secretion
   (B) decreased reabsorption of K^+ by the loop of Henle
   (C) decreased reabsorption of K^+ by the proximal tubule
   (D) increased glomerular filtration rate
   (E) increased secretion of K^+ by the distal and collecting tubules

2. Damage to Wernicke's area in the dominant hemisphere is likely to make a person unable to

   (A) hear high-frequency sounds
   (B) observe the details of a beautiful scene
   (C) perform complex mathematical functions
   (D) speak coherently

3. A heart murmur that is present during systole suggests which of the following conditions?
   (A) Aortic insufficiency
   (B) Aortic stenosis
   (C) Mitral stenosis

4. Dark adaptation is caused primarily by which of the following factors?

   (A) Decreasing concentrations of photopsin
   (B) Decreasing concentrations of retinal
   (C) Decreasing concentrations of scotopsin
   (D) Increasing concentrations of rhodopsin
   (E) Increasing concentrations of vitamin A

5. Which ion is most closely associated with fluid secretion?

   (A) Bicarbonate
   (B) Chloride
   (C) Phosphate
   (D) Potassium
   (E) Sodium

6. The renal-body fluid volume mechanism for regulating arterial pressure is important for

   (A) increasing arterial pressure during strenuous physical exercise
   (B) maintaining arterial pressure at a normal level over a period of weeks, months, or years
   (C) minimizing a decrease in arterial pressure immediately following severe hemorrhage
   (D) raising the pressure when a person stands suddenly after being in a supine position

7. High plasma levels of thyroxine are associated with which of the following effects?

   (A) Depleted fat stores
   (B) Exophthalmos
   (C) Increased body weight
   (D) Somnolence

8. Which of the following statements about the essential amino acids is correct?

   (A) They are all found in all dietary proteins
   (B) They are necessary to provide adequate amounts of ATP
   (C) They can be formed in the body
   (D) They must be present in the diet

9. The most important factor for regulating cerebral blood flow under normal conditions is the

   (A) degree of sympathetic stimulation of peripheral vasculature
   (B) rate of cerebral carbon dioxide formation
   (C) rate of cerebral oxygen consumption
   (D) rate of release of adenosine from the cerebrum
   (E) rate of release of potassium from the cerebrum

10. A sudden loud sound is more likely to damage the cochlea than a loud sound that develops slowly because

    (A) a sudden sound carries more energy
    (B) the basilar fibers are sensitive to sudden sounds but adapt to slowly developing sounds
    (C) the fluid pressure in the scala tympani decreases as a sound becomes louder
    (D) the tympanic membrane becomes flaccid as a sound becomes louder
    (E) there is a latent period before the attenuation reflex can occur

11. Which one of the following anticoagulants is preferred by the blood bank for blood storage?

    (A) Antivitamin K agents
    (B) Citrate
    (C) Coumarin
    (D) Heparin
    (E) Oxalate

12. The P_{O_2} in the arterial blood is normally lower than that of the alveolar gas primarily because

    (A) blood moves through the lungs too rapidly
    (B) oxygen requires a pressure gradient to diffuse into blood
    (C) some portions of the lungs are perfused but not ventilated
    (D) the lungs use oxygen

13. Which of the following is a characteristic result of sympathetic stimulation?

    (A) Decreased blood pressure
    (B) Increased peristalsis
    (C) Pupillary constriction
    (D) Slowing of the heart
    (E) Sweating

14. Arteriosclerosis is associated with an increase in pulse pressure because

(A) compliance of the arterial tree is decreased
(B) stroke volume is decreased
(C) total peripheral resistance is decreased
(D) vascular conductance is increased

Questions 15–17

A 35-year-old man weighing 60 kg has an extracellular fluid volume of 12.8 L, a blood volume of 4.3 L, a hematocrit of 40%, and 57% of his body weight is water. Answer the following questions based on this information.

15. The man's intracellular fluid volume is approximately

(A) 17.1 L
(B) 19.6 L
(C) 21.4 L
(D) 23.5 L
(E) 25.6 L

16. The man's plasma volume is approximately

(A) 2.0 L
(B) 2.3 L
(C) 2.6 L
(D) 3.0 L
(E) 3.3 L

17. The interstitial fluid volume is approximately

(A) 6.4 L
(B) 8.4 L
(C) 10.2 L
(D) 11.3 L
(E) 12.0 L

18. The blood vessels of the systemic circulation responsible for most of the resistance to blood flow in the circulation are the

(A) aorta and large arteries
(B) arterioles
(C) capillaries
(D) venae cavae and large veins
(E) venules

19. A 25-year-old woman presents at your office with an enlarged, swollen neck, which you suspect to be caused by an enlarged thyroid gland (endemic goiter). Which of the following conditions would you expect this patient to have?

    (A) High plasma levels of thyroid-stimulating hormone
    (B) High plasma levels of thyroxine
    (C) Low production of thyroglobulin
    (D) Mental retardation

20. The T wave of the normal electrocardiogram is caused by

    (A) atrial depolarization
    (B) atrial repolarization
    (C) ventricular depolarization
    (D) ventricular repolarization

21. Penile erection is caused primarily by

    (A) contraction of the bulbocavernosus muscle
    (B) parasympathetically induced dilation of the arterioles
    (C) reflex sympathetic constriction of the arterioles
    (D) reflex parasympathetic constriction of the venules
    (E) sympathetic induced constriction of the veins

22. At which time in the cardiac cycle does the period of isometric contraction occur?

    (A) During the P wave of the cardiac cycle
    (B) During the maximum ventricular ejection
    (C) During the second heart sound
    (D) When the aortic valve is closed

23. The formation of proteins on the ribosomes is a process called

    (A) replication
    (B) transcription
    (C) transduction
    (D) translation

24. Sympathetic stimulation of which vessels causes the greatest increase in total peripheral resistance?

    (**A**) Arteries
    (**B**) Arterioles
    (**C**) Capillaries
    (**D**) Veins
    (**E**) Venules

25. Which of the following stimulates milk production by breasts?

    (**A**) Dopamine
    (**B**) Estrogen
    (**C**) Oxytocin
    (**D**) Prolactin

26. Which of the following factors most likely affects myocardial blood flow to the greatest extent under normal conditions?

    (**A**) Degree of parasympathetic stimulation of coronary vessels
    (**B**) Degree of sympathetic stimulation of coronary vessels
    (**C**) Myocardial carbon dioxide concentration
    (**D**) Rate of release of adenosine from the myocardium
    (**E**) Rate of release of potassium from the myocardium

27. Suppose you have a patient who has a gradual loss of functional nephrons due to kidney disease, which reduced the glomerular filtration rate (GFR) from 100 to 50 ml/min. Which of the following changes would you expect to find under steady-state conditions after the decline in GFR, when compared with the patient's condition before the loss of kidney function?

    (**A**) A doubling of plasma creatinine concentration
    (**B**) A 50% decrease in urinary creatinine clearance
    (**C**) A 50% reduction in urinary creatinine excretion
    (**D**) A doubling of plasma creatinine concentration and a 50% decrease in urinary creatinine clearance
    (**E**) A doubling of plasma creatine concentration, a 50% decrease in urinary creatinine clearance, and a 50% reduction in urinary creatinine excretion

28. The number of chromosomes present in a normal mature sperm cell is

(**A**) 21
(**B**) 23
(**C**) 46
(**D**) 92

29. Autoregulation of tissue blood flow in response to an increase in arterial pressure occurs as a result of

(**A**) a decrease in tissue metabolism
(**B**) a decrease in vascular resistance
(**C**) an initial decrease in vascular wall tension
(**D**) excess delivery of nutrients such as oxygen to the tissues

30. Almost all of the active thyroid hormone entering the circulation is in the form of

(**A**) long-acting thyroid stimulator
(**B**) thyroglobulin
(**C**) thyrotropin
(**D**) thyroxine
(**E**) triiodothyronine

31. The cell membrane is least permeable to which of the following substances?

(**A**) Carbon dioxide
(**B**) Ethanol
(**C**) Oxygen
(**D**) Sodium
(**E**) Water

32. At which period is the rate of estrogen secretion highest?

(**A**) 3 days before ovulation
(**B**) 1 day after ovulation
(**C**) During menstruation
(**D**) 1 day before menstruation

33. Reabsorption of fluid by the renal peritubular capillaries can be increased by

(**A**) decreased filtration fraction
(**B**) decreased plasma colloid osmotic pressure
(**C**) decreased plasma protein concentration
(**D**) efferent arteriolar constriction

34. Which one of the following would result in an increase in tissue blood flow?

(**A**) A decrease in tissue carbon dioxide concentration
(**B**) A decrease in tissue lactic acid concentration
(**C**) An increase in tissue adenosine concentration
(**D**) An increase in tissue oxygen concentration

35. Which of the following hormones increases hepatic glycogenolysis and promotes gluconeogenesis?

(**A**) Cortisol
(**B**) Glucagon
(**C**) Growth hormone
(**D**) Insulin

36. Net fluid loss from blood capillaries occurs mainly by

(**A**) absorption
(**B**) diffusion
(**C**) filtration
(**D**) osmosis

37. In the cardiac cycle, closure of the atrioventricular (A-V) valves occurs at the same time as the

(**A**) beginning of diastole
(**B**) end of isovolumic relaxation
(**C**) first heart sound
(**D**) T complex in the electrocardiogram

38. Metabolic acidosis with partial respiratory compensation is characterized by which of the following in arterial blood?

(**A**) Normal P_{CO_2}, low pH, and low HCO_3^-
(**B**) Normal P_{CO_2}, normal pH, and low HCO_3^-
(**C**) Low P_{CO_2}, low pH, and low HCO_3^-
(**D**) Low P_{CO_2}, low pH, and normal HCO_3^-
(**E**) Low P_{CO_2}, high pH, and high HCO_3^-

39. Elevated parathyroid hormone levels lead to

(**A**) decreased activity of osteoclasts
(**B**) decreased renal phosphate excretion
(**C**) increased formation of 1,25-dihydroxy-cholecalciferol
(**D**) increased renal excretion of calcium

40. The natural rate of rhythmic discharge is greatest in which part of the heart?

(**A**) Atria
(**B**) Atrioventricular node
(**C**) Purkinje fibers
(**D**) Sinoatrial node
(**E**) Ventricular myocardium

41. Which of the following statements about lysosomes is most correct?

(**A**) They contain digestive enzymes
(**B**) They participate in the formation of ATP
(**C**) They participate in protein synthesis
(**D**) They phagocytize bacteria in many diseases

Questions 42 and 43

The following test results were obtained on specimens obtained from a patient during a 24-hour period:

Urine flow rate: 2.0 ml/min
Urine inulin: 1.0 mg/ml
Plasma inulin: 0.01 mg/ml
Urine urea: 220 mmol/L
Plasma urea: 5 mmol/L

42. The GFR of this patient is

(**A**) 100 ml/min
(**B**) 125 ml/min
(**C**) 150 ml/min
(**D**) 175 ml/min
(**E**) 200 ml/min

43. The urea clearance of this patient is

   (A) 4.4 ml/min
   (B) 22 ml/min
   (C) 44 ml/min
   (D) 88 ml/min
   (E) 440 ml/min

44. Blood flow through the coronary arteries is markedly attenuated during which period?

   (A) Diastole
   (B) Exercise
   (C) Isovolumic relaxation
   (D) Systole

45. Infusion of hypertonic sodium chloride solution will

   (A) decrease both intracellular and extracellular fluid volumes
   (B) increase both intracellular and extracellular fluid volumes
   (C) increase extracellular osmolarity only
   (D) increase extracellular volume and decrease intracellular volume
   (E) increase intracellular osmolarity only

46. The resistance of a blood vessel is 16 PRU (peripheral resistance unit). Doubling the vessel radius would change the resistance to

   (A) 10 PRU
   (B) 8 PRU
   (C) 4 PRU
   (D) 2 PRU
   (E) 1 PRU

47. What is the probable structure of pores in the cell membrane?

    (A) A cylindrical hole through the membrane
    (B) A large polysaccharide molecule entrapped in the membrane
    (C) A phospholipid molecule entrapped in the membrane
    (D) A protein molecule in the membrane with a channel through it
    (E) A slit in the membrane

48. A person with a PR interval of 0.23 second indicates

    (A) atrial flutter
    (B) incomplete heart block
    (C) paroxysmal tachycardia
    (D) nothing unusual

49. Which of the following statements concerning movement across capillary walls is true?

    (A) Amino acid movement across capillary walls occurs by active transport
    (B) Glucose movement across capillary walls occurs mainly by pinocytosis
    (C) Lipid movement across capillary walls is limited to intercellular junctions
    (D) Oxygen movement across capillary walls occurs mainly by diffusion
    (E) Protein movement across capillary walls occurs to the same extent in all capillary beds

50. Which of the following hormones impairs hydrolysis of triglycerides to fatty acids?

    (A) Cortisol
    (B) Glucagon
    (C) Growth hormone
    (D) Insulin

51. The velocity of impulse transmission is slowest in the

    (A) atria
    (B) A-V node
    (C) Purkinje system
    (D) S-A node
    (E) ventricular myocardium

52. Which of the following substances is actively secreted into the renal tubules?

    (A) Amino acids
    (B) Chloride
    (C) Glucose
    (D) Potassium
    (E) Sodium

53. The tensile strength of bones is attributed to

    (A) collagen fibers
    (B) ground substance
    (C) hydroxyapatite crystals
    (D) organic matrix

54. Which of the following two substances are best suited to measure interstitial fluid volume?

    (A) ^{51}Cr red blood cells and ^{125}I-albumin
    (B) Heavy water and ^{125}I-albumin
    (C) Inulin and heavy water
    (D) Inulin and ^{125}I-albumin
    (E) Inulin and Na^{22}

55. Ejection of blood from the left ventricle begins when the

    (A) A-V valves open
    (B) A-V valves close
    (C) left ventricular pressure exceeds aortic pressure
    (D) left ventricular pressure exceeds left atrial pressure

56. Protein molecules enter most cells by which process?

    (A) Opsonization
    (B) Passive diffusion
    (C) Phagocytosis
    (D) Pinocytosis

57. Destruction of the supraoptic nuclei of the brain will produce which of the following changes in urinary volume and concentration? (Assume that fluid intake equals fluid loss.)

    (A) Decreased urinary volume and a very dilute urine
    (B) Decreased urinary volume and a concentrated urine
    (C) Increased urinary volume and a very dilute urine
    (D) Increased urinary volume and a concentrated urine

58. Which of the following is the active form of vitamin D?

    (A) Calcitonin
    (B) Cholecalciferol
    (C) 1,25-Dihydroxycholecalciferol
    (D) Parathyroid hormone

59. Which of the following Starling forces tends to move fluid from interstitial spaces into blood capillaries?

    (A) Capillary hydrostatic pressure
    (B) Interstitial fluid colloid osmotic pressure
    (C) Plasma colloid osmotic pressure
    (D) Subatmospheric interstitial fluid pressure

60. Under normal conditions, most of the energy used by cardiac muscle comes from metabolism of

    (A) fatty acids
    (B) glucose
    (C) ketoacids
    (D) lactate

61. Blockage of the hypothalamic-hypophyseal venous portal system would be expected to cause increased secretion of which hormone?

    (A) Adrenocorticotropic hormone (ACTH)
    (B) Follicle-stimulating hormone (FSH)
    (C) Growth hormone
    (D) Prolactin
    (E) Thyroid-stimulating hormone (TSH)

62. Oxidative phosphorylation occurs in which of the following organelles?

    (A) Golgi apparatus
    (B) Lysosome
    (C) Mitochondria
    (D) Nucleus
    (E) Ribosome

63. A decrease in the velocity of impulse conduction through the A-V node will usually cause

    (A) atrial fibrillation
    (B) disappearance of the T wave
    (C) increased heart rate
    (D) the PR interval to decrease
    (E) the PR interval to increase

64. Which of the following produces human chorionic gonadotropin?

    (A) The anterior pituitary gland
    (B) The corpus luteum
    (C) The follicle
    (D) The trophoblasts

65. Which of the following conditions is associated with Addison's disease?

    (A) High blood levels of cortisol
    (B) Hypertension
    (C) Hypoglycemia between meals
    (D) Increased metabolic rate

66. Which type of heart failure is most likely to be associated with pulmonary edema?

    (A) Heart failure resulting from an arteriovenous fistula
    (B) High cardiac output heart failure
    (C) Left heart failure without right heart failure
    (D) Left heart failure with right heart failure
    (E) Right heart failure without left heart failure

67. Which of the following changes will decrease the rate of diffusion of a substance?

    (**A**) An increase in the concentration gradient
    (**B**) An increase in membrane permeability
    (**C**) An increase in the molecular weight of the substance
    (**D**) An increase in temperature

68. The pressure at one end of an artery is 60 mmHg, the pressure at the other end of the artery is 20 mmHg, and the flow through the artery is 200 ml/min. What is the resistance of the artery expressed in the above units?

    (**A**) 0.05
    (**B**) 0.1
    (**C**) 0.2
    (**D**) 0.4
    (**E**) 0.6

69. Under normal conditions, for which of the following substances would you expect renal clearance to be the lowest?

    (**A**) Creatinine
    (**B**) Glucose
    (**C**) Sodium
    (**D**) Urea
    (**E**) Water

70. During a normal menstrual cycle, a large surge of follicle-stimulating hormone and luteinizing hormone occurs

    (**A**) 1 to 2 days before ovulation
    (**B**) 1 to 2 days following ovulation
    (**C**) 1 to 2 days before menstruation
    (**D**) during menstruation

71. During a normal reproductive life, the ovaries will expel approximately

    (**A**) 45 ova
    (**B**) 450 ova
    (**C**) 4500 ova
    (**D**) 450,000 ova

72. A man drinks 2 L of water to replenish the fluids lost by sweating during a period of exercise. When compared with the situation prior to the period of sweating, his

    (A) extracellular fluid will be hypertonic
    (B) intracellular fluid will be hypertonic
    (C) extracellular fluid volume will be greater
    (D) intracellular fluid volume will be greater
    (E) intracellular and extracellular fluid volumes will be unchanged

73. Which of the following is increased by both growth hormone and glucagon?

    (A) Blood glucose concentration
    (B) Gluconeogenesis
    (C) Glycogenolysis
    (D) Lipolysis

74. An increase in sympathetic stimulation of the peripheral vasculature will most likely

    (A) decrease arterial blood flow
    (B) decrease arterial resistance
    (C) decrease venous resistance
    (D) increase venous compliance

75. Which of the following is thought to participate in the storage of memories?

    (A) Glial DNA
    (B) Myelin
    (C) Neuronal DNA
    (D) Neuronal RNA
    (E) Synapses

76. The most important physiologic function of the lymphatic system is to

    (A) concentrate proteins in the lymph
    (B) create negative pressure in the free interstitial fluid
    (C) remove particulate materials from the interstitium
    (D) transport antigenic materials to lymph nodes
    (E) transport fluid and proteins away from the interstitium to the blood

77. The Frank-Starling law of the heart states that

    (A) blood entering the atria is pumped immediately into the ventricles
    (B) cardiac output is controlled entirely by the activity of the heart
    (C) heart rate controls cardiac output during exercise
    (D) the heart can pump a certain amount of blood and no more
    (E) within physiologic limits, the heart pumps all the blood that comes to it

78. The secretion of parathyroid hormone is controlled by the concentration of

    (A) calcium bound to citrate anions
    (B) calcium bound to plasma proteins
    (C) calcium inside of the bone matrix
    (D) extracellular ionized calcium

79. Respiratory acidosis with full metabolic compensation is characterized by which of the following changes in arterial blood?

    (A) Low P_{CO_2}, high pH, and normal HCO_3^-
    (B) Low P_{CO_2}, normal pH, and high HCO_3^-
    (C) High P_{CO_2}, low pH, and normal HCO_3^-
    (D) High P_{CO_2}, normal pH, and high HCO_3^-
    (E) Normal P_{CO_2}, low pH, and low HCO_3^-

80. If the baroreceptor reflexes are fully functional when upright posture is assumed

    (A) arterial pressure in the foot will decrease markedly
    (B) bradycardia will occur
    (C) cerebral blood flow will not change appreciably
    (D) the blood vessels of the arms will become vasodilated

81. Quantitatively, the most abundant anion in the extracellular fluid is

    (A) bicarbonate
    (B) chloride
    (C) phosphate
    (D) protein
    (E) sulfate

82. Which of the following substances causes vasoconstriction?

    (A) Adenosine
    (B) Angiotensin II
    (C) Carbon dioxide
    (D) Histamine
    (E) Hydrogen ion

83. Metabolic acidosis can be caused by which of the following conditions?

    (A) Diabetes mellitus
    (B) Hysterical hyperventilation
    (C) Muscular dystrophy
    (D) Vomiting of gastric contents

84. The first heart sound is associated with

    (A) closing of the aortic and pulmonary valves
    (B) closing of the A-V valves
    (C) inrushing of blood into the ventricles due to atrial contraction
    (D) inrushing of blood into the ventricles in the early to middle part of diastole
    (E) opening of the A-V valves

85. Simple diffusion and facilitated diffusion are similar because both

    (A) can operate in the absence of adenosine triphosphate (ATP)
    (B) can transport substances against an electrochemical gradient
    (C) display saturation kinetics
    (D) require a carrier mechanism for transport

86. Which of the following statements about cardiac muscle is correct?

    (A) It has a longer duration of contraction during tachycardia
    (B) It has a velocity of conduction of action potentials of 0.3 m/sec to 0.5 m/sec
    (C) It is not influenced by norepinephrine
    (D) It never contracts for more than 0.12 sec

87. Which of the following occurs during dark adaptation?

(**A**) The eyes adapt fully within a few seconds
(**B**) The pupillary aperture increases
(**C**) The quantity of rhodopsin in the rods decreases
(**D**) The rate of rhodopsin decomposition increases

88. The frequency at which a neuron can generate action potentials is limited by the

(**A**) activity of the Na^+-K^+ pump
(**B**) concentration gradient of Na^+ across the neuronal membrane
(**C**) physical size of the neuron terminal
(**D**) K^+ concentration in the neuron
(**E**) refractory period of the neuron

89. The rate of conduction of action potentials in Purkinje fibers is about

(**A**) 0.2 to 1.1 m/sec
(**B**) 1.5 to 4.0 m/sec
(**C**) 5.0 to 8.5 m/sec
(**D**) 9.0 to 12.5 m/sec
(**E**) 15.0 to 18.5 m/sec

90. Which of the following changes would tend to decrease GFR?

(**A**) Decreased hydrostatic pressure in Bowman's capsule
(**B**) Decreased plasma colloid osmotic pressure
(**C**) Increased afferent arteriolar resistance
(**D**) Increased glomerular capillary filtration coefficient

91. Cirrhosis of the liver is often associated with which of the following?

(**A**) Decreased fluid in the peritoneal cavity
(**B**) Decreased liver lymph flow
(**C**) Decreased resistance of the portal vasculature
(**D**) Increased portal venous pressure

92. Which of the following statements about glucagon is correct?

    (A) It decreases gluconeogenesis
    (B) It is secreted by alpha cells of the islets of Langerhans
    (C) It is secreted by beta cells of the islets of Langerhans
    (D) It helps correct hyperglycemia
    (E) It promotes glycogen storage by the liver

93. In the normal heart, the majority of blood enters the left ventricle

    (A) after the aortic valve opens
    (B) as a result of atrial contraction
    (C) during early diastole
    (D) during isovolumic relaxation

94. In the circulation of the developed fetus

    (A) most of the cardiac output bypasses the lungs
    (B) the cardiac output of the right and left ventricles is exactly the same
    (C) the left atrial pressure is greater than the right atrial pressure
    (D) the pulmonary arterial pressure is lower than the aortic pressure

95. When comparing intracellular and interstitial body fluids, both have similar

    (A) chloride ion concentrations
    (B) colloid osmotic pressures
    (C) potassium ion concentrations
    (D) sodium ion concentrations
    (E) total osmolarity

96. A scuba diver at a depth of 66 feet (~20 meters) has a minute respiratory volume of 10 L/min. Expressed as a sea level equivalent, the rate of air use from the tank is

    (A) 5 L/min
    (B) 10 L/min
    (C) 20 L/min
    (D) 30 L/min

97. The most important difference between interstitial fluid and plasma is the

(A) concentration of sodium
(B) osmolarity
(C) potassium concentration
(D) protein concentration

98. Which of the following substances uses a protein-carrier molecule to traverse the cell membrane?

(A) Alcohol
(B) Glucose
(C) Glycerol
(D) Oxygen
(E) Water

99. Given the following conditions, calculate the net pressure difference across the capillary wall:

Interstitial fluid hydrostatic pressure = −3 mmHg
Plasma colloid osmotic pressure = 28 mmHg
Capillary hydrostatic pressure = 17 mmHg
Interstitial fluid colloid osmotic pressure = 8 mmHg

(A) −2 mmHg
(B) −1 mmHg
(C) 0 mmHg
(D) 1 mmHg
(E) 2 mmHg

100. An ejection fraction of 60% suggests

(A) athletic training
(B) heart failure
(C) normal ejection

101. The compressional strength of bones is attributed to

(A) collagen fibers
(B) ground substance
(C) hydroxyapatite crystals
(D) organic matrix

102. Which of the following is the receptor organ that generates nerve impulses in response to sound?

(A) Organ of Corti
(B) Round window
(C) Scala tympani
(D) Scala vestibuli
(E) Vestibular apparatus

103. Which of the following hormones is secreted by the posterior pituitary gland?

(A) Follicle-stimulating hormone (FSH)
(B) Luteinizing hormone (LH)
(C) Prolactin
(D) Vasopressin

104. A swimmer breathing through a snorkel has a respiration rate of 10/min, a tidal volume of 550 ml, and an effective anatomic dead space of 250 ml. What is her alveolar ventilation rate?

(A) 2500 ml/min
(B) 3000 ml/min
(C) 3500 ml/min
(D) 4000 ml/min
(E) 4500 ml/min

105. At which site on a motor neuron are action potentials most likely to be initiated?

(A) Axon
(B) Axon hillock
(C) Dendrites
(D) Soma
(E) Synaptic terminal

106. Which of the following might be expected in a patient with rickets resulting from vitamin D deficiency?

(A) Decreased plasma parathyroid hormone concentration
(B) Low plasma concentration of 1,25-dihydroxycholecalciferol
(C) Normal plasma phosphate concentration
(D) Suppressed osteoclastic activity

107. The majority of carbon dioxide is carried in the blood as

    (**A**) bicarbonate ions
    (**B**) carbonic anhydrase
    (**C**) carbon dioxide bound to hemoglobin
    (**D**) carbon dioxide bound to plasma proteins
    (**E**) dissolved carbon dioxide

108. When a person is dehydrated, hypotonic fluid will be found in the

    (**A**) collecting duct
    (**B**) distal end of the ascending loop of Henle
    (**C**) glomerular filtrate
    (**D**) late distal convoluted tubule
    (**E**) proximal tubule

109. The breakdown of complex foodstuffs is accomplished by which of the following chemical reactions?

    (**A**) Dehydration
    (**B**) Hydrolysis
    (**C**) Neutralization
    (**D**) Oxidation
    (**E**) Reduction

110. Which of the following statements about skeletal muscle is true?

    (**A**) Action potential lasts as long as the contraction
    (**B**) Action potential lasts longer than the contraction
    (**C**) Action potential precedes the contraction
    (**D**) Contraction and action potential begin simultaneously
    (**E**) Contraction precedes the action potential

111. A 25-year-old man has the posterior lobe of the pituitary gland surgically removed because of a tumor. In the absence of any hormone replacement therapy, which of the following conditions would you expect to find after the surgery?

    (**A**) Impaired secretion of growth hormone
    (**B**) Impaired synthesis and secretion of adrenocortical hormones
    (**C**) Impaired synthesis and secretion of thyroid hormones
    (**D**) Impaired urine-concentrating ability

112. A difference between skeletal muscle and smooth muscle is that

   (A) skeletal muscle fibers are smaller
   (B) skeletal muscle fibers can depolarize
   (C) smooth muscle fibers lack myosin filaments
   (D) smooth muscle fibers lack sarcomeres

113. The clearance rate for a substance that is freely filtered but neither secreted nor reabsorbed by the kidney is equal to the

   (A) filtration fraction
   (B) glomerular filtration rate
   (C) renal plasma flow
   (D) urinary excretion rate of the substance

114. The energy for skeletal muscle contraction is derived from the

   (A) binding of calcium to troponin
   (B) cleavage of ATP by the myosin head
   (C) influx of sodium during the action potential
   (D) membrane Na^+-K^+ ATPase pump

115. Muscles of inspiration include which of the following?

   (A) Abdominal muscles and external intercostals
   (B) Diaphragm and abdominal muscles
   (C) Diaphragm and external intercostals
   (D) Diaphragm and internal intercostals
   (E) Internal and external intercostals

116. The opening of the A-V valves occurs at about the same time in the cardiac cycle as the

   (A) beginning of diastole
   (B) end of isovolumic contraction
   (C) first heart sound
   (D) QRS complex of the electrocardiogram

117. The conduction velocity of the action potential is fastest in which type of axon?

    (A) Large diameter, myelinated fibers
    (B) Large diameter, unmyelinated fibers
    (C) Small diameter, myelinated fibers
    (D) Small diameter, unmyelinated fibers

118. Which process transports amino acids across the luminal surface of the epithelium that lines the small intestine?

    (A) Cotransport with the chloride ion
    (B) Cotransport with the sodium ion
    (C) Primary active transport
    (D) Simple diffusion

119. Creatine phosphate is important for intracellular energetics because

    (A) it can energize directly all intracellular reactions
    (B) it cannot transfer energy interchangeably with ATP
    (C) it is much more abundant than ATP
    (D) sufficient amount is stored to support anaerobic metabolism for hours

120. Which of the following controls the rate of ventricular contraction following complete heart block?

    (A) A-V node
    (B) Ectopic pacemaker
    (C) Internodal pathways
    (D) S-A node

121. Vitamin B_{12} or folic acid deficiency causes

    (A) erythroblastic cells of the bone marrow to become smaller than normal
    (B) the adult red blood cell to be smaller than normal
    (C) the adult red blood cell to have a thicker membrane than normal
    (D) the adult red blood cell to have an ovoid shape

122. Myasthenia gravis is an autoimmune disease in which the immune system has produced antibodies against

(**A**) AChE
(**B**) ACh receptors on the muscle membrane
(**C**) calcium channels on the muscle membrane
(**D**) molecules of ACh
(**E**) potassium channels on the muscle membrane

123. Stored fat is usually transported from one part of the body to another in the form of

(**A**) cholesterol
(**B**) free fatty acids
(**C**) glycerol
(**D**) neutral fat
(**E**) triglycerides

124. Complex starches are digested mainly by enzymes secreted from the

(**A**) large intestine
(**B**) pancreas
(**C**) salivary glands
(**D**) small intestine
(**E**) stomach

125. Contraction of the ciliary muscle of the eye

(**A**) causes the lens to become flat
(**B**) causes the pupil to dilate
(**C**) decreases the curvature of the lens
(**D**) reduces tension on the ligaments of the eye

126. Which of the following conditions is often caused by lack of iron in the diet?

(**A**) Erythremia
(**B**) Hemorrhagic anemia
(**C**) Hypochromic anemia
(**D**) Pernicious anemia
(**E**) Polycythemia

127. The threshold for initiation of a neuron action potential is the voltage at which

    (A) ACh is released
    (B) activation gates close
    (C) hyperpolarization occurs
    (D) potassium conductance decreases
    (E) progressively more sodium channels open

128. During the course of febrile illness, shivering is usually associated with which of the following?

    (A) Decreasing body temperature
    (B) Dilation of skin blood vessels
    (C) Recent increase in the hypothalamic set-point temperature
    (D) Recent reduction in the hypothalamic set-point temperature

129. Depression of the normal coagulation system and excessive bleeding most commonly result from

    (A) gastrointestinal disease
    (B) heart disease
    (C) kidney disease
    (D) liver disease
    (E) pulmonary disease

130. A major function of surfactant is to increase the

    (A) alveolar surface tension
    (B) pulmonary compliance
    (C) tendency of the lungs to collapse
    (D) work of breathing

131. At which site does a typical motoneuron receive most synapses?

    (A) Axon
    (B) Axon hillock
    (C) Dendrites
    (D) Soma
    (E) Synaptic terminal

132. Failure to absorb vitamin B_{12} from the gastrointestinal tract results in a condition called

   (A) erythremia
   (B) hemorrhagic anemia
   (C) pernicious anemia
   (D) polycythemia
   (E) thalassemia

133. Which ion is most closely associated with fluid absorption?

   (A) Bicarbonate
   (B) Chloride
   (C) Phosphate
   (D) Potassium
   (E) Sodium

134. The color of a monochromatic light that stimulates the red cones about twice as much as the green cones is

   (A) blue
   (B) green
   (C) orange
   (D) red
   (E) yellow

135. The need for vitamin B_{12} and folic acid in the formation of red blood cells is related primarily to their effects on

   (A) absorption of iron from the gut
   (B) DNA synthesis in the bone marrow
   (C) hemoglobin formation in the red blood cell
   (D) synthesis and release of erythropoietin from the kidney

136. During smooth muscle contraction, calcium ions

   (A) bind to troponin
   (B) do not enter the cell
   (C) enter the cell from T tubules
   (D) increase myosin ATPase activity

137. During rest, oxygen tension is lowest in which of the following blood vessels?

(A) Aorta
(B) Coronary artery
(C) Coronary vein
(D) Pulmonary artery
(E) Pulmonary vein

138. Pushing on the footpads of an animal with a transected spinal cord causes the foot to

(A) move anteriorly
(B) move laterally
(C) move posteriorly
(D) thrust downward
(E) withdraw

139. When the core temperature is below the hypothalamic set-point temperature, which of the following occurs?

(A) Blood flow to skin increases
(B) Heat production increases
(C) Pilorelaxation occurs
(D) Sweating occurs

140. Which of the following is a characteristic result of parasympathetic stimulation?

(A) Dry mouth
(B) Hypertension
(C) Profuse sweating
(D) Slowing of the heart

141. Anemia is usually characterized by increased

(A) blood viscosity
(B) exercise performance
(C) hematocrit
(D) total peripheral resistance
(E) workload on the heart

142. When excess amounts of carbohydrates or proteins are consumed, they are stored in the body as

    (A) cholesterol
    (B) glucose
    (C) glycogen
    (D) protein
    (E) triglycerides

143. Synaptic vesicles are released at the skeletal muscle neuromuscular junction when

    (A) calcium enters the nerve terminal
    (B) the motor endplate hyperpolarizes
    (C) the motor endplate releases ACh
    (D) the nerve terminal releases ATP
    (E) the skeletal muscle shortens

144. Which of the following can produce the most potent effect in stimulating the respiratory center and thus increasing respiration?

    (A) Decreases in CO_2 tension in the cerebrospinal fluid
    (B) Decreases in the blood $[H^+]$
    (C) Decreases in the blood O_2 tension
    (D) Increases in the blood CO_2 tension
    (E) Increases in the blood $[H^+]$

145. Excitatory synaptic transmitter substances cause which ions preferentially to move through the postsynaptic membrane of a typical neuron?

    (A) Calcium ions
    (B) Chloride ions
    (C) Magnesium ions
    (D) Potassium ions
    (E) Sodium ions

146. A patient is admitted to the emergency department with increased depth and rate of ventilation. Laboratory evaluation of an arterial blood sample reveals the following data:

$[HCO_3^-]$ = 8 mmol/L; P_{CO_2} = 20 mmHg; pH = 7.22

On the basis of these findings, what do you suspect is the cause of this patient's hyperventilation?

(**A**) Metabolic acidosis
(**B**) Metabolic alkalosis
(**C**) Respiratory acidosis
(**D**) Respiratory alkalosis

147. The fluid that cushions the brain in the cranium is

(**A**) aqueous humor
(**B**) cerebrospinal fluid
(**C**) intraocular fluid
(**D**) vitreous humor

148. Which condition can lead to anemia characterized by destruction of circulating red blood cells?

(**A**) Bone marrow aplasia
(**B**) Living at high altitude
(**C**) Presence of hemoglobin S
(**D**) Total gastronomy

149. Which of the following is stimulated by secretin?

(**A**) Gastric acid secretion
(**B**) Gastric emptying
(**C**) Intestinal motility
(**D**) Release of pancreatic bicarbonate

150. The Na^+-K^+ pump is an example of

(**A**) an ion channel
(**B**) carrier-mediated cotransport
(**C**) facilitated diffusion
(**D**) primary active transport
(**E**) vesicular transport

151. When the environmental temperature is greater than the body temperature, the only available mechanism of heat loss from the body is

   (A) conduction
   (B) convection
   (C) evaporation
   (D) forced convection
   (E) radiation

152. Which of the following substances or cell components can usually be found in a mature red blood cell?

   (A) ATP
   (B) DNA
   (C) Mitochondria
   (D) RNA

153. An abnormally low arterial oxygen tension is often caused by

   (A) decreased hematocrit
   (B) dilation of the respiratory passageways
   (C) inadequate hemoglobin in the blood
   (D) pulmonary edema

154. The nerve fibers of the pyramidal tract

   (A) carry sensory information to the pyramidal cells
   (B) most often cross to the contralateral side
   (C) originate in the cerebellum
   (D) rarely synapse with interneurons in the spinal cord

155. The tidal volume can never become greater than the

   (A) anatomic dead space
   (B) functional residual capacity
   (C) inspiratory capacity
   (D) residual volume
   (E) vital capacity

156. A positive ion will diffuse across a cell membrane

(**A**) down its electrochemical gradient
(**B**) down its partial pressure gradient
(**C**) to the side having a greater negativity
(**D**) to the side having a lower concentration

157. The rate of gas diffusion across the pulmonary membrane is

(**A**) inversely proportional to the solubility of the gas
(**B**) inversely proportional to the diffusion constant for the gas
(**C**) proportional to the membrane area
(**D**) proportional to the membrane thickness
(**E**) proportional to the molecular weight of the gas

158. Which portion of the brain is most likely to be malfunctioning when the voluntary movements of a person are jerky but there is no tremor when the person is resting?

(**A**) Basal ganglia
(**B**) Cerebellum
(**C**) Hypothalamus
(**D**) Premotor cortex
(**E**) Reticular formation

159. Which of the following substances causes the gallbladder to contract?

(**A**) Cholecystokinin
(**B**) Gastrin inhibitory polypeptide (GIP)
(**C**) Secretin
(**D**) Vasoactive intestinal polypeptide (VIP)

160. The spinal nerve endings of the neurons whose cell bodies are located in the raphe magnus nucleus release

(**A**) endorphin
(**B**) enkephalin
(**C**) glycine
(**D**) serotonin
(**E**) substance P

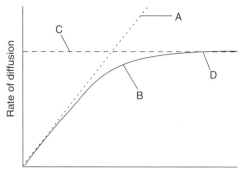

161. Which of the following statements concerning diffusion is correct (refer to the figure)?

(A) Line A depicts facilitated diffusion
(B) At point B the concentration gradient is maximal
(C) The position of line C changes with time
(D) All protein carriers are occupied at point D

162. Smooth muscle cells are innervated by which type of neuron?

(A) Alpha motoneurons
(B) Gamma motoneurons
(C) Interneurons
(D) Postganglionic autonomic neurons
(E) Preganglionic autonomic neurons

163. Which of the following statements about the transverse tubules is correct?

(A) They are better developed in thin muscle fibers
(B) They are invaginations of the sarcoplasmic reticulum
(C) They conduct depolarizations to the muscle cell interior
(D) They contain intracellular fluid

164. Which of the following functions is associated with the macula of the utricle?

(A) It conducts sound vibrations to the cochlea
(B) It is a receptor organ of the vestibular apparatus
(C) It protects the cochlea from loud sounds
(D) It is nonfunctional in persons with hearing deficits

165. The participation of calcium in the contraction of skeletal muscle is facilitated by or associated with the

   (A) active transport of calcium out of longitudinal tubules
   (B) binding of calcium to the myosin heads
   (C) release of calcium from longitudinal tubules
   (D) uptake of calcium by T tubules

166. During resting conditions, what amount of oxygen is transported to the average tissue in each 100 ml of blood?

   (A) 2 ml O_2/100 ml blood
   (B) 5 ml O_2/100 ml blood
   (C) 10 ml O_2/100 ml blood
   (D) 20 ml O_2/100 ml blood
   (E) 30 ml O_2/100 ml blood

167. Of the following taste buds, which helps protect against the ingestion of certain plant poisons?

   (A) Bitter taste buds
   (B) Salt taste buds
   (C) Sour taste buds
   (D) Sweet taste buds

168. Pepsinogen is activated by

   (A) hydrochloric acid and pepsin
   (B) cholecystokinin
   (C) chymotrypsin
   (D) gastrin and pepsin
   (E) trypsin and acid pH

169. Oxygen tension is greatest in which of the following blood vessels?

   (A) Aorta
   (B) Coronary artery
   (C) Coronary vein
   (D) Pulmonary artery
   (E) Pulmonary venules

170. Nervous control of muscular activity is typified by which type of neuronal circuitry?

    (A) Converging circuit
    (B) Diverging circuit
    (C) Integrative circuit
    (D) Parallel circuit
    (E) Reverberating circuit

171. After its release from the skeletal muscle neuromuscular junction, ACh

    (A) activates presynaptic potassium channels
    (B) causes postsynaptic depolarization
    (C) enters the sarcoplasmic reticulum
    (D) is triggered by AChE
    (E) suppresses norepinephrine secretion

172. Gallstones are composed mainly of which substance?

    (A) Bile salts
    (B) Bilirubin
    (C) Calcium
    (D) Cholesterol
    (E) Lecithin

173. Which of the following proteins is involved in skeletal muscle contraction but not in smooth muscle contraction?

    (A) ATPase
    (B) Calmodulin
    (C) Tropomyosin
    (D) Troponin

174. About 70% of the carbon dioxide is transported to the lungs

    (A) in chemical combination with albumin
    (B) in the dissolved state in the water of the plasma and cells
    (C) in the form of bicarbonate ions
    (D) in the form of carbaminohemoglobin
    (E) in the form of carbonic acid

175. The neuronal circuit with the greatest potential for producing a long-lasting output is the

   (A) converging circuit with multiple outputs
   (B) diverging circuit with multiple inputs
   (C) integrative circuit
   (D) parallel circuit
   (E) reverberating circuit

176. Parkinson's disease is usually caused by damage to which of the following structures?

   (A) Caudate nucleus
   (B) Globus pallidus
   (C) Putamen
   (D) Substantia nigra
   (E) Subthalamus nucleus

177. From which region of the CNS do emotions and complex behaviors arise?

   (A) Basal ganglia
   (B) Cerebellum
   (C) Limbic system
   (D) Reticular activating system
   (E) Spinal cord

178. In skeletal muscle, the region near the Z disk is composed mainly of

   (A) actin
   (B) myosin
   (C) overlapping actin and myosin

179. The flexor reflex is integrated in the

   (A) primary motor cortex
   (B) reticular formation
   (C) spinal cord
   (D) thalamus

180. Factors contributing to the initiation of parturition include each of the following EXCEPT

    (A) increased pressure on the cervix
    (B) increased progesterone secretion relative to estrogen secretion
    (C) increased secretion of oxytocin
    (D) increased stretch of the uterus

181. A 30-year-old healthy man reads that a diet high in potassium protects against cardiovascular disease and decides to take potassium supplements. Increasing his potassium intake would cause each of the following conditions EXCEPT

    (A) increased aldosterone release from the adrenal cortex
    (B) increased excretion of potassium in the urine
    (C) increased tubular secretion of potassium
    (D) renal retention of sodium

182. Each of the following hormones mediates its major effects without actually entering the target cell EXCEPT

    (A) cortisol
    (B) insulin
    (C) glucagon
    (D) growth hormone
    (E) parathyroid hormone

183. Increased pressure in the carotid sinus causes each of the following effects EXCEPT

    (A) increased sympathetic stimulation of the heart
    (B) increased vagal stimulation of the heart
    (C) reflex dilation of the peripheral blood vessels
    (D) reflex slowing of the heart rate

184. Mucus is secreted by each of the following organs EXCEPT the

    (A) duodenum
    (B) ileum
    (C) large intestine
    (D) pancreas
    (E) stomach

185. Each of the following statements concerning the determinants of glomerular filtration rate (GFR) and renal blood flow (RBF) is correct EXCEPT

(A) an increase in RBF, even with little change in glomerular hydrostatic pressure, increases GFR

(B) constriction of the afferent arteriole decreases both RBF and GFR

(C) constriction of the efferent arteriole decreases RBF and slightly increases GFR

(D) in a normal kidney, an increase in systemic arterial pressure from 100 to 150 mmHg increases GFR severalfold

186. Each of the following conditions causes an increase in arterial pulse pressure EXCEPT

(A) aortic valve insufficiency

(B) arteriosclerosis of aorta

(C) arteriovenous shunt

(D) mitral valve stenosis

(E) patent ductus arteriosus

187. Cortisol can cause each of the following metabolic effects EXCEPT

(A) blood glucose concentration to increase

(B) fat to be used for energy

(C) inflammation to be suppressed

(D) lysosomal membranes to become unstable

(E) proteins to be degraded in many tissues

188. Each of the following structures plays an important role in protein synthesis EXCEPT

(A) lysosomes

(B) messenger RNA

(C) ribosomal RNA

(D) ribosomes

(E) transfer RNA

189. High plasma levels of thyroxine can cause each of the following effects EXCEPT

(A) decreased body weight
(B) increased cardiac output
(C) increased heart rate
(D) increased metabolic rate
(E) increased plasma triglyceride concentration

190. Each of the following modalities of sensation is detected by free nerve endings EXCEPT

(A) crude touch
(B) itch sensations
(C) high-frequency vibration
(D) pain
(E) tickle sensations

191. Each of the following statements about the peripheral circulation is correct EXCEPT

(A) blood flow velocity in the capillaries is greater than in the large veins
(B) increased sympathetic nerve stimulation tends to constrict the small arterioles
(C) reduced oxygen tension in the tissues tends to relax precapillary sphincters
(D) total surface area of the capillaries is much greater than that of the large veins

192. Which of the following anticoagulants will NOT prevent coagulation when placed in a blood sample outside the body?

(A) Citrates
(B) Coumarin
(C) Heparin
(D) Oxalate

193. Which of the following conditions would LEAST likely be associated with Graves' disease?

(A) Goiter
(B) Increased sweating
(C) Increased metabolic rate
(D) Increased thyroid-stimulating hormone secretion
(E) Weight loss

194. High-altitude acclimatization may be facilitated by each of the following changes EXCEPT

(A) growth of new blood vessels
(B) growth of new skeletal muscle fibers
(C) increased alveolar ventilation
(D) increased production of red blood cells

195. Each of the following statements about protein synthesis is correct EXCEPT

(A) adenine is a purine base in DNA
(B) cytosine is a purine base in DNA
(C) guanine is a purine base in DNA
(D) thymine is a pyrimidine base in DNA

196. It is likely that growth hormone either directly or indirectly causes growth or contributes to growth by each of the following actions EXCEPT

(A) activating the RNA translation process
(B) decreasing somatomedin release from the liver
(C) increasing DNA replication
(D) increasing the rate of DNA transcription

197. Each of the following statements about the endoplasmic reticulum of the cell is correct EXCEPT

(A) it is involved with protein synthesis
(B) it often connects with the nuclear membrane
(C) it often has ribosomes attached to the outer surface
(D) it shuttles DNA between the cytoplasm and nucleus

198. Circus movements in the heart can be caused either directly or indirectly by each of the following EXCEPT

(A) atrial or ventricular dilation
(B) chronic mitral stenosis
(C) damage to the Purkinje system
(D) decreased myocardial refractory period
(E) increased myocardial conduction velocity

199. Hypothyroidism is often associated with each of the following physiologic effects EXCEPT

(A) decreased metabolic rate
(B) increased respiratory rate
(C) lethargy
(D) myxedema

200. The stomach does NOT digest itself because

(A) acid is completely neutralized by food
(B) acid is not secreted between meals
(C) gastric mucosal cells are not digestible
(D) gastric mucosal cells transport hydrogen ions out of the gastric mucosa

201. The appearance of large QRS complexes between normal beats NOT preceded by P waves indicates

(A) atrial flutter
(B) increased conduction velocity in Purkinje fibers
(C) partial atrioventricular block
(D) premature ventricular contraction

202. Important effects of testosterone include each of the following EXCEPT

(A) descent of the testes into the scrotum
(B) formation of the fetal penis
(C) increased muscle development
(D) increased thickness of the skin
(E) initiation of ejaculation

203. Which of the following would NOT be expected to occur during strenuous physical exercise?

(A) Large decrease in pulmonary vascular resistance
(B) Large increase in pulmonary arterial pressure
(C) Large increase in pulmonary blood flow
(D) Pulmonary capillary distention
(E) Pulmonary capillary recruitment

204. The final products of carbohydrate digestion include each of the following EXCEPT

(**A**) fructose
(**B**) galactose
(**C**) glucose
(**D**) sucrose

205. During chronic respiratory acidosis, each of the following effects will occur in a person with normal kidneys EXCEPT

(**A**) almost all of the filtered bicarbonate will be reabsorbed by the kidney
(**B**) glutamine uptake by the kidney will be enhanced
(**C**) hydrogen ion secretion in the distal nephron will be enhanced
(**D**) the production of ammonia by the kidney will increase
(**E**) the urinary pH will be increased

206. The repolarization of a neuron action potential is associated with each of the following EXCEPT

(**A**) closure of sodium channels in the cell membrane
(**B**) decreased potassium permeability of the cell membrane
(**C**) loss of positive charges from inside the cell
(**D**) outward diffusion of potassium ions
(**E**) return of the membrane potential toward its resting value

207. Estrogen can cause each of the following effects EXCEPT

(**A**) deposition of fat in the breasts
(**B**) enlargement of the external genitalia
(**C**) growth of hair on the pubic region
(**D**) skin to develop a soft texture

208. Widespread discharge of the sympathetic nervous system will cause each of the following effects EXCEPT

    (**A**) decreased blood glucose concentration
    (**B**) dilation of the pupils
    (**C**) increased basal metabolic rate
    (**D**) increased heart rate
    (**E**) increased myocardial contractility

209. Extracellular edema may result from each of the following EXCEPT

    (**A**) increased capillary permeability
    (**B**) increased capillary pressure
    (**C**) increased interstitial fluid colloid osmotic pressure
    (**D**) increased plasma colloid osmotic pressure
    (**E**) lymphatic blockage

210. Cardiopulmonary changes in the infant immediately upon birth include each of the following EXCEPT

    (**A**) decreased pulmonary arterial pressure
    (**B**) decreased pulmonary vascular resistance
    (**C**) decreased total peripheral vascular resistance
    (**D**) increased systemic arterial pressure

211. Each of the following factors will cause lymph flow to increase EXCEPT

    (**A**) increased capillary hydrostatic pressure
    (**B**) increased interstitial fluid colloid osmotic pressure
    (**C**) increased permeability of the capillaries
    (**D**) increased plasma colloid osmotic pressure

212. A decrease in blood pressure at the level of the internal carotid artery would cause each of the following changes EXCEPT

    (**A**) decreased nerve impulses from carotid sinus nerves
    (**B**) increased heart rate
    (**C**) increased total peripheral resistance
    (**D**) increased parasympathetic stimulation of the heart
    (**E**) increased strength of heart muscle contraction

213. Renal compensation for metabolic acidosis involves each of the following factors EXCEPT

    (A) activation of the tubular ammonia buffer system
    (B) increased excretion of NH_4^+ in the urine
    (C) increased filtration of bicarbonate
    (D) increased renal production of HCO_3^-

214. Increased plasma parathyroid hormone concentration tends to increase each of the following EXCEPT

    (A) absorption of calcium from the gastrointestinal tract
    (B) absorption of calcium from the renal tubules
    (C) absorption of phosphate from bone
    (D) extracellular phosphate concentration

215. When the core temperature of the body falls below the hypothalamic set-point temperature, each of the following changes occurs EXCEPT

    (A) basal metabolic rate increases
    (B) blood vessels of the skin dilate
    (C) heat production increases within minutes
    (D) the person feels cold

216. Each of the following statements about the renal medulla is correct EXCEPT

    (A) blood flow through the vasa recta is very slow, compared to blood flow through peritubular capillaries of cortical nephrons
    (B) countercurrent flow in the vasa recta minimizes solute loss from the medulla of the kidney
    (C) the thick ascending limb of the loop of Henle is highly permeable to water
    (D) there is net movement of water out of the descending limb of the loop of Henle

217. Thyroid-stimulating hormone stimulates thyroid function in many ways, but it does NOT increase

    (A) iodination of tyrosine
    (B) iodine uptake from the blood
    (C) rate of synthesis of thyroglobulin
    (D) size of the thyroid gland
    (E) synthesis of thyroxine-binding globulin

218. Hemostasis is normally accomplished by or associated with each of the following EXCEPT

(A) blood coagulation
(B) dilation of local blood vessels
(C) formation of a platelet plug
(D) polymerization of plasma fibrin molecules

219. In a patient with Conn's syndrome (primary aldosteronism), you would expect to find each of the following under steady-state conditions EXCEPT

(A) hypokalemia
(B) mild hypertension
(C) modest increases in extracellular fluid volume
(D) reduced urinary sodium excretion

220. Progesterone can cause each of the following effects EXCEPT

(A) development of lobules and alveoli of the breasts
(B) milk secretion by the alveoli of the breasts
(C) mucosal secretion of the fallopian tubes
(D) secretory changes in the endometrium

221. Peristalsis may be initiated or affected directly by each of the following EXCEPT

(A) distention of the gut wall
(B) inhibitors of striated muscle contraction
(C) irritation of the mucosa
(D) the composition of the chyme
(E) the myenteric plexus

222. Which of the following statements concerning ACh is NOT correct? The release of ACh at the neuromuscular junction

(A) always causes the muscle fiber to contract
(B) is followed by rapid destruction of ACh
(C) increases sodium movement into the muscle fiber
(D) produces an endplate potential

223. Which of the following changes would NOT occur as a result of dehydration?

    (A) Decreased permeability of the collecting ducts to water
    (B) Increased plasma sodium concentration
    (C) Increased secretion of antidiuretic hormone
    (D) Increased solute concentration in the renal medulla

224. Each of the following transport pathways through cell membranes is highly selective for specific substances EXCEPT

    (A) active transport via carrier proteins
    (B) facilitated diffusion via carrier proteins
    (C) simple diffusion through lipid bilayer
    (D) simple diffusion through protein channels

225. Stimulation of visceral pain often results from each of the following EXCEPT

    (A) a pin prick
    (B) chemical irritation of tissues
    (C) overdistension of tissues
    (D) smooth muscle spasm
    (E) tissue ischemia

226. The stomach secretes each of the following substances EXCEPT

    (A) chyme
    (B) gastrin
    (C) hydrochloric acid
    (D) intrinsic factor
    (E) pepsinogen

227. Increased amounts of erythropoietin might be released from the kidney EXCEPT when the

    (A) arterial P_{O_2} and arterial O_2 content are reduced
    (B) arterial P_{O_2} is low and the saturation of hemoglobin with O_2 is much reduced
    (C) arterial P_{O_2} is normal and the arterial O_2 content is reduced
    (D) tissue P_{O_2} and renal blood flow are both increased

228. When compared with intracellular fluid, each of the following substances is found in higher concentrations in the extracellular fluid EXCEPT

(A) calcium
(B) oxygen
(C) potassium
(D) sodium

229. Each of the following is increased by growth hormone EXCEPT

(A) blood free fatty acid concentration
(B) blood glucose concentration
(C) metabolism of carbohydrates
(D) protein synthesis
(E) storage of proteins in cells

230. Each of the following occurs when the core temperature of a person increases to more than 106°F (41°C) EXCEPT

(A) heat production increases
(B) sweating increases greatly
(C) the CNS may begin to malfunction
(D) the person may become poikilothermic

231. In which of the following conditions would oxygen therapy be LEAST beneficial?

(A) Anemia
(B) Emphysema
(C) Pneumonia
(D) Pulmonary edema

232. Which neurons of the autonomic nervous system usually do NOT release ACh?

(A) Postganglionic parasympathetic neurons
(B) Postganglionic sympathetic neurons
(C) Preganglionic parasympathetic neurons
(D) Preganglionic sympathetic neurons

233. Which couple CANNOT be the genetic parents of a child with blood group AB?

(A) Mother AA, father BB
(B) Mother OB, father AA
(C) Mother AB, father OO
(D) Mother OA, father OB
(E) Mother BB, father AB

234. Emptying of the stomach is inhibited by each of the following EXCEPT

(A) alkaline chyme in the duodenum
(B) hyperosmolar chyme in the duodenum
(C) irritation of the duodenal mucosa
(D) fatty acids in the chyme of the duodenum

235. Untreated diabetes mellitus is often associated with each of the following conditions EXCEPT

(A) decreased intracellular glucose concentration in some tissues
(B) increased fat metabolism
(C) increased plasma cholesterol concentration
(D) metabolic alkalosis

236. Increasing alveolar ventilation about threefold during resting conditions causes each of the following effects EXCEPT

(A) dizziness
(B) decreased arterial carbon dioxide content
(C) increased arterial pH
(D) increased arterial oxygen content

237. The sympathetic and parasympathetic nervous systems have opposite effects on each of the following EXCEPT

(A) contraction of the skeletal muscle
(B) contraction of the urinary bladder
(C) heart rate
(D) motility of the gastrointestinal tract

238. The process of swallowing involves each of the following EXCEPT

   (**A**) closure of the glottis
   (**B**) esophageal peristalsis
   (**C**) involuntary movements of the tongue against the palate
   (**D**) involuntary relaxation of the upper esophageal sphincter
   (**E**) transmission of action potentials from the pharyngeal region to the brain stem

239. In most instances of erythroblastosis fetalis, each of the following occurs EXCEPT

   (**A**) the fetus is Rh negative
   (**B**) the mother is Rh negative
   (**C**) the father is Rh positive
   (**D**) many of the fetal red blood cells are nucleated

240. The resting membrane potential of a cell is established by or maintained by each of the following EXCEPT

   (**A**) a net inward movement of positive ions
   (**B**) ATP
   (**C**) inward movement of K^+
   (**D**) outward movement of Na^+
   (**E**) the Na^+-K^+ pump

241. Vasodilation of the efferent arterioles of the kidney causes each of the following changes EXCEPT

   (**A**) decreased glomerular capillary hydrostatic pressure
   (**B**) decreased glomerular filtration rate
   (**C**) decreased peritubular capillary hydrostatic pressure
   (**D**) increased renal blood flow

242. Carbon dioxide is carried in the blood in each of the following ways EXCEPT

   (**A**) as carbamino compounds
   (**B**) as dissolved gas
   (**C**) as bicarbonate
   (**D**) in combination with hemoglobin
   (**E**) as carbonic anhydrase

243. The overall response of a neuronal pool characterized as a converging circuit depends on each of the following factors EXCEPT

    (A) the basic excitability of the neurons in the pool
    (B) the number of excitatory impulses entering the pool
    (C) the number of inhibitory impulses entering the pool
    (D) the total number of impulses entering the pool
    (E) whether or not there might be some diverging circuits also in the pool

244. Each of the following statements about cholecystokinin is true EXCEPT

    (A) fat and proteins in the chyme stimulate its release
    (B) it helps regulate contraction of the gallbladder
    (C) it inhibits gastric emptying
    (D) it stimulates insulin release
    (E) it stimulates the secretion of pancreatic enzymes

245. A transfusion reaction may lead to or cause each of the following conditions EXCEPT

    (A) fever
    (B) hemolysis
    (C) oliguria
    (D) polycythemia
    (E) uremia

246. Visual acuity is greatest in the retinal fovea for each of the following reasons EXCEPT

    (A) blood vessels and ganglion cells do not cover foveal cones
    (B) many cones are innervated by a single optic nerve fiber
    (C) only cones are present in the fovea
    (D) the cones are smaller in the fovea

247. Which of the following statements concerning the Na$^+$-K$^+$ pump is NOT correct? If the Na$^+$-K$^+$ pump were suddenly poisoned,

    (A) no further transmission of nerve impulses could occur
    (B) the intracellular potassium concentration would decrease
    (C) the intracellular sodium concentration would increase
    (D) the resting membrane potential would become less negative

248. The pupillary light reflex might be abolished by discrete damage of each of the following structures EXCEPT the

    (A) fovea centralis
    (B) optic nerve
    (C) pretectal nuclei
    (D) third cranial nerve

249. The strength of contraction of an entire skeletal muscle is dependent on each of the following factors EXCEPT

    (A) frequency of contraction of each muscle fiber
    (B) frequency of slow waves
    (C) number of active crossbridges in each muscle fiber
    (D) number of muscle fibers that contract simultaneously

250. A patient with a duodenal ulcer would be expected to exhibit each of the following signs EXCEPT

    (A) increased gastric H$^+$ secretion
    (B) increased pepsin secretion
    (C) increased serum gastrin levels
    (D) primary defect in the mucosa

DIRECTIONS: Each set of matching questions in this section consists of a list of three to twenty-six options followed by several numbered items. For each numbered item, select the ONE lettered option that is most closely associated with it. Each lettered option may be selected once, more than once, or not at all.

Questions 251–253

(**A**) Aldosterone
(**B**) Angiotensin II
(**C**) Antidiuretic hormone
(**D**) Atrial natriuretic peptide

Match each function listed below with the appropriate hormone.

251. Increases urea and water permeability in the collecting duct

252. Increases potassium secretion in the cortical collecting tubule

253. Increases urinary excretion of sodium

Questions 254–256

(**A**) Eosinophil
(**B**) Lymphocyte
(**C**) Macrophage
(**D**) Megakaryocyte
(**E**) Neutrophil

For each characteristic or function, choose the type of cell with which it is usually associated.

254. First line of defense against bacterial invasion from the environment

255. Combats parasitic infections

256. The most numerous of the white blood cells

Questions 257–261

	Pco$_2$ (mmHg)	[HCO$_3^-$] (mmol/L)	pH
(A)	29	22.0	7.50
(B)	33	32.0	7.61
(C)	35	17.5	7.32
(D)	40	25.0	7.41
(E)	60	37.5	7.42

The information above shows the arterial blood acid-base data for five individuals, designated by the letters A to E. For each of the following descriptions of acid-base status, choose the individual with the appropriate acid-base data.

257. Normal

258. Partially compensated metabolic acidosis

259. Fully compensated respiratory acidosis

260. Uncompensated respiratory alkalosis

261. Combined respiratory and metabolic alkalosis

Questions 262–265

(A) Arteries
(B) Arterioles
(C) Capillaries
(D) Veins
(E) Venules

For each characteristic listed below, match the type of blood vessel that is usually associated with it.

262. Slowest velocity of blood flow

263. Lowest hydrostatic pressure

264. Lowest resistance to blood flow

265. Largest surface area

Questions 266–269

(**A**) Golgi tendon apparatus
(**B**) Joint receptor
(**C**) Meissner's corpuscle
(**D**) Muscle spindle
(**E**) Pacinian corpuscle

For each function described, select the type of sensory receptor with which it is usually associated.

266. Detects the degree of tension in the muscles

267. Detects the muscle length or rate of change of muscle length

268. Detects rapid changes in mechanical deformation

269. Responds specifically to light touch

Questions 270–272

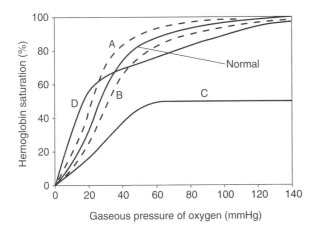

Refer to the figure when answering the following three questions. The curve indicated as "normal" shows the loading characteristics of normal human hemoglobin as the PO_2 increases. Select the appropriate curve for each of the following situations.

270. Increased temperature

271. Increased blood pH

272. Carbon monoxide poisoning

Questions 273–276

	[K+] (mmol/L)	[Na+] (mmol/L)	[HCO₃⁻] (mmol/L)	Plasma renin activity (× Normal)
Normal	4.0	142	24	1.0
(A)	2.9	124	12	0.2
(B)	2.9	146	30	0.1
(C)	5.2	130	19	6.0
(D)	4.5	152	26	2.0

The table shows plasma laboratory values from a normal subject and from four patients, designated by the letters A to D. Select that patient whose plasma laboratory values best match the stated disorder.

273. Primary aldosteronism (excessive aldosterone secretion)

274. Diabetes insipidus

275. Addison's disease (adrenal insufficiency)

276. Syndrome of inappropriate ADH (excessive ADH secretion)

Questions 277–279

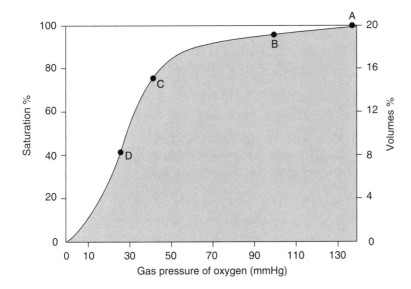

The figure shows an oxygen-hemoglobin dissociation curve for normal human hemoglobin. Select the point on the curve that corresponds to each of the following questions.

277. Systemic arterial blood

278. Mixed venous blood

279. Mixed venous blood during exercise

Questions 280 and 281

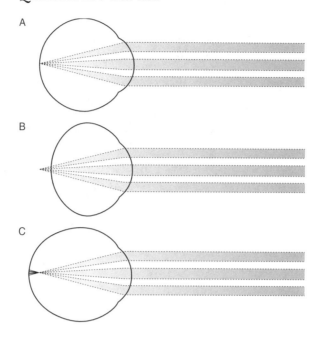

The figure shows parallel light rays that enter the eye and focus on the retina (case A), behind the retina (case B), and in front of the retina (case C). Select the appropriate clinical condition that represents the case shown.

280. Hyperopia

281. Myopia

Questions 282 and 283

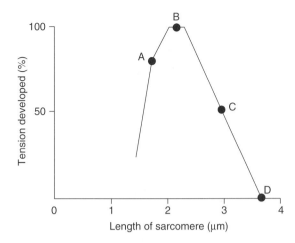

The figure shows a length-tension diagram for a single sarcomere. Select the point on the curve that corresponds to each of the following situations.

282. Maximum overlap between actin filaments and myosin crossbridges

283. Partial overlap between actin and myosin filaments

Questions 284 and 285

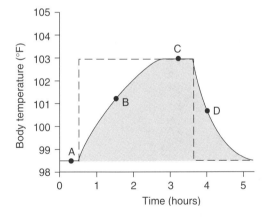

The figure shows the effect of changing the hypothalamic set-point temperature (*dashed line*) on body temperature (*solid line*). Select the point on the curve that corresponds to each of the following situations.

284. The individual feels cold

285. Sweating and vasodilation of the skin blood vessels occur

Questions 286–288

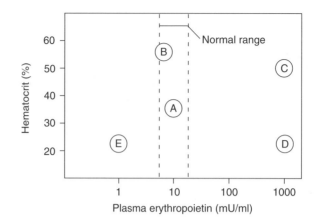

The figure shows the relationship between hematocrit and plasma erythropoietin levels in various states or diseases. Select the area that corresponds to each of the following clinical conditions.

286. Chronic obstructive lung disease

287. Polycythemia vera

288. Chronic renal disease

Questions 289 and 290

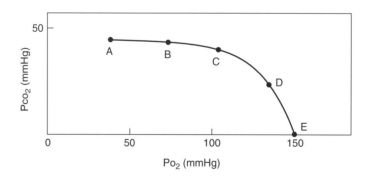

The figure shows the normal Po_2-Pco_2 ventilation/perfusion $\dot{V}_A/\dot{Q}$ diagram. Select the point on the curve that corresponds to each of the following situations.

289. A lung unit with normal blood flow but no ventilation

290. A lung unit at the apex of the lung following hemorrhage

Questions 291–293

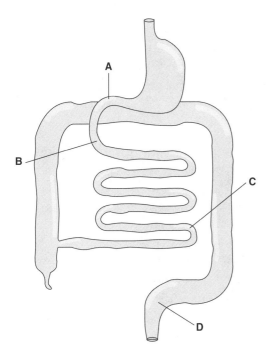

The figure shows characteristic points of obstruction in the gastrointestinal tract. Select the location that is most likely to correspond to each of the following situations.

291. History of peptic ulceration and persistent acidic vomitus

292. Metabolic acidosis and basic vomitus that becomes fecal in character

293. Moderate vomiting and extreme constipation

Questions 294–296

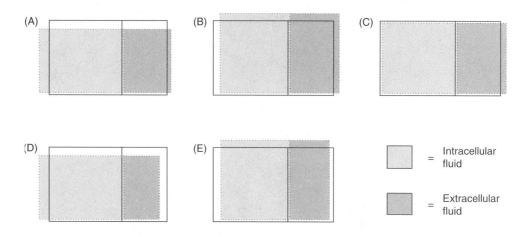

Each diagram represents a state of abnormal hydration. In each diagram, the normal state (*solid lines*) is superimposed on the abnormal state (*dashed lines*) to illustrate the shifts in the volumes (width of rectangles) and total osmolalities (height of rectangles) of the extracellular fluid and intracellular fluid compartments. Each of the numbered phrases that follows describes a condition that leads to abnormal hydration. Match each phrase to the lettered diagram that represents the appropriate state of abnormal hydration.

294. Syndrome of inappropriate ADH secretion (SIADH, excess ADH secretion)

295. Infusion of isotonic saline

296. Infusion of hypertonic saline

Questions 297–300

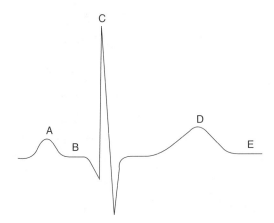

Match each of the events of the cardiac cycle with the correct lettered point shown on the electrocardiogram.

297. The blood pressure in the aorta is at its lowest value

298. The atria are contracting

299. The A-V valve (mitral valve) closes

300. The atria and ventricles are completely repolarized

Questions 301–305

(**A**) 1,25-Dihydroxycholecalciferol
(**B**) Glucagon
(**C**) Insulin
(**D**) Parathyroid hormone
(**E**) Thyroxine

Match each of the following descriptions with the appropriate hormone.

301. Increases the activity of osteoclasts to cause resorption of bone

302. The active form of this hormone is formed in the kidney

303. Reduces lipolysis, promotes storage of triglycerides in fat tissue, and stimulates glucose uptake and utilization by skeletal muscle

304. Stimulates glycogenolysis by the liver

305. A deficiency of this hormone produces an edematous state called myxedema

PHYSIOLOGY ANSWERS AND DISCUSSION

1—E (Chapter 2) The major cause of increased renal excretion of K^+ in a person on a diet high in K^+ is increased secretion of K^+ by the distal and collecting tubules.

2—C (Chapter 7) Wernicke's area is a general interpretative area that has a critical role in the higher levels of brain function that are commonly referred to as intelligence. When it is damaged in the left hemisphere of a right-handed person, the person may lose the ability to put together sensory information to determine the overall meaning. However, the person will still hear perfectly well and may even be able to read certain words.

3—B (Chapter 5) In aortic stenosis, the blood is expelled through only a small opening in the aortic valve, causing the ventricular pressure to become extremely high, sometimes as high as 300 mmHg. The blood is thus ejected from the ventricle at a tremendous velocity, causing severe turbulence of the blood and vibration of the walls of the aorta.

4—D (Chapter 7) When a person spends a long period of time in darkness, only very small amounts of rhodopsin are split, while the metabolic systems of the rods are continually building more rhodopsin. Consequently, rhodopsin collects in a very high concentration after a period of time and greatly increases the sensitivity of the retina to light.

5—B (Chapter 8) Chloride is most closely associated with fluid secretion.

6—B (Chapter 5) The renal-body fluid volume mechanism is slow to act, but it is a powerful mechanism for regulating arterial pressure extremely accurately over long periods of time. The most rapid control of arterial pressure is achieved by the cardiovascular nervous reflexes.

7—A (Chapter 10) Excess thyroxine mobilizes lipids from fat tissue, depleting fat stores. Exophthalmos occurs in many individuals with hyperthyroidism, but the condition is not caused by the high levels of thyroxine in the plasma. Instead, exophthalmos is thought to result from an autoimmune process in which antibodies directed against the extraocular muscles cause the muscles to weaken and the eyeball to protrude.

8—D (Chapter 9) The essential amino acids must be present in the diet, but not all dietary proteins contain all 10 of the essential amino acids. The other 10 nonessential amino acids are essential for the formation of proteins, but it is not *essential* that they be present in the diet.

9—B (Chapter 5) Blood flow through the brain is controlled mainly in response to carbon dioxide released from the brain tissue.

10—E (Chapter 7) When a loud sound is transmitted into the CNS, an attenuation reflex occurs after a latent period of 40 to 80 milliseconds. The reflex involves the contraction of two muscles that pull the malleus and stapes toward each other, thereby causing the entire ossicular system to develop a high degree of rigidity. In turn, the ossicular conduction of low-frequency sounds to the cochlea can be reduced by as much as 30 to 40 decibels. Since loud sounds are usually low-frequency sounds, the attenuation reflex can protect the cochlea from damage caused by loud sounds when they develop slowly.

11—B (Chapter 3) The citrate ion combines with calcium in the blood to form an unionized calcium compound. The lack of ionic calcium prevents coagulation. Following a single transfusion, the citrate ion is removed from the blood within a few minutes by the liver without any dire consequences. Oxalate anticoagulants work in a similar manner, but oxalate is toxic to the body.

12—C (Chapter 6) Blood that perfuses nonventilated portions of the lungs cannot be oxygenated. This blood mixes with fully oxygenated blood from ventilated alveoli, decreasing the arterial P_{O_2} below the alveolar value. Although oxygen must diffuse down a partial pressure gradient from the alveoli into the blood, the capillary P_{O_2} becomes virtually identical to the alveolar P_{O_2} when the red blood cell is about one-third of the way through the capillary.

13—E (Chapter 7) Sweating is a characteristic result of sympathetic stimulation as are dilation of the pupils, increased heart rate, decreased peristalsis, and increased blood pressure.

14—A (Chapter 5) Increased pulse pressure can occur either because of an increase in stroke volume or a decrease in compliance of the arterial tree. In atherosclerosis, fatty lesions develop on the inside surfaces of the arterial walls, leading to calcifications that make the arteries lose their distensibility and therefore cause decreased compliance.

15—C (Chapter 2) The total body water is equal to 60 kg × 0.57, or 34.2 L. The intracellular fluid volume is equal to the total body water minus the extracellular fluid volume, or 34.2–12.8 = 21.4 L.

16—C (Chapter 2) Because 60% of the blood volume is plasma (hematocrit = 40%), the plasma volume is equal to 4.3 × 0.6, or 2.6 L.

17—C (Chapter 2) The interstitial fluid volume is equal to the extracellular fluid volume (12.8 L) minus the plasma volume (2.6 L) = 10.2 L.

18—B (Chapter 5) The blood vessels of the systemic circulation responsible for most of the resistance to blood flow in the circulation are the arterioles.

19—A (Chapter 10) Lack of iodine in the diet prevents the production of both thyroxine (T_4) and triiodothyronine (T_3) but does not stop the formation of thyroglobulin. No hormone is thus available to inhibit secretion of thyroid-stimulating hormone (TSH), allowing the anterior pituitary to secrete large quantities of TSH. The TSH causes the thyroid cells to secrete excessive amounts of thyroglobulin into the follicles, and the gland then grows larger and larger.

20—D (Chapter 5) The T wave in the normal electrocardiogram is an upward deflection representing ventricular repolarization and occurs during the latter half of ventricular systole.

21—B (Chapter 10) Penile erection and erection of the female clitoris and associated erectile tissues occur by parasympathetically mediated dilation of arterioles.

22—D (Chapter 5) The period of isometric contraction occurs when the valves of the heart are closed so that during this period contraction in the ventricles is occurring but there is no emptying. This period is called isometric contraction because tension is increased in the muscle but no shortening of the fibers is occurring.

23—D (Chapter 1) Translation is the process by which polypeptides are synthesized on the ribosome from the genetic code contained in the mRNA. Transcription is the process by which genetic information contained in the DNA is transferred to RNA as a complementary sequence of bases.

24—B (Chapter 5) The arterioles are often called the resistance vessels of the body because they account for about half the resistance of the entire systemic circulation. The arteriolar wall has a thick smooth muscle coat in relation to the size of the vessel, and it is innervated richly by the autonomic nervous system.

25—D (Chapter 10) Prolactin stimulates milk production by the breasts. During pregnancy, the mother's anterior pituitary produces increased amounts of prolactin, increasing to about 10-fold the normal rate of secretion. Prolactin is required to cause the final development of the breasts and also to cause them to secrete milk.

26—D (Chapter 5) Low levels of oxygen in the myocardium cause the local tissues to release adenosine, which in turn acts to dilate the local blood vessels.

27—D (Chapter 2) A 50% reduction in glomerular filtration rate would cause a 50% decrease in urinary creatinine clearance, because creatinine is neither reabsorbed nor secreted to any significant extent by the renal tubules. The transient reduction in urinary creatinine excretion would raise the plasma creatinine concentration, which would return the filtered load and urinary excretion rate of creatinine toward normal under steady-state conditions. Therefore, under steady-state conditions, urinary creatinine excretion would be normal, but at the expense of an elevated plasma creatinine concentration.

28—B (Chapter 10) There are 23 chromosomes present in a normal mature sperm cell.

29—D (Chapter 5) Increased arterial pressure tends to increase flow to the tissues and therefore delivery of nutrients, such as oxygen, while increasing the removal of vasodilator waste products of metabolism, such as adenosine. The increased delivery of oxygen tends to cause vasoconstriction, as does the increased removal of vasodilator waste products of metabolism.

30—D (Chapter 10) Almost all of the active thyroid hormone entering the circulation is in the form of thyroxine.

31—D (Chapter 1) Lipid-soluble substances such as oxygen, ethanol, and carbon dioxide can dissolve directly in the lipid bilayer and diffuse rapidly through the cell membrane. Although water is highly insoluble in the membrane lipids, it penetrates the cell membrane very rapidly, moving through the protein channels as well as directly through the lipid bilayer. Charged molecules such as sodium, potassium, and chloride penetrate the lipid bilayer about 1 million times less rapidly than does water.

32—A (Chapter 10) Estrogen secretion is highest during the period of 3 days before ovulation.

33—D (Chapter 2) Constriction of efferent arterioles decreases peritubular capillary hydrostatic pressure allowing greater amounts of fluid to be reabsorbed by the peritubular capillaries. Also, increased efferent arteriolar resistance reduces renal plasma flow while increasing glomerular filtration rate. This raises the filtration fraction and the concentration of protein in peritubular capillaries, thereby causing greater amounts of fluid to be reabsorbed. Decreased plasma protein concentration, decreased plasma colloid osmotic pressure, and decreased filtration fraction all lead to decreased peritubular capillary reabsorption.

34—C (Chapter 5) Adenosine accumulates in the tissues when metabolic rate is increased and causes vasodilation, thereby increasing tissue blood flow to match the metabolic rate.

35—B (Chapter 10) Glucagon increases hepatic glycogenolysis and promotes gluconeogenesis.

36—C (Chapter 2) It is important to distinguish between filtration and diffusion through the capillary wall. Diffusion occurs in both directions, whereas filtration is the net movement of fluid out of the capillaries. The rate of diffusion of water through all the capillary membranes of the entire body is about 240,000 ml/min, whereas the rate of filtration at the arterial ends of all the capillaries is only about 16 ml/min.

37—C (Chapter 5) When listening with a stethoscope, the "lub" sound is associated with closure of the atrioventricular (A-V) valves at the beginning of systole, and the "dub" sound is associated with closure of the semilunar valves at the end of systole. Because the cardiac cycle is considered to start with the beginning of systole, the "lub" sound is called the *first heart sound,* and the "dub" sound is called the *second heart sound.*

38—C (Chapter 2) The high [H$^+$] in metabolic acidosis causes increased pulmonary ventilation, which removes CO_2 from the body fluids and leads to a reduction toward normal in the [H$^+$]. However, the respiratory system cannot fully compensate for metabolic acidosis. Thus respiratory compensation is characterized by low levels of CO_2 and $NaHCO_3$ as well as a low pH in the extracellular fluids.

39—C (Chapter 10) Parathyroid hormone increases the tubular reabsorption of calcium in the ascending limbs of the loops of Henle, distal tubules, and collecting ducts, and at the same time it decreases reabsorption of phosphate in the proximal tubules. Parathyroid hormone also enhances calcium and phosphate reabsorption from the intestines by increasing the formation of 1,25-dihydroxycholecalciferol from vitamin D in the kidneys, and increases absorption of calcium and phosphate from bone by stimulating both the proliferation and activity of the osteoclasts.

40—D (Chapter 5) The sinoatrial node, or S-A node, has the highest rate of automatic discharge of any tissue in the heart. This makes the S-A node the pacemaker of the heart because it excites other potentially self-excitatory tissues before self-excitation can actually occur.

41—A (Chapter 1) The cell phagocytizes bacteria and other particulate substances, and digestive enzymes from lysosomes digest the phagocytized particles.

42—E (Chapter 2) The glomerular filtration rate (GFR) is equal to the inulin clearance because inulin, a low-molecular-weight polysaccharide, is filtered freely into Bowman's capsule but is not reabsorbed or secreted. The clearance (C) of any substance can be calculated as follows: $C = (U \times V)/P$, where U and P are the urine and plasma concentrations of the substance, respectively, and V is the urine flow rate. Thus, GFR = $(1.0 \times 2.0)/0.01 = 200$ ml/min.

43—D (Chapter 2) Urea clearance = $(0.220 \times 2.0)/0.005 = 88$ ml/min.

44—D (Chapter 5) The strong compression of the ventricular muscle around the coronary blood vessels causes the coronary blood flow to fall to a low value during systole. During diastole, the ventricular muscle relaxes completely, allowing the blood to flow rapidly throughout diastole.

45—D (Chapter 2) Infusion of hypertonic sodium chloride solution increases both intracellular and extracellular osmolarity while increasing extracellular fluid volume and decreasing intracellular fluid volume by causing osmosis of water out of the cells into the extracellular compartment.

46—E (Chapter 5) As the radius of a vessel changes, the resistance of the vessel changes in inverse proportion to the fourth power of the radius. Therefore, increasing the radius of a vessel to twice the initial radius would decrease the vessel resistance to one-sixteenth of the initial resistance.

47—D (Chapter 1) Large globular protein molecules, protruding through the lipid bilayer membrane, are believed to serve as minute pores in the cell membrane that allow water and water-soluble substances, especially ions, to move through the membrane.

48—B (Chapter 5) The duration of time between the beginning of the P wave and the beginning of the QRS complex is the interval between the beginning of atrial depolarization and the beginning of ventricular depolarization. This interval, called the PR interval, is approximately 0.16 sec during normal, resting conditions. If the PR interval increases above approximately 0.20 sec in a heart beating at a normal rate, the patient is said to have *first-degree incomplete heart block*.

49—D (Chapter 2) Oxygen is lipid soluble and can therefore move by diffusion through all portions of the lipid capillary wall. Glucose moves primarily through aqueous pores in the capillary wall. The movement of proteins across the capillary wall is greatest in the liver, where large gaps exist between endothelial cells, and is lowest in the brain, where adjacent endothelial cells form very tight junctions. Amino acid movement across the capillary occurs mainly by diffusion.

50—D (Chapter 10) Insulin inhibits the action of hormone-sensitive lipase, an enzyme that causes hydrolysis of the triglycerides already stored in the fat cells.

51—B (Chapter 5) Impulse transmission through the atrioventricular (A-V) node is slower than through the ventricular myocardium, atria, Purkinje system, and sinoatrial (S-A) node. This slow velocity delays the impulse from reaching the ventricles, so that the atria contract 0.1 to 0.2 sec ahead of the ventricles.

52—D (Chapter 2) Potassium is secreted by the principal cells of the cortical collecting tubule. All of the other substances listed (amino acids, chloride, glucose, and sodium) are *reabsorbed* by the renal tubules.

53—A (Chapter 10) The collagen fibers of bone have great tensile strength, which means that they can bear great longitudinal stress without tearing apart.

54—D (Chapter 2) Interstitial fluid volume cannot be measured directly, but it is calculated as the difference between extracellular fluid volume (which is comprised of interstitial fluid volume and plasma volume) and the plasma volume. Extracellular fluid volume can be measured from inulin space and plasma volume can be measured from ^{125}I-albumin space.

55—C (Chapter 5) When the left ventricular pressure rises above the diastolic pressure in the aorta (about 80 mmHg), the aortic valve opens, and blood is ejected from the ventricle.

56—D (Chapter 1) Pinocytosis means imbibition of liquids by cells. It is a receptor-mediated, energy-requiring process that is especially important for transporting protein molecules through the brush border of the proximal tubular epithelium, where about 30 grams of plasma protein are absorbed daily. Phagocytosis means the ingestion of large particles, such as bacteria, cells, and portions of degenerating tissue. Unlike pinocytosis, only certain cells have the capability of phagocytosis.

57—C (Chapter 2) Destruction of the supraoptic nuclei of the brain removes the source of antidiuretic hormone and therefore makes the late distal tubule and collecting tubules impermeable to water. For this reason, water fails to be reabsorbed in these parts of the tubules, causing a large volume of diluted urine.

58—C (Chapter 10) Vitamin D_3 is cholecalciferol. Vitamin D_3 is converted to 25-hydroxycholecalciferol in the liver and this, in turn, is converted in the kidney to the active form, 1,25-dihydroxycholecalciferol.

59—C (Chapter 2) When the interstitial fluid hydrostatic pressure is negative, or subatmospheric, only the plasma colloid osmotic pressure promotes fluid transfer into blood capillaries. However, during the edematous state the interstitial fluid pressure can become positive, and thus oppose fluid loss from capillaries. Interstitial fluid hydrostatic pressure is thought to be subatmospheric during normal conditions in such tissues as skin, subcutis, skeletal muscle, and lung; positive pressures are usually found in brain, kidney, heart, and liver.

60—A (Chapter 5) Under resting conditions, the cardiac muscle derives approximately 70% of its energy from the metabolism of fatty acids instead of carbohydrates. Anaerobic glycolysis is called upon during exercise for extra amounts of energy, but this can supply very little energy in relation to the large amounts of energy required by the heart.

61—D (Chapter 10) The hypothalamic-hypophyseal venous portal system carries prolactin-inhibitory hormone from the hypothalamus to the anterior pituitary gland. In the absence of this hormone, prolactin secretion increases to about three times normal levels.

62—C (Chapter 1) The mitochondria are often called the powerhouses of the cell because as much as 95% of the adenosine triphosphate (ATP) used by a cell is formed in the mitochondria by oxidative phosphorylation.

63—E (Chapter 5) Damage to the A-V node resulting from ischemia or inflammation may increase the PR interval from a normal value of about 0.16 sec to as much as 0.25 to 0.40 sec. If conduction through the A-V node becomes severely impaired, complete block of impulses may occur, in which case the atria and ventricles will beat independently.

64—D (Chapter 10) Trophoblasts produce human chorionic gonadotropin.

65—C (Chapter 10) Addison's disease results from failure of the adrenal cortices to produce adrenocortical hormones. Loss of cortisol secretion makes it impossible to maintain normal blood glucose concentration between meals because glucose cannot be synthesized in significant quantities by gluconeogenesis.

66—C (Chapter 5) In left heart failure, large amounts of blood are pumped from the systemic circulation into the lungs, leading to pulmonary edema. The amount of blood transferred into the lungs is obviously dependent on the pumping capabilities of the right heart.

67—C (Chapter 2) The rate of diffusion of a substance is proportional to the difference in concentration of the diffusing substance between the two sides of the membrane, the temperature of the solution, the permeability of the membrane and, in the case of ions, the electrical potential difference between the two sides of the membrane. Increased molecular weight *decreases* the rate of diffusion.

68—C (Chapter 5) Resistance is equal to the pressure difference divided by the flow: $(60 - 20)/200 = 0.2$ mmHg/ml/min.

69—B (Chapter 2) Under normal conditions, the renal clearance of glucose is zero, since glucose is completely reabsorbed in the renal tubules and not excreted.

70—A (Chapter 10) A large surge of follicle-stimulating hormone (FSH) and luteinizing hormone (LH) occurs 1 to 2 days before ovulation.

71—B (Chapter 10) During a woman's reproductive life, about 450 of the primordial follicles grow into vesicular follicles and ovulate, which corresponds to 1 ova per month for about 37 years.

72—D (Chapter 2) Sweating causes fluid loss and salt loss, mainly from the extracellular compartment. Replenishing the fluid volume (without replacement of the salt) will restore the total body water, but extracellular fluid volume will be reduced because of osmosis of water from the extracellular compartment into the cell compartment.

73—A (Chapter 10) Glucagon increases blood glucose concentration, mainly by causing glycogenolysis and gluconeogenesis in the liver. Growth hormone increases blood glucose concentration by decreasing glucose utilization and uptake by the cells. Also, growth hormone increases the mobilization of fatty acids from adipose tissue and increases the use of fatty acids for energy.

74—A (Chapter 5) Sympathetic stimulation of the peripheral arterioles causes the vessels to constrict, which increases their resistance and decreases the flow of blood. When the veins are stimulated by the sympathetic nervous system, they become *less* compliant.

75—E (Chapter 7) Long-term memory is believed to result from structural changes at the synapses that augment or attenuate signal conduction.

76—E (Chapter 2) The lymphatic system performs all of the functions listed as choices (i.e., concentrate proteins in the lymph, create negative pressure in the free interstitial fluid, remove particulate materials from the interstitium, and transport antigenic materials to lymph nodes), but removal of fluid and proteins (especially proteins) is the *most important* physiologic function. In the absence of the lymphatic system, the interstitial fluid protein concentration would increase greatly, resulting in widespread extracellular edema.

77—E (Chapter 5) The Frank-Starling law of the heart states that the more the heart is filled during diastole, the greater will be the amount of blood pumped into the aorta; that is, within physiologic limits, the heart can pump all the blood that comes to it without excessive damming of blood in the veins. The ability of stretched muscle to contract with greater force is characteristic of all striated muscles. Stretching the muscle is believed to increase interdigitation of the actin and myosin filaments and thus to facilitate a more forceful contraction.

78—D (Chapter 10) The secretion of parathyroid hormone (PTH) is controlled by the concentration of extracellular ionized calcium.

79—D (Chapter 2) With chronic respiratory acidosis, the P_{CO_2} is increased because of impaired pulmonary ventilation. However, increased renal secretion of H^+ and generation of new HCO_3^- cause an increase in plasma HCO_3^-, which returns pH toward normal.

80—C (Chapter 5) The arterial baroreceptors play an important role in preventing excessive reductions in blood pressure and therefore cerebral blood flow when an upright posture is assumed. The baroreceptor reflex results in vasoconstriction of the blood vessels in the peripheral circulation, increased heart rate, and increased contractility of the heart.

81—B (Chapter 2) The most abundant anion in the extracellular fluid is chloride.

82—B (Chapter 5) Angiotensin II is one of the most powerful vasoconstrictor substances in the body.

83—A (Chapter 2) In untreated diabetes mellitus, there is an excess of free fatty acids in the plasma because of increased hydrolysis of stored triglycerides. This leads to an increased hepatic oxidation of fatty acids and thus the formation of acetoacetic acid. The concentration of acetoacetic acid often rises very high and causes severe acidosis.

84—B (Chapter 5) The first heart sound is caused by closure of the atrioventricular (A-V) valves, which occurs when the ventricles contract. The second heart sound occurs when the aortic and pulmonary valves close at the end of systole. The third heart sound sometimes occurs at the end of the first third of diastole and is believed to be caused by blood flowing with a rumbling motion into the almost-filled ventricles.

85—A (Chapter 2) Both simple and facilitated diffusion rely on an electrochemical gradient as the energy source for movement of a molecule. However, facilitated diffusion also utilizes a carrier mechanism for transport, unlike simple diffusion.

86—B (Chapter 5) Cardiac muscle has a velocity of conduction of action potentials of 0.3 m/sec to 0.5 m/sec.

87—B (Chapter 7) Increasing the pupillary aperture can cause an immediate 30-fold adaptation by increasing the amount of light that enters the eye. When a person spends a long period of time in darkness, only very small amounts of rhodopsin are split while the metabolic systems of the rods are continually building more and more rhodopsin. Consequently, rhodopsin collects in very high concentration after a period of time and greatly increases the sensitivity of the retina. Full adaptation to darkness requires as long as 1 hour.

88—E (Chapter 4) The refractory period of the action potential is the duration of time between the beginning of depolarization and the end of repolarization. In large nerve fibers, the repolarization process follows about 1/2500 sec after the depolarization process. Therefore, as many as 2500 nerve impulses (action potentials) can be transmitted along a large nerve fiber per second. On the other hand, the process of repolarization is so slow in some smooth muscle fibers that only one impulse can be transmitted every few seconds.

89—B (Chapter 5) The rate of conduction of action potentials in Purkinje fibers is about 6 times greater than the velocity in normal cardiac muscle, or 1.5 to 4.0 m/sec.

90—C (Chapter 2) Increased afferent arteriolar resistance reduces glomerular hydrostatic pressure and therefore decreases glomerular filtration rate (GFR). Increased glomerular capillary filtration coefficient increases GFR, as does decreased hydrostatic pressure in Bowman's capsule or decreased colloid osmotic pressure, both of which increase the net filtration force in the glomerular capillaries.

91—D (Chapter 5) The resistance to blood flow through the portal vasculature is increased in the cirrhotic liver, and this increases the portal venous pressure, leakage of fluid out of the capillaries, and fluid accumulation in the peritoneal cavity.

92—B (Chapter 10) Glucagon is secreted by the alpha cells of the pancreas and, within minutes, helps correct hypoglycemia by increasing the rate of breakdown of glycogen by the liver. It acts more slowly to promote gluconeogenesis.

93—C (Chapter 5) Nearly 75% of the blood enters the ventricles during the first third of diastole, called the *period of rapid filling of the ventricles*. Atrial contraction during the last third of diastole accounts for only about 25% of the filling of the ventricles.

94—A (Chapter 10) Most of the blood pumped by the right ventricle into the pulmonary artery bypasses the lungs through the ductus arteriosus. In the developed fetus, the right and left hearts pump in parallel rather than in series as occurs after birth. The left ventricle pumps only about 30% of the combined ventricular output.

95—E (Chapter 2) The concentrations of protein, potassium, sodium, and chloride ions are markedly different in intracellular and extracellular fluids. However, the total osmolarity of intracellular and extracellular (including interstitial) fluids are similar because the cell membrane is highly permeable to water.

96—D (Chapter 6) At 66 feet (~ 20 meters) below sea level, the pressure is 3 atmospheres, 1 atmosphere of pressure for each 33 feet (~ 10 meters) of water plus 1 atmosphere caused by the air above the water. Therefore, a 10-L volume of air at 66 feet is equivalent to a 30-L volume at sea level.

97—D (Chapter 2) The capillary membrane, which separates the interstitial fluid and plasma, is highly permeable to virtually all substances in the plasma except for the proteins. Therefore, the osmolarity and concentration of almost all solutes, except proteins, are similar in interstitial fluid and plasma.

98—B (Chapter 1) Glucose and amino acids cross cell membranes by facilitated diffusion, which is also called carrier-mediated diffusion. In the case of glucose, the carrier molecule can also transport several other monosaccharides that have structures similar to glucose, including mannose, galactose, xylose, and arabinose.

99—C (Chapter 2) The net hydrostatic pressure across the capillary wall is calculated as follows: Capillary hydrostatic pressure (17) – interstitial fluid hydrostatic pressure (– 3) – plasma colloid osmotic pressure (28) + interstitial fluid colloid osmotic pressure (8) = 0 mmHg.

100—C (Chapter 5) The fraction of the *end diastolic volume* that is ejected during systole is called the *ejection fraction*. During diastole, the volume of each ventricle increases to about 110 to 120 ml. During systole, about 70 ml of this blood is ejected. The 40 to 50 ml of blood that remains in each ventricle is called the *end systolic volume;* therefore an ejection fraction of 60% is normal.

101—C (Chapter 10) The major crystalline salt of the bone, hydroxy-apatite, is composed mainly of calcium and phosphate. The great compressional strength of hydroxyapatite crystals, combined with the tensile strength of collagen fibers, provides a type of construction not unlike reinforced concrete, with the sand, mortar, and rock providing the compressional strength and the iron rods providing the tensile strength.

102—A (Chapter 7) The actual sensory receptors of the organ of Corti are hair cells that synapse with a network of cochlear nerve endings. Bending the hair cells in one direction depolarizes the hair cells, and bending them in the opposite direction hyperpolarizes them. The more forceful the vibration and subsequent bending of the hair cells the greater the rate of nerve impulses. The loudness of the sound is determined by the rate of impulse transmission from the hair cells into the brain by way of the cochlear nerve.

103—D (Chapter 10) Only vasopressin (also called antidiuretic hormone, or ADH) and oxytocin are secreted by the posterior pituitary.

104—B (Chapter 6) Alveolar ventilation = respiration rate × (tidal volume – anatomic dead space volume). Therefore, alveolar ventilation = 10 × (550 – 250) = 3000 ml/min.

105—B (Chapter 7) Action potentials are most likely to be initiated at axon hillocks.

106—B (Chapter 10) In vitamin D deficiency, extra amounts of parathyroid hormone are released, which maintains the plasma calcium concentration at a value only slightly lower than normal. However, plasma phosphate concentration falls to extremely low levels because the increased parathyroid activity increases the excretion of phosphates in the urine.

107—A (Chapter 6) About 70% of the carbon dioxide in the blood is carried in the form of bicarbonate ions. Dissolved carbon dioxide entering red blood cells reacts very rapidly with water because the red blood cells contain carbonic anhydrase, an enzyme that catalyzes the reaction between water and carbon dioxide to form carbonic acid. The carbonic acid then dissociates into hydrogen and bicarbonate ions.

108—B (Chapter 2) Fluid in the distal end of the ascending loop of Henle is hypotonic regardless of the state of hydration because of the active reabsorption of sodium chloride but not water, which is impermeable in this tubular segment. In hydration, the late distal convoluted tubule and the collecting tubule become progressively more hypertonic in the presence of antidiuretic hormone (ADH), which increases water permeability of these tubular segments. The glomerular filtrate and proximal tubule have the same tonicity as plasma.

109—B (Chapter 8) The breakdown of complex foodstuffs such as carbohydrates, fats, and proteins is accomplished by the same basic process of hydrolysis. The only difference is that different enzymes are required to promote the reactions for each type of food.

110—C (Chapter 4) The contraction of a muscle fiber begins a few milliseconds after the upstroke of the action potential, and it usually lasts about 100 times longer than the action potential.

111—D (Chapters 2 and 10) The posterior pituitary is the site of the release of antidiuretic hormone (ADH), which is necessary for normal reabsorption of water by the distal tubules and collecting ducts. In the absence of ADH, a large volume of diluted urine is formed, resulting in a condition called diabetes insipidus.

112—D (Chapter 4) Smooth muscle contains both actin and myosin filaments but the physical organization of these filaments differs from that of skeletal muscle. Instead of being arranged in sarcomeres, the actin filaments radiate from "dense bodies" that are either attached to the cell membrane or held in place within the cell by a scaffold of structural proteins. The actin filaments, which radiate from two dense bodies, overlap a single myosin filament located midway between the dense bodies so that when contraction occurs the dense bodies move toward each other.

113—B (Chapter 2) If a substance passes through the glomerular membrane with perfect ease, the glomerular filtrate contains virtually the same concentration of the substance as does the plasma, and if the substance is neither secreted nor reabsorbed by the tubules, all of the filtered substance continues on into the urine. Therefore, the plasma clearance of the substance is equal to the glomerular filtration rate (GFR). Inulin clearance is commonly used as a measure of GFR because it is freely filtered, but neither secreted nor reabsorbed by the kidney tubules.

114—B (Chapter 4) The myosin head functions as an ATPase enzyme, allowing it to cleave ATP and to use the energy derived from the high-energy phosphate bond to energize the contraction process.

115—C (Chapter 6) Contraction of the diaphragm elongates the thoracic cavity, causing the lungs to expand. Contraction of the external intercostals causes the ribs to rotate upward. This increases the volume of the thoracic cavity, causing the lungs to expand. The abdominal muscles and internal intercostals are muscles of expiration.

116—A (Chapter 5) At the end of systole, the ventricular muscle relaxes, allowing the ventricular pressure to decrease greatly. When the ventricular pressure falls below the atrial pressure, the A-V valves open, and blood from the atria flows into the ventricles.

117—A (Chapter 7) The myelin sheaths of the myelinated neuron virtually eliminate ion flow through the neuronal membrane except at the juncture of two adjacent Schwann cells that make up the myelin sheaths. These breaks in myelination, called nodes of Ranvier, occur about every 1 to 3 mm along the length of an axon and allow the ions to move with ease between the extracellular fluid and the axon. Therefore, the depolarization process can occur only at the nodes, causing the action potential to jump relatively large distances from node to node. This mechanism increases the velocity of nerve transmission as much as 5-fold to 50-fold.

118—B (Chapter 8) Cotransport with the sodium ion is the process that transports amino acids across the luminal surface of the epithelium that lines the small intestine.

119—C (Chapter 9) Creatine phosphate (also called phosphocreatine) is the most abundant store of high-energy phosphate bonds in the cells. Although it cannot energize directly many of the intracellular reactions, it can transfer energy interchangeably with ATP; for example, the slightest decrease in ATP levels leads to rapid transfer of energy from creatine phosphate to replenish the ATP stores. When extra amounts of ATP become available, the stores of creatine phosphate are replenished. During anaerobic metabolism the stores of creatine phosphate are used up within a few minutes.

120—B (Chapter 5) Ectopic pacemakers frequently develop in the Purkinje system when the transmission of impulses from the sino-atrial (S-A) node is blocked along the conduction pathway to the ventricular muscle, causing the atria to beat at a normal rate and the ventricles to beat at a much lower rate, between approximately 15 and 40 beats per minute.

121—D (Chapter 3) Vitamin B_{12} and folic acid are essential for the synthesis of DNA. Lack of either vitamin leads to decreased DNA and thus failure of nuclear maturation and division. The erythroblastic cells of the bone marrow become larger than normal and are called *megaloblasts*. The adult red blood cell has a flimsy membrane and is often large, with an ovoid shape rather than the usual biconcave disk shape. Therefore, vitamin B_{12} or folic acid deficiency causes *maturation failure* in the process of erythropoiesis.

122—B (Chapter 4) The endplate potentials are often too small to stimulate the muscle fibers and the result is muscle weakness. In severe cases, the patient dies following paralysis of the respiratory muscles. When treated with a drug, such as neostigmine, which can inactivate the enzyme that normally inactivates ACh, the levels of ACh build and thereby enhance neuromuscular transmission.

123—B (Chapter 9) Fat is transported from one part of the body to another almost entirely in the form of free fatty acids. On leaving the fat cells, the free fatty acids ionize strongly in the plasma and immediately combine with albumin. The concentration of free fatty acid in plasma under resting conditions is about 15 mg/dl; there is approximately 0.45 g of fatty acids in the entire circulatory system.

124—B (Chapter 8) Pancreatic secretion contains a large quantity of α-amylase, a digestive enzyme that hydrolyzes starches, glycogen, and most other carbohydrates to form disaccharides and a few trisaccharides.

125—D (Chapter 7) The ciliary muscle encircles the lens and is attached to the lens by ligaments. When an object moves closer to the eye, the ciliary muscle contracts, which decreases the tension on the ligaments. The elastic lens can then change passively from a relatively flat shape to a round shape, allowing the image to be focused on the retina.

126—C (Chapter 3) Iron is a primary nutritive factor necessary for the formation of hemoglobin. Iron is present in the diet in only very small quantities and even then is rather poorly absorbed from the gastrointestinal tract. Therefore, many persons fail to form sufficient quantities of hemoglobin to fill the red blood cells as they are being produced. This causes *hypochromic anemia,* in which the number of cells may be normal but the amount of hemoglobin in each cell is far below normal.

127—E (Chapter 4) Any event that leads to the opening of sodium channels can lead to the development of an action potential if the rising voltage itself is sufficient to cause still more voltage-gated sodium channels to open and, thus, a positive feedback vicious cycle to begin. A sudden increase in membrane potential from a typical resting value of –90 mV to a typical threshold value of –65 mV will usually initiate an action potential in a motor neuron.

128—C (Chapter 9) During the course of febrile illness, pyrogens are released from degenerating tissues or secreted by toxic bacteria. These pyrogens increase the set-point temperature of the hypothalamus to a value greater than the normal body temperature. All of the mechanisms for raising body temperature—shivering, constriction of skin blood vessels, and so on—are then brought into play, and within a few hours the body temperature rises to the set-point temperature.

129—D (Chapter 3) Most of the coagulation factors are formed in the liver. Therefore, hepatitis, cirrhosis, and other diseases of the liver can depress the normal coagulation system, causing a person to bleed excessively. Another cause of decreased coagulation factor production by the liver is vitamin K deficiency. Vitamin K is necessary for the formation of prothrombin and factors VII, IX, and X.

130—B (Chapter 6) Surfactant is a *surface active substance* that decreases the surface tension of the alveolar fluid. Decreased surface tension causes increased pulmonary compliance, which decreases the work required to expand the lungs.

131—C (Chapter 7) The typical motoneuron receives excitatory and inhibitory inputs from other neurons; approximately 80% to 95% of them are located on the dendrites and only 5% to 20% are located on the soma.

132—C (Chapter 3) Pernicious anemia is caused by a vitamin B_{12} deficiency. This usually results from failure to absorb vitamin B_{12} from the gastrointestinal tract rather than from lack of vitamin B_{12} in the diet.

133—E (Chapter 8) The sodium ion is most closely associated with fluid absorption.

134—C (Chapter 7) The color orange has a wavelength closer to that of red light than to that of green light.

135—B (Chapter 3) Both vitamin B_{12} and folic acid are required for DNA synthesis in all cells. Because tissues producing red blood cells grow more rapidly than most other tissues, deficiencies in these vitamins inhibit especially the rate of production of red blood cells.

136—D (Chapter 4) When calcium ions bind to calmodulin, the calmodulin-calcium combination then joins with and activates myosin kinase, a phosphorylating enzyme. A portion of the myosin head becomes phosphorylated in response to the myosin kinase, allowing the head to bind with the actin filament. The smooth muscle membrane contains many voltage-gated calcium channels. Thus, unlike skeletal muscle, the flow of calcium ions into the interior of the smooth muscle cell is mainly responsible for the generation of action potentials. However, it is also true that smooth muscle contraction is very often initiated not by action potentials but by factors that act directly on the contractile machinery.

137—C (Chapter 6) Even during rest, about 70% of the oxygen in the arterial blood is removed as it passes through the coronary circulation, causing the coronary venous oxygen tension to be very low, at about 20 mmHg. Mixed venous blood in the pulmonary artery has an oxygen tension of about 40 mmHg.

138—D (Chapter 7) Pressure on the bottom of the feet causes the extensor muscles of the legs to tighten, helping to support the weight of the body. This reflex, called the positive supportive reflex, is believed to be integrated entirely in the spinal cord.

139—B (Chapter 9) When the body is too cool (below the hypothalamic set-point temperature), heat production increases, piloerection occurs, and the blood vessels in the skin constrict.

140—D (Chapter 7) The characteristic result of parasympathetic stimulation is slowing of the heart. Parasympathetic stimulation can cause limited sweating on the palms of the hands and copious secretion of the salivary glands.

141—E (Chapter 3) In anemia, the total peripheral resistance is decreased by two factors: a reduced viscosity of the blood and vasodilation resulting from decreased oxygenation of the tissues. The increase in cardiac output caused by the decrease in peripheral resistance can provide the tissues with adequate amounts of oxygen during resting conditions. However, when an anemic person begins to exercise, the heart may not be capable of pumping the extra amounts of blood needed by the tissues.

142—E (Chapter 9) When more food energy is consumed than is used each day, the excess is stored in the fat depots (triglycerides) of the body.

143—A (Chapter 4) When an action potential spreads over the terminal end of a motoneuron, voltage-gated calcium channels open, causing calcium ions to diffuse into the neuron terminal. The calcium ions cause ACh vesicles within the neuron terminal to fuse with the neural membrane and empty their contents into the synaptic trough by the process of exocytosis.

144—D (Chapter 6) The blood-brain barrier is almost totally impermeable to hydrogen ions so that increases in the blood hydrogen ion concentration have relatively little effect on the hydrogen ion concentration in the vicinity of the respiratory center. Carbon dioxide, on the other hand, permeates the blood-brain barrier with ease and immediately reacts with water to form hydrogen ions. Thus, more hydrogen ions are released in the respiratory center when the blood carbon dioxide concentration increases than when the blood hydrogen ion concentration increases. Unlike the effects of carbon dioxide and hydrogen ion concentration, oxygen lack does not directly stimulate the respiratory center. Instead, it excites special nerve receptors called *chemoreceptors,* which are located in minute *carotid* and *aortic bodies* that lie, respectively, in the carotid bifurcations and along the aorta.

145—E (Chapter 7) The movement of sodium ions to the inside of the membrane causes the neuron to depolarize and therefore to be stimulated.

146—A (Chapter 2) An arterial pH of 7.22 indicates that the patient has acidosis. The next issue is whether the acidosis is metabolic or respiratory. If the patient has respiratory acidosis, the P_{CO_2} would be increased and the plasma bicarbonate would be increased. The low plasma bicarbonate of 8 mmol/L indicates metabolic acidosis, and the low P_{CO_2} of 20 mmHg indicates that there is some degree of respiratory compensation for the acidosis.

147—B (Chapter 7) The cerebrospinal fluid (CSF) is found in the cisterns around the brain, in the ventricles of the brain, and in the subarachnoid space around both the brain and the spinal cord. Because the brain and CSF have approximately the same specific gravity, the brain simply floats in the fluid.

148—C (Chapter 3) Hemoglobin S is an abnormal type of hemoglobin caused by abnormal composition of the beta chains. When hemoglobin S is exposed to low concentrations of oxygen, it precipitates into long crystals inside the red blood cell, causing the cell to be shaped like a sickle rather than a biconcave disk. The red blood cells become very fragile and rupture during passage through the microcirculation, especially in the spleen. The life span of the cells is so short that serious anemia results in a condition called sickle cell anemia.

149—D (Chapter 8) Secretin is released by the presence of acid in the duodenum. The actions of secretin that function to decrease the amount of acid in the duodenum include inhibition of gastric acid secretion, inhibition of gastric emptying, and release of pancreatic bicarbonate. For these reasons, secretin has been nicknamed nature's antacid.

150—D (Chapter 4) In primary active transport, the energy required to transport a substance against its electrochemical gradient is derived directly from ATP or some other high-energy phosphate compound. The Na^+-K^+ pump transports sodium ions out of the cell at the same time that potassium ions are transported into the cell, and it is responsible for maintaining the sodium and potassium concentration differences across the cell membrane as well as for establishing a negative electrical potential inside the cells. The pump is present in all living cells of the body.

151—C (Chapter 9) Evaporation will occur as long as the relative humidity is below 100%. For each gram of water that evaporates from the body surface, 580 calories of heat are lost from the body.

152—A (Chapter 3) Mature red blood cells of mammals do not have a nucleus, mitochondria, or endoplasmic reticulum. They do, however, contain enzymes that are capable of metabolizing glucose and forming small amounts of ATP, both of which serve the red blood cells in several important ways.

153—D (Chapter 6) Hematocrit and hemoglobin concentration can vary widely without significantly affecting the arterial oxygen tension because oxygen diffuses along a partial pressure gradient from the alveoli into the blood until the gradient is virtually dissipated. This event is nearly independent of the oxygen-carrying capacity of the blood, as exemplified by the fact that arterial oxygen tension is usually normal during anemia and polycythemia. In contrast, the hemoglobin concentration and hematocrit have a great effect on arterial oxygen content. Pulmonary edema can decrease the arterial oxygen tension in two ways: by increasing the thickness of the pulmonary membrane through which the oxygen must diffuse, and by decreasing the surface area available for oxygen diffusion when the edema fluid actually fills the alveoli.

154—B (Chapter 7) The fibers of the pyramidal tract begin at pyramidal cells in the primary motor cortex. They usually cross to the contralateral side at the juncture of the medulla and spinal cord, and most of them synapse with interneurons in the spinal cord gray matter.

155—E (Chapter 6) The tidal volume can never be greater than the vital capacity because the vital capacity is the maximum amount of air that can be expired after the deepest possible breath has been taken.

156—A (Chapter 4) Small ions such as Na^+, K^+, and Cl^- diffuse across a cell membrane in accordance with both chemical and electrical gradients. Assume that a cell contains only K^+ and that the concentration of K^+ is the same on both sides of the cell membrane. When a negative charge is applied to the inside of the cell, the positively charged K^+ will continue to diffuse into the cell even against a K^+ concentration gradient, until electrochemical balance is achieved. The electrical potential that prevents net diffusion of an ion in either direction through a cell membrane is called the *Nernst potential* for that ion.

157—C (Chapter 6) The rate of gas diffusion across the pulmonary membrane is proportional to the membrane area; it is *inversely* proportional to membrane thickness. The total surface area of the pulmonary membrane is 50 to 100 m^2 and its thickness averages about 0.5μm. The rate of diffusion is also proportional to the diffusion constant for the gas, which is proportional to the solubility of the gas and inversely proportional to the molecular weight of the gas. When compared with oxygen, carbon dioxide has a much greater solubility but a similar molecular weight, and it diffuses through the pulmonary membrane far more easily than does oxygen.

158—B (Chapter 7) When the voluntary movements of a person are jerky but there is no tremor when the person is resting, the condition is known as an *intention tremor*. Such a tremor is usually caused by failure of the cerebellum to dampen the motor movements. Damage to certain of the basal ganglia will also cause a tremor, but this tremor usually continues as long as the person is awake.

159—A (Chapter 8) Cholecystokinin will cause the gallbladder to contract. Gastrin in high concentrations also can cause the gallbladder to contract, but the effect may not have physiologic relevance.

160—D (Chapter 7) Serotonin is the transmitter substance that is released by the spinal nerve endings of the neurons whose cell bodies are located in the raphe magnus nucleus. These neurons are part of a pain control system called an analgesia system.

161—D (Chapter 2) The rate of simple diffusion increases proportionately with the concentration of the diffusing substances, as shown by line A. Facilitated diffusion, also called carrier-mediated diffusion, indicated by line B, differs from simple diffusion in the following way: The rate of facilitated diffusion approaches a maximum as the concentration of the substance increases. This maximum rate (point *D*) is dictated by the rate at which carrier-protein molecules can move the diffusing substances across the membrane and also by the total number of protein carriers in the membrane.

162—D (Chapter 7) Each sympathetic pathway from the spinal cord comprises a preganglionic neuron and a postganglionic neuron. Postganglionic neurons can originate in one of the sympathetic chain ganglia or in the prevertebral ganglion, and from there each neuron travels to a tissue where it terminates. In contrast, the preganglionic neurons of the parasympathetic nervous system travel all the way to the target organ, where they synapse with relatively short postganglionic neurons.

163—C (Chapter 4) Because the transverse tubules (T tubules) are invaginations of the outer cell membrane, their interior is open to the extracellular fluid. They function to transmit depolarizations rapidly to the depths of the cell interior, and they appear to be especially well developed in thick muscle fibers.

164—B (Chapter 7) The macula of the utricle functions to help maintain equilibrium in the upright position and is independent of the hearing apparatus.

165—C (Chapter 4) The sarcoplasmic reticulum is composed of longitudinal tubules that terminate in large chambers called terminal cisternae. When an action potential of an adjacent T tubule causes current to flow through the cisternae, calcium is released from the cisternae and longitudinal tubules. The calcium ions then diffuse to adjacent myofibrils, where they bind strongly with troponin; this, in turn, elicits the muscle contraction. The muscle contraction will continue as long as the calcium ions remain in high concentration in the sarcoplasmic fluid. However, a continually active calcium pump located in the walls of the sarcoplasmic reticulum pumps calcium ions out of the sarcoplasmic fluid, returning it to the vesicular cavities of the reticulum.

166—B (Chapter 6) In a normal person, each 100 ml of blood contains about 15 grams of hemoglobin, and each gram of hemoglobin can bind with about 1.34 ml of oxygen when it is 100% saturated (15 × 1.34 = 20 ml O_2/100 ml blood). The hemoglobin in venous blood leaving the peripheral tissues is about 75% saturated with oxygen so that the amount of oxygen transported by hemoglobin in venous blood is about 15 ml O_2/100 ml blood. Therefore, about 5 ml of O_2 is normally transported to the tissues in each 100 ml of blood.

167—A (Chapter 7) Many naturally occurring plant poisons have a bitter taste, causing animals to reject them.

168—A (Chapter 8) Pepsinogen is an inactive precursor of the proteolytic enzyme pepsin. Following secretion from the peptic and mucous cells of the gastric glands, the pepsinogen comes into contact with hydrochloric acid and previously formed pepsin, which split the pepsinogen to form active pepsin.

169—E (Chapter 6) About 2% of the blood that enters the aorta has passed through the bronchial circulation, which is not exposed to the pulmonary air and thus has an oxygen tension similar to that of venous blood—about 40 mmHg. The bronchiolar blood mixes with the blood in the pulmonary veins, causing the oxygen tension to decrease from about 104 mmHg to about 95 mmHg.

170—B (Chapter 7) The diverging circuit is typified by the nervous control of muscular activity. Stimulation of a single large neuron in the motor cortex might lead to stimulation of several thousand fibers.

171—B (Chapter 4) When ACh is released into the synaptic trough, it attaches to ACh-gated ion channels on the postsynaptic membrane, causing them to open. The net effect of opening the channels is to allow large numbers of sodium ions to pour inside the fiber. This creates a local potential inside the fiber that initiates an action potential at the muscle membrane.

172—D (Chapter 8) Bile salts are formed in the hepatic cells from cholesterol, and in the process of secreting the bile salts, about 1 to 2 g of cholesterol are also secreted daily. Under abnormal conditions, the cholesterol may precipitate, causing cholesterol gallstones to develop in the gallbladder. Some causes of gallstone formation include excess absorption of water or bile acids from bile, excess cholesterol in bile, and inflammation of gallbladder epithelium.

173—D (Chapter 4) The mechanism of smooth muscle contraction differs somewhat from that of skeletal muscle. In the place of troponin (in skeletal muscle), the smooth muscle contains a regulatory protein called calmodulin. When calmodulin reacts with calcium ions, it initiates contraction by activating myosin crossbridges.

174—C (Chapter 6) Dissolved CO_2 reacts with water inside red blood cells to form *carbonic acid*. This reaction is catalyzed by a protein enzyme in the red cells called *carbonic anhydrase*. Most of the carbonic acid immediately dissociates into bicarbonate ions and hydrogen ions; in turn, the hydrogen ions combine with hemoglobin. Approximately 23% of the CO_2 produced in the tissues combines directly with hemoglobin to form *carbaminohemoglobin,* and an additional 7% is transported in the dissolved state in the water of the plasma and cells.

175—E (Chapter 7) In a reverberating circuit, the output of a neuronal circuit feeds back to reexcite the same circuit, making it possible for the circuit to discharge repetitively for a very long time.

176—D (Chapter 7) Parkinson's disease almost invariably results from widespread destruction of areas in the substantia nigra that send dopamine-secreting nerve fibers to the caudate nucleus and putamen.

177—C (Chapter 7) Several structures of the limbic system, including the hypothalamus, are particularly concerned with the affective nature of sensory sensations, that is, whether the sensations are pleasant or unpleasant. Electrical stimulation of certain areas evokes pleasant sensations, whereas electrical stimulation of other areas causes pain, fear, defense or escape reactions, and other unpleasant sensations.

178—A (Chapter 4) The ends of the actin filaments are attached to the Z disks, and they extend in both directions to interdigitate with the myosin filaments. The Z disk passes from myofibril to myofibril, attaching the myofibrils together all the way across a muscle fiber. The portion of a myofibril that lies between two Z disks is called a sarcomere.

179—C (Chapter 7) The flexor reflex is integrated in the spinal cord. This reflex, also called the withdrawal reflex, is elicited most powerfully by the stimulation of pain endings, which causes the limb to withdraw from the stimulus.

180—B (Chapter 10) Contraction of the uterus and initiation of parturition are inhibited by progesterone and stimulated by estrogens.

181—D (Chapter 2) Although increased potassium intake stimulates aldosterone secretion, which increases tubular secretion and urinary excretion of potassium, it also has a natriuretic effect that opposes the sodium-retaining action of aldosterone. The net result is either no change or a transient increase in urinary sodium excretion.

182—A (Chapter 10) The general model for the action of steroid hormones such as cortisol is the following: The steroid hormone enters the target cell, is transported to the nucleus with a receptor protein molecule, and finally interacts with one or more specific parts of the genome to initiate or increase synthesis of a specific protein. This mechanism is different from the action of most protein hormones, which rely on a second-messenger system, often the cAMP system, to mediate their intracellular effects.

183—D (Chapter 5) Increased pressure in the carotid sinus activates the baroreceptor reflect, which in turn elicits mechanisms that tend to decrease blood pressure: Vagal stimulation of the heart decreases heart rate, and decreased sympathetic stimulation of the entire cardiovascular system decreases both heart rate and contractility and produces widespread vasodilation.

184—D (Chapter 8) Mucus is secreted by all segments of the gastrointestinal tract. Mucus is an excellent lubricant, and it protects the mucosa from the intestinal juices as well as from different types of food. The pancreatic juice contains enzymes for digestion of fats, proteins, and carbohydrates as well as bicarbonate ions that serve to neutralize acid emptied into the duodenum from the stomach.

185—D (Chapter 2) In normal kidneys, autoregulatory mechanisms maintain a relatively constant glomerular filtration rate (GFR) and renal blood flow in the range of systemic arterial pressures from 80 to about 160 mmHg. Constriction of the afferent arterioles reduces glomerular hydrostatic pressure and therefore GFR as well as renal blood flow. Constriction of the efferent arterioles, however, decreases renal blood flow while increasing glomerular hydrostatic pressure and therefore increasing GFR. An increase in renal blood flow, even with little change in glomerular hydrostatic pressure, can increase GFR by decreasing filtration fraction and therefore the colloid osmotic pressure of the glomerular capillaries, which increases the net filtration force.

186—D (Chapter 5) The pulse pressure—the difference between the systolic and diastolic pressures—is increased when the amount of blood that is ejected during systole (stroke volume) increases or when the distensibility of the arterial tree decreases. Mitral valve stenosis would tend to reduce filling of the ventricles and therefore reduce stroke volume.

187—D (Chapter 10) Cortisol suppresses inflammation in part by stabilizing lysosomal membranes.

188—A (Chapter 1) The lysosomes contain enzymes that are used by the cell for digestive purposes.

189—E (Chapter 10) Thyroxine decreases the plasma concentrations of triglycerides, cholesterol, and phospholipids, but it increases the level of free fatty acids.

190—C (Chapter 7) High-frequency vibration is detected by pacinian corpuscles, which are specialized sensory receptors. Other modalities of sensation detected by free nerve endings are heavy pressure and probably warmth and cold.

191—A (Chapter 5) Blood flow velocity is inversely related to surface area. Since capillaries have the greatest surface area, they have the lowest blood flow velocity of any vascular segment.

192—B (Chapter 3) A coumarin such as warfarin will not prevent coagulation of a blood sample *outside* the body, but it does prevent coagulation of blood in the body. Warfarin causes this effect by competing with vitamin K for reactive sites in the enzymatic processes for formation of prothrombin as well as factors VII, IX, and X in the liver.

193—D (Chapter 10) Most individuals with Graves' disease (hyperthyroidism) have an enlarged thyroid gland with each cell secreting more than normal amounts of thyroid hormone; however, the plasma levels of thyroid-stimulating hormone (TSH) are found to be less than normal and often zero. Instead, thyroid-stimulating antibodies, designated as TSAb, which bind to the same membrane receptors as TSH, are found in the plasma, and the high levels of thyroid hormone in turn suppress anterior pituitary formation of TSH.

194—B (Chapter 6) High-altitude hypoxia decreases the size of the skeletal muscle fibers, which facilitates acclimatization by decreasing the distances between the capillaries.

195—B (Chapter 1) Cytosine is a *pyrimidine* base, rather than a purine base. DNA is composed of multiple units of the sugar ribose, phosphoric acid, and four nitrogenous bases, including two purines (adenine and guanine) and two pyrimidines (thymine and cytosine).

196—B (Chapter 10) Growth hormone causes the liver to release several small proteins called somatomedins that function to increase the growth of bones and other peripheral diseases.

197—D (Chapter 1) The large surface area of the endoplasmic reticulum and the numerous associated enzymes provide the machinery for many metabolic functions of the cell. The granular endoplasmic reticulum has many ribosomes attached that function in the synthesis of proteins. Electron micrographs show that the internal portion of the endoplasmic reticulum is connected with the space between the two membranes of the double nuclear membrane.

198—E (Chapter 5) One definition of circus movement is the continuous movement of an impulse around the heart. When the impulse arrives back at the starting point, the tissue is reexcited, causing another impulse to travel around the heart because the tissue is not in a refractory state. Therefore, almost any factor that increases the path length around the atria or ventricles, or decreases the refractory period of the heart tissue, can promote circus movement.

199—B (Chapter 10) The physiologic effects of hypothyroidism are the same regardless of the cause. They include fatigue, extreme muscular weakness, decreased heart rate, decreased cardiac output, decreased metabolic rate, myxedema, decreased blood volume, sometimes increased body weight, constipation, *decreased* respiratory rate, and mental sluggishness.

200—D (Chapter 8) Transporting hydrogen ions out of the gastric mucosa decreases the local hydrogen ion concentration, and this is believed to offer some protection against autodigestion of the stomach. The gastric mucosa also is protected by a thick layer of mucus.

201—D (Chapter 5) If the ventricular contraction is *not* preceded by a P wave, this means that atrial depolarization and contraction did not precede the ventricular contraction associated with the QRS complex. Premature ventricular contractions (PVCs) usually result from *ectopic foci* in the heart that emit abnormal impulses during the cardiac cycle. Although some PVCs originate from cigarette smoking, coffee, or emotional stress, a large share of PVCs originate from infarcted or ischemic areas of the heart.

202—E (Chapter 10) Testosterone has several important effects, including descent of the testes into the scrotum, formation of the fetal penis, increased muscle development, and increased thickness of the skin. It is not associated with the initiation of ejaculation.

203—B (Chapter 5) The pulmonary blood flow (cardiac output) can increase greatly without causing much increase in pulmonary arterial pressure because the pulmonary resistance decreases almost as much as the pulmonary flow increases. This decrease in resistance is caused by recruitment of previously closed pulmonary capillaries as well as distention of the capillaries.

204—D (Chapter 8) Sucrose is a disaccharide that is digested in the small intestine to make a molecule of glucose and a molecule of fructose. Essentially all carbohydrates are absorbed from the gut as monosaccharides.

205—E (Chapter 2) The urinary pH in normal kidneys is *decreased* in response to acidosis. In respiratory acidosis, the increased P_{CO_2} stimulates secretion of hydrogen ions that react with the filtered bicarbonate to cause almost complete bicarbonate reabsorption. The excess hydrogen ions in the renal tubule combine with urinary buffers, especially ammonia, and are excreted as buffer salts. Glutamine uptake by the kidney and ammonia production are increased in acidosis, providing more urinary buffers.

206—B (Chapter 4) The potassium permeability of the cell membrane increases, allowing rapid loss of potassium ions to the exterior.

207—C (Chapter 10) Hair develops in the pubic region and in the axillae after puberty, but androgens formed by the adrenal glands are mainly responsible for this.

208—A (Chapter 7) Adrenergic stimulation causes the liver to release glucose, thereby increasing the blood glucose concentration.

209—D (Chapter 2) Increased capillary permeability, increased capillary pressure, and increased interstitial fluid colloid osmotic pressure would increase fluid movement into the interstitial spaces. Increased plasma colloid osmotic pressure, however, would oppose fluid movement from the capillaries into the interstitial compartment.

210—C (Chapter 10) Loss of blood flow through the placenta increases the total peripheral resistance, and expansion of the lungs pulls open the pulmonary vessels, decreasing the pulmonary vascular resistance.

211—D (Chapter 2) The factors that promote fluid loss from capillaries cause lymph flow to increase because the lymphatic vascular system transports the extravasated fluid back to the blood vascular system. Increased plasma colloid osmotic pressure opposes fluid loss from the capillaries.

212—D (Chapter 5) Decreased blood pressure in the internal carotid artery reduces the stretch of arterial baroreceptors and therefore decreases the rate of nerve impulses of the carotid sinus nerves. This increases the sympathetic outflow from the vasomotor centers, causing increased heart rate, increased peripheral vascular resistance, and increased strength of heart muscle contraction.

213—C (Chapter 2) In metabolic acidosis, there is a reduction in plasma $[HCO_3^-]$ and in the filtered load of HCO_3^-, which causes a relative excess of H^+ in the renal tubules. The H^+ that does not titrate with HCO_3^- combines with urinary buffers, especially ammonia, and is excreted as NH_4^+. This results in increased renal production of new HCO_3^- and compensates for the initial reduction in plasma $[HCO_3^-]$.

214—D (Chapter 10) Although parathyroid hormone increases the absorption of both calcium and phosphate from bone, the extracellular phosphate concentration often actually decreases because parathyroid hormone increases the rate of excretion of phosphate by the kidneys.

215—B (Chapter 9) When the hypothalamus becomes too cold, heat loss is decreased by constriction of skin blood vessels, and heat production is increased by shivering and by an epinephrine-induced increase in metabolic rate. These mechanisms help return the body temperature to the hypothalamic set-point temperature.

216—C (Chapter 2) The thick ascending limb of the loop of Henle is highly *impermeable* to water, although it actively transports sodium chloride from the tubular fluid to the interstitial fluid. The descending loop of Henle, however, is highly permeable to water. The vasa rectae help to minimize solute loss from the renal medulla because of the very slow blood flow and the countercurrent flow pattern of these vessels.

217—E (Chapter 10) Thyroid-stimulating hormone has no known effect on the synthesis of thyroxine-binding globulin.

218—B (Chapter 3) When a vessel is severed, the wall of the vessel contracts or spasms immediately, and this decreases the loss of blood from the injured vessel.

219—D (Chapter 2) Excess aldosterone secretion results in increased tubular secretion of potassium as well as a net movement of potassium into the cells, causing marked hypokalemia. Aldosterone also causes transient decreases in urinary sodium excretion, modest expansion of extracellular fluid volume, and mild hypertension. However, within a few days the kidneys "escape" from sodium retention, and sodium balance is reestablished under steady-state conditions. Therefore, sodium excretion returns to normal under steady-state conditions.

220—B (Chapter 10) Progesterone does not cause the alveoli to actually secrete milk. However, progesterone promotes development of lobules and alveoli of the breasts.

221—B (Chapter 8) The propulsive movement of the gastrointestinal tract results from smooth muscle contraction. Smooth muscle is not striated.

222—A (Chapter 4) When insufficient amounts of ACh are released, the endplate potential may be too small to initiate an action potential, in which case the muscle fiber will *not* contract.

223—A (Chapter 2) Dehydration, or loss of water but not solute, causes increased plasma osmolarity, which stimulates secretion of antidiuretic hormone (ADH). ADH, in turn, *increases* the permeability of the collecting ducts to water, thereby increasing renal tubular water reabsorption.

224—C (Chapter 1) Active transport via carrier proteins, facilitated diffusion via carrier proteins, and simple diffusion through protein channels are all highly selective for specific substances. Simple diffusion through lipid bilayer is not highly selective for specific substances; any lipid-soluble substance can rapidly penetrate the lipid bilayer.

225—A (Chapter 7) Visceral pain is usually caused by stimulation of pain endings over a wide area; for example, the organs of the abdomen are usually insensitive to a pin prick or even a sharp knife cut.

226—A (Chapter 8) Chyme is the semifluid material produced by gastric digestion of food. Gastrin is secreted from gastrin cells (G cells) located in the antrum of the stomach. Hydrochloric acid and intrinsic factor are both secreted by parietal (or oxyntic) cells, and pepsinogen is secreted by peptic (or chief) cells, all located in the body and fundus of the stomach.

227—D (Chapter 3) Any condition that sufficiently decreases the renal microvascular or tissue Po_2 will cause increased amounts of erythropoietin to be released from the kidneys. Such conditions include anemia and CO poisoning, in which arterial Po_2 is normal and arterial O_2 content is reduced; hypoxemia, in which the hemoglobin is not fully saturated and arterial Po_2 and arterial O_2 content are low; and others.

228—C (Chapter 4) All cell membranes contain sodium-potassium pumps that transport sodium to the outside of the cell and potassium to the inside. The result is relatively high intracellular potassium concentration and relatively low intracellular sodium concentration.

229—C (Chapter 10) Growth hormone decreases carbohydrate metabolism.

230—B (Chapter 9) When the hypothalamus becomes overheated, its temperature-regulating capabilities are greatly diminished, sweating may cease to occur, and the person may become poikilothermic. The rise in body temperature will perpetuate itself by increasing the metabolic rate. When the core temperature rises above 106°F to 108°F, the person is likely to develop heat stroke.

231—A (Chapter 6) Oxygen therapy is of little value in anemia because plenty of oxygen is already available in the alveoli and the hemoglobin in the blood has fully equilibrated with the alveolar oxygen. Instead, the problem in anemia is too little hemoglobin in the blood, which reduces the amount of oxygen that is carried in the blood. Nevertheless, oxygen therapy can cause small amounts of extra oxygen to be transported in the dissolved state although the amount transported by the hemoglobin is barely altered.

232—B (Chapter 7) The preganglionic neurons of the parasympathetic and sympathetic nervous systems are cholinergic; that is, they release ACh at the terminal synapse. The postganglionic neurons of the parasympathetic nervous system are all cholinergic. On the other hand, most of the postganglionic sympathetic neurons are adrenergic; that is, they release norepinephrine at the effector organ. Notable exceptions include postganglionic cholinergic fibers to the sweat glands, piloerector muscles, and a few blood vessels.

233—C (Chapter 3) The O-A-B blood groups are determined by two genes, one on each pair of chromosomes. Because one gene comes from each parent, a mother with blood group AB and a father with blood group OO can produce offspring with blood groups OB or OA, but not blood group AB.

234—A (Chapter 8) Acidic chyme in the duodenum inhibits gastric emptying by way of a neural reflex.

235—A (Chapter 10) Diabetes mellitus is associated with decreased glucose uptake and utilization in most cells of the body. As a result, cell utilization of glucose falls and utilization of fats is increased, releasing acetoacetic acid into the plasma more rapidly than it can be taken up and oxidized by the cells. As a result, the patient develops *acidosis*.

236—D (Chapter 6) During resting conditions, arterial blood is almost fully saturated with oxygen. Therefore, it is not possible to increase significantly the arterial oxygen content by increasing alveolar ventilation.

237—A (Chapter 7) The sympathetic and parasympathetic nervous systems have opposite effects on virtually every tissue and organ that they innervate. Skeletal muscle fibers are not innervated by the autonomic nervous system.

238—C (Chapter 8) Swallowing is initiated voluntarily when the tongue forces a bolus of food upward and backward against the palate. Once the food is forced into the pharynx, the swallowing process becomes involuntary.

239—A (Chapter 3) In most instances of erythroblastosis fetalis, the mother is Rh negative and the father is Rh positive. The infant has inherited the Rh-positive characteristic from the father, and the mother has developed anti-Rh agglutinins, which have diffused through the placenta into the fetus to cause red blood cell agglutination. Erythroblasts are nucleated erythrocytes.

240—A (Chapter 4) The rate of sodium movement out of the cell exceeds the rate of potassium movement into the cell. This produces a net outward movement of positive charges, which contributes to negativity inside the cell.

241—C (Chapter 2) Decreased efferent arteriolar resistance *increases* peritubular capillary hydrostatic pressure while reducing glomerular capillary hydrostatic pressure and glomerular filtration rate. At the same time, decreased efferent arteriolar resistance increases renal blood flow.

242—E (Chapter 6) Carbon dioxide is transported in the blood as dissolved gas (10%), as bicarbonate (70%), and as carbamino compounds (20%). Carbamino compounds are formed by the combination of carbon dioxide with terminal amine groups of blood proteins. The most important protein for transporting carbon dioxide in the carbamino form is hemoglobin. Carbonic anhydrase catalyzes the reaction between carbon dioxide and water to form carbonic acid.

243—D (Chapter 7) A converging circuit is one that, after receiving incoming signals from several sources, determines the level of reaction that will occur. That is, impulses "converge" into the pool, some from inhibitory nerves, some from excitatory nerves, some from peripheral nerves, and some from parts of the brain. From the list of possible factors that can affect the output from the neuronal pool, one can readily understand that basic differences in the anatomic organization of different neuronal pools can give thousands of different responses to incoming signals.

244—D (Chapter 8) Fat and proteins in the chyme stimulate the release of cholecystokinin. Cholecystokinin helps regulate the contraction of the gallbladder; it inhibits gastric emptying; and it stimulates the secretion of pancreatic enzymes. In addition to these functions, cholecystokinin also has trophic effects. It stimulates DNA synthesis and growth of the exocrine pancreas, but it has little or no trophic influence on the small intestine or stomach.

245—D (Chapter 3) A transfusion reaction usually causes hemolysis with the release of large amounts of hemoglobin into the plasma. Because the hemoglobin can enter the renal tubules through the normal glomerular membrane, the tubular load of hemoglobin increases greatly. When the tubular load of hemoglobin surpasses the reabsorption capabilities of the proximal tubules, the excess hemoglobin often precipitates in the nephron, causing oliguria or anuria with subsequent uremia. The untreated patient may die within 1 to 2 weeks from uremia, rather than from the acute effects of the transfusion reaction. Polycythemia is a condition characterized by an excessive increase in hematocrit.

246—B (Chapter 7) The fovea is located in the very center of the retina and has an area of only 0.4 mm. Cones provide greater visual acuity than rods primarily because the ratio of cones to optic nerve fibers is close to one; that is, nearly every cone is innervated by a single optic nerve fiber.

247—A (Chapter 4) Nerve metabolism can be blocked for minutes to hours without greatly affecting the transmission of nerve impulses. This is because the action potential is a physical phenomenon dependent only on ionic gradient across the cell membrane, and only a minute number of ions move across the cell membrane with each action potential.

248—A (Chapter 7) Because the entire retina is likely to be involved in the reflex, discrete damage to the fovea would have little effect on the reflex.

249—B (Chapter 4) Slow waves are rhythmic changes in membrane potential that can sometimes initiate action potentials in smooth muscles.

250—D (Chapter 8) Duodenal ulcers result from excessive secretion of acid and pepsin, and duodenal ulcer patients have higher than normal serum gastrin levels in response to a meal. However, unlike gastric ulcers, duodenal ulcers do not result from a defect in the mucosa itself.

251—C (Chapter 2) Antidiuretic hormone, released by the posterior pituitary, increases water and urea permeability in the collecting ducts and therefore increases water reabsorption.

252—A (Chapter 2) Aldosterone is produced by the adrenal cortex and stimulates sodium reabsorption and potassium secretion in the cortical collecting tubules.

253—D (Chapter 2) Atrial natriuretic peptide is released by the cardiac atria in response to increased stretch, as occurs with increased blood volume, and reduces renal tubular sodium reabsorption, thereby increasing urinary sodium excretion.

254—C (Chapter 3) Macrophages most often provide the first line of defense against exogenous bacterial invasion from the environment because macrophages reside in great abundance in the tissues most likely to be exposed to bacteria, such as the lung alveoli, lymph nodes, and subcutaneous tissues. When other tissues of the body become infected, the concentration of neutrophils increases rapidly; then, several days later, the macrophages begin to replace the neutrophils.

255—A (Chapter 3) Eosinophils are produced in large numbers in persons with parasitic infections. The parasites are usually too large to be phagocytized, but the eosinophils attach themselves to the surface and release lethal substances that can kill many of the parasites. Large numbers of eosinophils also appear in the blood in allergic conditions and may help to detoxify toxins that are released by allergic reactions.

256—E (Chapter 3) Neutrophils represent about 60% of the total white cells in the blood. They are highly motile and highly phagocytic, and they are attracted out of the blood into tissue areas where tissue destruction is occurring by a process called *chemotaxis*, which means attraction by the destruction products from the damaged tissues. Once in the tissue area, the neutrophils phagocytize bacteria and small amounts of dead tissue debris.

257—D (Chapter 2) The individual has a normal acid-base status (with pH 7.41, HCO_3^- of 25.0 mmol/L, and P_{CO_2} of 40 mmHg).

258—C (Chapter 2) In metabolic acidosis, the primary disturbance is a decrease in extracellular $[HCO_3^-]$, which is the same as an increase in $[H^+]$ and a decrease in pH. Partial respiratory compensation causes a decrease in P_{CO_2}.

259—E (Chapter 2) In fully compensated respiratory acidosis, the plasma pH and $[H^+]$ are returned to nearly normal in the face of increased P_{CO_2} because of the compensatory rise in $[HCO_3^-]$ that occurs as a result of increased renal production of HCO_3^-.

260—A (Chapter 2) In respiratory alkalosis, the primary disturbance is a decrease in P_{CO_2}, which reduces $[H^+]$ and increases pH. In uncompensated respiratory alkalosis, plasma $[HCO_3^-]$ is near normal.

261—B (Chapter 2) In combined respiratory and metabolic alkalosis, there are two primary disturbances: decreased P_{CO_2} because of hyperventilation and increased plasma $[HCO_3^-]$ either because of ingestion of bases or loss of H^+ (for example, vomiting of gastric acid). This leads to marked decrease in plasma $[H^+]$.

262—C (Chapter 5) The capillaries have the greatest cross-sectional area and thus the slowest velocity of blood flow.

263—D (Chapter 5) Veins must have the lowest hydrostatic pressure so that blood can flow down a pressure gradient from the arteries to the veins.

264—A (Chapter 5) One might suspect that the large veins would have the lowest resistance, but they are usually partially collapsed, which increases their resistance above the arterial value.

265—C (Chapter 5) The total surface area of all the capillaries is many times greater than that of the other vessels.

266—A (Chapter 7) The Golgi tendon apparatus is located at the juncture of the muscle and tendon. Several muscle fibers are connected in series with the tendon organ, and the organ is stimulated when the tension of these muscle fibers increases. Stimulation of the Golgi tendon apparatus produces reflex inhibition of muscle contraction and thereby prevents the muscle from developing too much tension.

267—D (Chapter 7) The muscle spindles, distributed throughout the belly of a muscle, send information to the central nervous system regarding muscle length and rate of change of muscle length. Stimulation of the muscle spindles causes the muscle to contract and thereby prevents it from being overstretched. The knee-jerk test is based on stimulation of the muscle spindles.

268—E (Chapter 7) The pacinian corpuscle is a rapidly adapting mechanoreceptor, located in the skin and deep fascial tissues, that is capable of detecting rapid movements of the tissues such as vibrations.

269—C (Chapter 7) Meissner's corpuscles are found in nonhairy parts of the skin where touch sensations are highly developed, such as the lips and fingertips.

270—B (Chapter 6) The temperature of an exercising muscle often rises as much as 2°C to 3°C, which shifts the oxygen-hemoglobin curve of the muscle capillary blood to the right. This rightward shift allows oxygen to be released to the muscle at much lower oxygen pressures than would otherwise be possible.

271—A (Chapter 6) An increase in blood pH causes a leftward shift of the oxygen-hemoglobin dissociation curve; a rightward shift occurs when the blood becomes slightly acidic.

272—C (Chapter 6) Carbon monoxide interferes with the transport of oxygen because it combines with hemoglobin with about 250-fold as much tenacity as oxygen. Therefore, relatively small amounts of carbon monoxide can tie up a large portion of the hemoglobin, making it unavailable for oxygen transport.

273—B (Chapters 2 and 10) Primary aldosteronism is associated with excessive secretion of aldosterone by the adrenal glands. This stimulates sodium secretion by the renal tubules and loss of potassium by the kidneys, and shifts potassium from the extracellular fluid into the cells, thereby decreasing plasma potassium concentration. Aldosterone also stimulates sodium and bicarbonate reabsorption by the renal tubules, thereby increasing plasma sodium and bicarbonate concentration. The sodium retention and increased arterial pressure decrease renin secretion by the kidneys and therefore reduce plasma renin activity.

274—D (Chapters 2 and 10) Diabetes insipidus can be caused either by inadequate secretion of antidiuretic hormone (ADH) or by the inability of the kidney to respond to ADH appropriately. In either case, diabetes insipidus results in decreased water reabsorption by the renal tubules, increased water excretion, and dehydration. This, in turn, markedly increases extracellular fluid concentration of sodium while causing smaller increases in potassium and bicarbonate concentration because the concentrations of these substances are also regulated by other control mechanisms. Dehydration also causes increased renin secretion and therefore increased plasma renin activity.

275—C (Chapters 2 and 10) Addison's disease is associated with inadequate formation of aldosterone, causing decreased sodium and bicarbonate reabsorption and decreased potassium secretion by the renal tubules. This causes a loss of sodium bicarbonate in the urine and renal retention of potassium, contributing to increases in plasma potassium concentration while decreasing slightly plasma concentrations of sodium and bicarbonate. The loss of sodium and tendency toward volume depletion stimulate secretion of renin by the kidney, causing an increase in plasma renin activity.

276—A (Chapters 2 and 10) Excessive ADH secretion causes increased renal tubular reabsorption of water and dilution of extracellular electrolytes, thereby reducing the plasma concentration of potassium, sodium, and bicarbonate. The fluid retention and extracellular fluid volume expansion also lead to a decreased renin secretion.

277—B (Chapter 6) In systemic arterial blood, the Po_2 averages 95 to 100 mmHg, the hemoglobin is almost 100% saturated with oxygen, and the oxygen content is nearly 20 ml O_2/100 ml blood.

278—C (Chapter 6) In mixed venous blood (taken from the pulmonary artery), the Po_2 is about 40 mmHg, the hemoglobin is about 75% saturated with oxygen, and the oxygen content is about 15 ml O_2/100 ml blood.

279—D (Chapter 6) Muscles consume greater amounts of oxygen during exercise; this decreases the oxygen pressure, oxygen content, and hemoglobin saturation with oxygen in the mixed venous blood.

280—B (Chapter 7) Hyperopia, also called farsightedness, is due to an eyeball that is too short, or less often, a lens system that is too weak when the ciliary muscle is relaxed. Vision can be corrected by adding refractive power with a convex lens.

281—C (Chapter 7) Myopia, or nearsightedness, is usually due to an eyeball that is too long, but it can sometimes be caused by too much refractive power of the lens system of the eye when the ciliary muscle is relaxed. Myopia can be corrected by placing a concave spherical lens in front of the eye.

282—B (Chapter 4) When the actin filaments have overlapped all of the myosin crossbridges, the activated sarcomere develops full tension. At point D on the curve, the actin filaments have pulled all the way to the end of the myosin filament with no overlap at all and, therefore, zero tension is now developed by the activated sarcomere.

283—C (Chapter 4) Because the actin filaments have overlapped all of the myosin crossbridges at point B, and because there is no overlap at all at point D, it is understandable that actin and myosin filaments overlap partially at point C. The development of tension is low at point A because the Z disks of the sarcomere abut the ends of the myosin filament.

284—B (Chapter 9) When the hypothalamic set-point temperature is increased, the person experiences chills and feels cold although the body temperature may already be above normal. Temperature-increasing mechanisms such as skin vasoconstriction and increased heat production continue to raise the body temperature until it reaches the set-point values of 103°F.

285—D (Chapter 9) Temperature-decreasing mechanisms are initiated when the hypothalamic set-point temperature is decreased. The rate of heat loss is increased by vasodilation of the skin blood vessels, which causes excess amounts of heat to be lost to the surroundings, and by sweating over the entire body, which causes evaporative heat loss. The person feels hot as the body temperature is decreasing, returning to normal.

286—C (Chapter 3) Any lung disease that results in tissue hypoxia stimulates the secretion of erythropoietin, which in turn increases the production of red blood cells. Prolonged cardiac failure can produce similar increases in both erythropoietin and hematocrit.

287—B (Chapter 3) Polycythemia vera is a tumorous condition of the red blood cell-producing tissues of the body. The excess production of red blood cells can cause slight suppression of erythropoietin secretion by the kidneys. Excess production of white blood cells and platelets also may occur.

288—E (Chapter 3) When the kidneys are destroyed by disease or are removed from a person, the result is virtually always anemia because 80% to 90% of all erythropoietin is secreted from the kidneys. The 10% to 20% of the normal erythropoietin formed in other tissues, primarily the liver, is not sufficient to produce the numbers of red blood cells needed by the body.

289—A (Chapter 6) $\dot{V}A/\dot{Q}$ equals zero when the lung unit has blood flow but no ventilation. Air trapped in the alveolus comes to equilibrium with the O_2 and CO_2 in the alveolar blood because the gases diffuse through the alveolar membrane. The PO_2 and PCO_2 of alveolar air equal that of mixed venous blood—that is, 40 mmHg and 45 mmHg, respectively.

290—E (Chapter 6) The capillaries at the top of the lung are pressed flat when the arterial and venous pressures are both greater than the alveolar pressure. This zone 1 condition does not occur during normal conditions, but it can occur when the pulmonary artery pressure is decreased, such as a result of hemorrhage. $\dot{V}A/\dot{Q}$ equals infinity when the lung unit has ventilation but no blood flow. The inspired air loses no O_2 and gains no CO_2 because there is no capillary blood flow. The PO_2 and PCO_2 of alveolar air equal that of inspired air—that is, 150 mmHg and 0 mmHg, respectively.

291—A (Chapter 8) An obstruction at the pylorus often results from fibrotic constriction after peptic ulceration. Persistent vomiting of stomach contents causes excessive loss of hydrogen ions and metabolic alkalosis.

292—C (Chapter 8) Patients with obstruction near the lower end of the small intestine tend to vomit more basic substances than acidic substances, which may result in acidosis. In addition, after a few days of obstruction, the vomitus may become fecal in nature.

293—D (Chapter 8) An obstruction near the distal end of the large intestine causes feces to accumulate in the colon. Vomiting eventually occurs when it becomes impossible for additional chyme to move from the small intestine into the large intestine.

294—A (Chapter 2) Excess ADH secretion causes renal retention of water, which would decrease the osmolarity of intracellular and extracellular fluid while increasing the volume of both compartments.

295—C (Chapter 2) Infusion of isotonic saline, which has the same osmolarity as both intracellular and extracellular fluid, causes no significant change in the osmolarity of these fluid compartments. Since sodium does not readily cross the cell membrane, most of the infused sodium and water remain in the extracellular fluid.

296—B (Chapter 2) Infusion of hypertonic saline, which has an osmolarity greater than plasma, initially causes an increase in extracellular fluid osmolarity, which in turn causes osmosis of water from the cells into the extracellular compartment. The net result, after osmotic equilibrium, is marked expansion of the extracellular compartment, decreased intracellular volume, and increased osmolarity of both intracellular and extracellular fluids.

297—C (Chapter 5) The pressure in the aorta is at its lowest value just prior to contraction of the ventricles, which occurs after ventricular depolarization, at point C, the QRS complex.

298—B (Chapter 5) The atria contract just after the P wave, at point A, which is caused by atrial depolarization.

299—C (Chapter 5) The AV valve (mitral valve) closes at the peak of ventricular depolarization, at point C. When the left ventricle contracts it ejects blood into the aorta, not back into the left atrium, because of the closure of the mitral valve.

300—E (Chapter 5) The ventricles depolarize during the QRS complex (point C) and then repolarize during the T wave (point D). The atria depolarize during the P wave (point A) and then repolarize. Therefore, both atria and ventricles are fully repolarized after the completion of the T wave, or at point E.

301—D (Chapter 10) Parathyroid hormone (PTH) secretion is stimulated by decreased plasma calcium concentration. The increased PTH stimulates activity of osteoclasts that secrete hydrogen ions and enzymes to cause breakdown (resorption) of bone. This leads to the movement of calcium from the bone into the extracellular fluid, helping to return extracellular fluid calcium concentration toward normal.

302—A (Chapters 2 and 10) Vitamin D_3 is formed in the skin and converted to 25-hydroxycholecalciferol in the liver. This is then converted in the kidneys to 1,25-dihydroxycholecalciferol, the active form of vitamin D.

303—C (Chapter 10) Insulin is perhaps the most important hormone in the body for maintaining normal blood glucose concentration. After a meal, increased insulin secretion promotes storage and utilization of glucose by skeletal muscle. Insulin also reduces the rate of fatty acid released (lipolysis) from fat tissue by depression of hormone-sensitive lipase.

304—B (Chapter 10) The actions of glucagon are almost exactly opposite those of insulin. Glucagon has a hyperglycemic action mainly by stimulating liver glycogenolysis.

305—E (Chapter 10) A deficiency of the thyroid hormone thyroxine greatly inhibits many functional systems in the body, which causes a reduction in the metabolic rate as well as a collection of mucinous fluid in the tissue spaces between the cells. This, then, creates an edematous state called myxedema.

Physiology
Must-Know Topics

The following are must-know topics discussed in this review. It would be useful for you to formulate outlines on these subjects since knowledge of the related material will be key to your understanding of the subject and material and for performing well on the examination.

Basic Organization of the Body, Homeostasis, and Cell Function

- The functional systems of the body and how each helps to maintain homeostasis

- The major parts of the cell, including the different organelles and their overall functions

- The principal mechanisms by which energy is released in the cell and the role of mitochondria in energy release

- The function of the endoplasmic reticulum

- The functions of mRNA, transfer RNA, and ribosomal RNA

- The process of mitosis and how cellular reproduction occurs

(continued)

Body Fluids

- The distribution of body fluids between the extracellular and intracellular compartments

- The differences between the ionic compositions of extracellular and intracellular fluids

- The differences between active and passive transport, between primary and secondary active transport, and between simple and facilitated diffusion

- The mechanisms for cotransport of glucose and amino acids with sodium and countertransport of sodium and hydrogen ions in the cell membrane

- The changes in intracellular and extracellular fluid volumes and osmolarity after intravenous administration of isotonic, hypertonic, and hypotonic fluids

- The forces that determine fluid movement across the capillary membrane

- The different factors that can cause interstitial fluid edema

Kidneys and Acid-Base Regulation

- The overall relationship between glomerular filtration, tubular reabsorption, and tubular secretion in urine formation

- The basic anatomy of the nephron, including vascular and tubular elements

- The effects on glomerular filtration rate of changes in the following:

- glomerular capillary filtration coefficient

- glomerular hydrostatic pressure

- afferent arteriolar resistance

- efferent arteriolar resistance

- How water and solutes are reabsorbed from the tubules into the peritubular capillaries

- The differences between the functions of the following tubular segments:

  - proximal tubules

  - descending and ascending segments of the loop of Henle

  - distal and collecting tubules

- The different tubular effects of the following hormones: aldosterone, angiotensin II, ADH, and atrial natriuretic peptide

- How renal clearance is used to calculate glomerular filtration, renal plasma flow, and tubular reabsorption

- The factors that determine whether the kidneys will form a concentrated or a dilute urine

- The role of the osmoreceptor-ADH-thirst systems in controlling extracellular fluid sodium concentration and osmolarity

- The mechanisms that cause K^+ secretion in the distal and collecting tubules and the factors that regulate K^+ secretion in these tubular segments

(continued)

- The three primary lines of defense against changes in body fluid $[H^+]$

- How the HCO_3^- buffer system is regulated

- How the respiratory system regulates $[H^+]$

- The mechanisms by which the kidneys secrete H^+, reabsorb HCO_3^-, and generate new HCO_3^-

- The characteristic changes in plasma pH, $[H^+]$, P_{CO_2}, and plasma $[HCO_3^-]$ in the following acid-base disturbances: respiratory acidosis, respiratory alkalosis, metabolic acidosis, and metabolic alkalosis

Blood, Hemostasis, and Immunity

- How red blood cell concentration is regulated; role of hypoxia, erythropoietin

- Roles of vitamin B_{12} and folic acid in red blood cell production

- Cause of various anemias: hypochromic anemia, aplastic anemia, megaloblastic anemia, hemolytic anemia

- Blood concentration and function of white blood cells: neutrophils, eosinophils, monocytes, lymphocytes

- Fours lines of defense against infection; role of macrophage and neutrophils

- Basic mechanism for initiation of blood clotting; difference between extrinsic and intrinsic pathways

- Cause of excessive bleeding in hemophilia, thrombocytopenia, prothrombin deficiency, and liver disease

- Difference between acquired and passive immunity, and between cellular and humoral immunity

- Three major types of sensitized T cells and function of each

- Basis for the four basic blood groups; genotypes, agglutinogens, agglutinins of each

- Why mismatching of blood groups causes transfusion reactions; effects of a transfusion mismatch

- How Rh blood types differ from A-B-O blood groups in causing transfusion reactions

Nerve and Muscle

- Cause of negative resting membrane potential; Nernst equation

- Cause of action potential: involvement of Na^+ and K^+ in depolarization and repolarization

- How an action potential spreads along a nerve membrane; the all-or-nothing law; what limits frequency of action potential; energy used by action potential

- Dynamics of ACh release and destruction at neuromuscular junction

(continued)

- Excitation-contraction coupling in skeletal muscle; role of ATP, calcium, transverse tubules, troponin complex

- Mechanism of muscle contraction: role of actin and myosin; sliding filament theory; walk-along theory

- Length-tension relationship of skeletal and cardiac muscle

- Major structural and functional differences between cardiac muscle, skeletal muscle, and smooth muscle

The Heart

- The basic mechanisms for control of rhythmicity in the heart, including the "pacemaker" function of the S-A node

- The significance of the junctional fibers in the Purkinje system

- The characteristics of the normal electrocardiogram

- The pressure changes in the heart during the cardiac cycle and the relationship to the ECG, the heart sounds, and ventricular volume changes

- The Frank-Starling law of the heart

- The types of autonomic nerve fibers that supply the heart and the function of each type

The Circulation and Its Regulation

- The approximate percentage of the total blood volume distributed in the systemic circulation, the pulmonary circulation, and the heart

- The determinants of blood flow through a blood vessel

- The factors that determine the resistance to blood flow in a blood vessel

- The meaning of vascular compliance and how it is related to distensibility of the vasculature

- The meaning of the following terms: stroke volume, diastolic pressure, systolic pressure, and pulse pressure

- The relationship between blood flow velocity and total cross-sectional area of blood vessels

- The local mechanisms by which blood flow is regulated in tissues

- The cardiovascular autonomic reflexes and their role in short-term blood pressure regulation

- The effects of each of the following hormones on blood pressure: angiotensin II, norepinephrine, aldosterone, vasopressin

- The renal-body fluid feedback mechanism for long-term regulation of mean arterial pressure

- The primary mechanisms that regulate pumping action of the heart

- The factors that regulate venous return

(continued)

- The differences that occur in the circulation in left heart failure compared to right heart failure

- The major classifications of circulatory shock, and the main causes of each type of shock

Respiratory System

- Muscles of inspiration and expiration: those used during rest; those used during exercise or restrictive and obstructive lung disease

- Alveolar and pleural pressures during normal breathing

- Pulmonary volumes and capacities: which can be measured; which must be calculated; equation for each

- Ventilation equations: minute respiratory volume and alveolar ventilation; types of dead space air

- Lung compliance: how it is measured; relation to lung elasticity; effect of restrictive and obstructive lung disease

- How surfactant reduces work of breathing; helps to prevent atelectasis

- Factors that affect airway resistance

- Concentrations and partial pressures of respiratory gases in room air, alveolus, blood; why partial pressure is used

- Determinants of gas diffusion across pulmonary membrane; how these are affected by emphysema, edema

- Definition of diffusing capacity; why CO_2 diffuses through body tissues more rapidly than O_2

- Oxygen-hemoglobin dissociation curve: factors that change P_{50}; importance of sigmoid shape, in tissues, in lung

- Normal values of P_{O_2} and P_{CO_2} and content in blood

- CO_2 transport in blood: quantitative importance of each mechanism

- Blood flow zones: relation between arterial, venous, and alveolar pressure in each zone; determinants of blood flow in each zone

- Ventilation-perfusion mismatch: causes; values at different levels in lung

- Central respiratory centers: three main groups of neurons and function of each; Hering-Breuer reflex

- Mechanism by which CO_2, blood pH, and O_2 lack increase pulmonary ventilation; involvement of central and peripheral chemoreceptors

- Different causes of hypoxia and which can benefit from oxygen therapy; mechanisms of acclimatization to hypoxia

- Obstructive and restrictive diseases: causes, examples, pulmonary function test results

- Maximum expiratory flow curve: cause of dynamic compression, diagnosis of lung disease

- Toxic effects of high oxygen pressure and high nitrogen pressure on the body

- Decompression sickness, its cause and treatment

(continued)

Central Nervous System

- Main parts of the nervous system for controlling body functions; basic mechanism of reflex arc

- Function of the synapse: examples of excitatory and inhibitory transmitters and how each works

- Mechanism of excitatory postsynaptic potential and inhibitory postsynaptic potential

- Types of summation at synapses

- Basis of diverging circuit, converging circuit, repetitive firing circuit; example of system that utilizes each

- Possible mechanism by which thoughts occur in the CNS

- Possible mechanism of memory; portion of brain most concerned with memory

Somatic Sensory System

- Modalities of sensation and basic types of sensory receptors

- Pathways for transmission of somatic sensations into the CNS; major differences between pathways and sensations conducted in each

- Types of nerve fibers; classification systems; characteristic of each type

- The "purpose" of pain; stimuli that cause pain; how the brain controls the sensitivity to pain

- Types of stimuli that cause visceral pain; mechanism of referred pain

- Mechanism (and purpose) of various cord reflexes: stretch reflex, withdrawal reflex, crossed-extensor reflex, positive supportive reflex, walking reflexes

- How equilibrium is maintained: functions of macula of the utricle, semicircular canals, semicircular ducts

- Functions of somatic, auditory, and visual sensory association areas

- Organization of motor cortex; portions of musculature represented to greatest extent in cortex

- Basic differences between pyramidal pathways and extrapyramidal pathways

- Function of premotor cortex in control of muscular movements

- Functions of basal ganglia in the control of muscular movements

- How cerebellum damps muscular movement; why cerebellar dysfunction causes ataxic movements

- Hormones secreted by sympathetic and parasympathetic nerve endings

- Effect of sympathetic and parasympathetic stimulation on various organs and tissues in body and which receptors are involved

(continued)

Special Senses

- How lens system of eye functions as camera

- Mechanism of accommodation: function of ciliary muscle and its control by central nervous system

- Control of pupillary diameter; how it alters depth of focus; mechanism of pupillary light reflex

- Common errors of refraction and how they are corrected

- Functions of rods and cones; importance of foveal region of retina

- Operation of rhodopsin-retinal cycle of rods; mechanism of dark and light adaptation

- Mechanism of color vision

- Mechanism of sound transmission from tympanic membrane to cochlea

- Function of cochlea; how resonance occurs; how frequency of sound is determined; how loudness is determined

- Pathway for conduction of taste impulses into brain

- Function of taste buds and four primary taste sensations

Gastrointestinal Tract

- Gastrointestinal hormones, paracrines, neurocrines: their actions, stimulus for secretion, site of action

- Major types of gastrointestinal movements and their purpose; mechanism of peristalsis; function of slow waves

- Phases of gastric secretion: percent of total response, initiating stimulus, stimulus at parietal cell

- Function and control of salivary secretion, pancreatic secretion, intestinal secretion, bile secretion

- Mechanisms of digestion and absorption of carbohydrates, fats, and proteins; importance of intestinal villi

- Mechanism of absorption of water and ions

- Cause of peptic ulcers, duodenal ulcers, diarrhea, cholera, pernicious anemia, achalasia, megaesophagus

Metabolism and Energy

- Glucose storage and release: blood glucose-buffering function of glycogen; glycogenesis; glycogenolysis; effect of insulin

- Mechanism of ATP production: importance of glycolysis; oxidative phosphorylation, chemiosmotic mechanism

- Fat transport, storage, and energy production; functions of phospholipids and cholesterol

- Amino acid transport, storage, metabolic rate, energy production

- Regulation of food intake: causes of obesity

(continued)

- Starvation: sequence of carbohydrate, fat, and protein consumption

- Principal function(s) of each vitamin

- Mechanisms of heat production and heat loss; control of body temperature

- Cause of fever and sequence of events in febrile illness

Endocrine Control Systems

- The general mechanisms of action for hormones that have receptors located on the cell membrane compared to hormones that interact with specific intracellular receptors

- How cAMP acts as an intracellular hormonal second messenger

- The six hormones released by the hypothalamus that control function of the anterior pituitary gland, as well as their actions

- The major hormones secreted by the anterior pituitary gland

- The functions of adrenocorticotrophic hormone (ACTH)

- The major functions of thyroid-stimulating hormone (TSH)

- The two principal gonadotrophic hormones and their major actions

- The two principal hormones secreted by the posterior pituitary gland and their major actions

- The three major types of steroid hormones secreted by the adrenal cortex and their principal actions

- The basic effects of thyroxine and triiodothyronine on cell function

- The effects of insulin on carbohydrate and fat metabolism

- The effects of glucagon on carbohydrate metabolism

- The two types of diabetes mellitus and their primary causes

- The principal steps in the formation of bone

- The mechanisms by which parathyroid hormone regulates plasma calcium concentration

- The physiologic functions of testosterone

- How testicular function is controlled by the anterior pituitary gland

- The principal effects of estrogens and progesterone on the uterine endometrium

- The events of the ovarian cycle during the female sexual month and the associated hormonal changes

- The effects of estrogens and progesterone on the tissues of the body other than the sex organs

- The role of the hypothalamus and anterior pituitary in regulating the female sexual cycle

- The importance of human chorionic gonadotropin for the continuation of pregnancy

(continued)

- The mechanisms of parturition

- The changes that occur in the infant's respiration and circulation immediately after birth

- The factors that cause growth of the mother's breasts and then milk secretion following birth of the infant

- The function of oxytocin in lactation

Index

Rypins' Intensive Reviews

Series Editor: Edward D. Frohlich, MD

Behavioral Science

Internal Medicine

Surgery

Psychiatry and Behavioral Medicine

Pharmacology

Pediatrics

Physiology

FUTURE VOLUMES

Anatomy

Biochemistry

Microbiology and Immunobiology

Pathology

Obstetrics and Gynecology

Community Health